Teaching STEM in the Secondary School

This book looks at the purpose and pedagogy of STEM teaching and explores the ways in which STEM subjects can interact in the curriculum to enhance student understanding, achievement and motivation. By reaching outside their own classroom, teachers can collaborate across STEM subjects to enrich learning and help students relate school science, technology and maths to the wider world.

Packed with ideas and practical details for teachers of STEM subjects, the new revised edition of this book:

- considers what the STEM subjects contribute separately to the curriculum and how they relate to each other in the wider education of secondary school students;
- describes and evaluates different curriculum models for STEM;
- suggests ways in which a critical approach to the pedagogy of the classroom, laboratory and workshop can support and encourage all pupils to engage fully in STEM;
- addresses the practicalities of introducing, organising and sustaining STEM-related activities in the secondary school;
- looks to ways schools can manage and sustain STEM approaches in the long-term.

This new revised edition is essential reading for trainee and practising teachers, those engaged in further professional development and all who wish to make the learning of science, technology, engineering and mathematics an interesting, motivating and exciting experience for their students.

Prof. Frank Banks is Emeritus Professor in Teacher Education at The Open University in the UK where he directed the innovative on-line initial teacher education programme. Frank has worked as a school teacher of science, engineering and technology in different secondary high schools in England and in Wales.

Dr David Barlex is an acknowledged leader in design and technology education, curriculum design and curriculum materials development. A former school teacher and Senior Lecturer in Education at Brunel University, David's teacher development activity stems from his conviction that there should be a dynamic relationship between curriculum development and academic research.

Teaching STEM in the Secondary School

Helping Teachers Meet the Challenge

Second Edition

Frank Banks and David Barlex

Routledge
Taylor & Francis Group

LONDON AND NEW YORK

Second edition published 2021
by Routledge
2 Park Square, Milton Park, Abingdon, Oxon, OX14 4RN

and by Routledge
52 Vanderbilt Avenue, New York, NY 10017

Routledge is an imprint of the Taylor & Francis Group, an informa business

First edition published by Routledge 2014

British Library Cataloguing-in-Publication Data
A catalogue record for this book is available from the British Library

Library of Congress Cataloging-in-Publication Data
Names: Banks, Frank, 1953- author. | Barlex, David.
Title: Teaching STEM in the secondary school : helping teachers meet the challenge / Frank Banks and David Barlex.
Other titles: Teaching Science, Technology & Engineering, and Mathematics in the secondary school
Description: Second Edition. | New York : Routledge, 2020. | "First edition published by Routledge 2014"--T.p. verso. | Includes bibliographical references and index. |
Identifiers: LCCN 2020035124 (print) | LCCN 2020035125 (ebook) | ISBN 9780367330453 (Hardback) | ISBN 9780367330460 (Paperback) | ISBN 9780429317736 (eBook)
Subjects: LCSH: Science--Study and teaching (Secondary) | Technology--Study and teaching (Secondary) | Engineering--Study and teaching (Secondary) | Mathematics--Study and teaching (Secondary)
Classification: LCC Q181 .B26 2020 (print) | LCC Q181 (ebook) | DDC 507.1/2--dc23
LC record available at https://lccn.loc.gov/2020035124
LC ebook record available at https://lccn.loc.gov/2020035125

ISBN: 978-0-367-33045-3 (hbk)
ISBN: 978-0-367-33046-0 (pbk)
ISBN: 978-0-429-31773-6 (ebk)

Typeset in Bembo
by SPi Global, India

Contents

Foreword

It is very encouraging that Routledge have asked the authors to write this second edition of their excellent book, because it indicates that there is a readership, particularly among secondary school teachers, who value the latest thinking on the role of STEM subjects in the curriculum and the importance of integrating STEM across the whole of secondary school education if we are to prepare the next generation for the many challenges facing humanity in the twenty-first century.

Of course, not everyone enjoys science or feels confident about mathematics, but the idea that studying STEM subjects is only about learning scientific facts or solving algebraic equations – or indeed to prepare the next generation of technologists and engineers – is way off the mark. While it is still the case, as we enter the third decade of the century, that science, technology, engineering and mathematics are mostly taught as separate subjects, the problem can often be more serious than that because pupils do not appreciate the interconnectedness of what they are learning or how these subjects overlap in the real world, which is becoming increasingly broad and interdisciplinary. When we talk today about STEM-based industries, we mean much more than just those in the engineering and technology sector, as it encompasses more broadly almost all aspects of modern life. Thus, if school science exams do little more than test pupils' abilities to recall facts then they will be ill-prepared to enter the modern workforce.

In an age when all information is available, literally at our fingertips, it is vital that STEM teaching also focuses on the necessary skills needed to navigate through a world where problem-solving and rational thinking are more important than ever. In addition, STEM teaching must encompass softer skills, often acquired in group-based work, such as communication skills or an appreciation of the values and ethics of the application of new scientific knowledge in technology and an appreciation of the role of scientific evidence.

The 2020 coronavirus pandemic brought to the fore a number of issues relating to the role of STEM in everyday life and the crucial importance of well-informed citizens who understand how our scientific knowledge of the world around us develops and evolves, particularly when it comes to knowing what to believe and whom to trust, or the willingness to change one's views in the light of new data or evidence.

This is now even more important than when the first edition of this book was published, with the accelerating increase in the effects of the negative influence of social media on society's views and opinions and the devaluing of the role of reason, debate and rational argument. Young people may have grown up in a world of instant access, echo chambers and polarising ideologies, but that doesn't make them any better at understanding what and whom to trust. Teaching pupils the necessary skills to cope with the modern world will involve an integration of STEM into all areas of their education.

Computer science is one area of increasing importance in the teaching of STEM, not only because of the role that computing, digital literacy and IT play in everyday life or the range of careers that require these skills, but because of its use as a tool in the teaching of all subjects across the curriculum. We are now seeing Technology Enhanced Learning (TEL), whereby teachers are making use of a wide range of computer-based learning techniques and technologies, really coming to the fore. The authors, for example, highlight the move away from IT packages such as Word and Excel to the use of the BBC micro:bit or the Raspberry Pi for coding and problem solving.

Of course, there is more than one way to teach STEM or enhance the learning experience of STEM subjects, and the authors draw on their wealth of experience, both nationally and internationally, to lay out the different models for STEM teaching. They explore some of the latest ideas from around the world to highlight how STEM education is evolving and where it needs to go next.

This new edition is therefore both inspiring and timely.

JIM AL-KHALILI, JUNE 2020

Jim Al-Khalili OBE FRS is Distinguished Professor of Physics at the University of Surrey, where he has taught continuously for 28 years. He is widely known as a populariser of science through his writing and many TV and radio broadcasts. He sits on the Royal Society Education Committee and was one of the authors of its *Vision for Science and Mathematics Education* report in 2014.

Preface

At the start of our teaching careers, we trained to be science teachers; Frank a physics teacher and David a chemistry teacher. As happened in those days, we were soon required to teach all the sciences to pupils up to the age of 14. While teaching in comprehensive schools, both of us then became interested in and enjoyed teaching technology too, and ultimately moved into higher education with responsibility for training technology teachers. David then concentrated on curriculum development directing the Nuffield Design & Technology and Young Foresight projects. Frank was, in turn, in charge of both design & technology and science initial teacher education (PGCE) courses at The Open University. Both of us have an interest in the professional development of teachers.

Given our background and interests, it is not surprising that we were intrigued by the rise of STEM as a potentially unifying concept across the related yet different disciplines of science, mathematics and technology, which could be used to mutually enhance pupils' learning in these subjects. We saw that it was not easy for teachers to capitalise on the STEM potential despite successive initiatives and exhortations across many years for them to do so. In this second edition we have taken advantage of the opportunity to update what has been happening in STEM education worldwide, but our motivation for writing it remains the same; to explore the advantages for teachers from mathematics, science and design & technology in 'looking sideways' in their school's curriculum to see what is happening in the STEM subjects other than their own. We suggest that such a view will stimulate conversations that are the first and vital step in enhancing and developing synergy in pupils' learning across the STEM subjects. We hope that we have been realistic in appreciating the difficulties in such work, yet have provided sufficient argument, guidance and examples to give those working in secondary schools the confidence to have those essential conversations and turn the emerging ideas in to action – action that will result in improved learning for pupils and more rewarding teaching for teachers.

Working together, we have been able to critically support each other in drafting this fully revised second edition. Frank took the lead on writing Chapters 1, 2, 6, 9 and 10; and David led on Chapters 3, 4, 5, 7, 8 and 12. Chapter 11 on 'Looking at Stem education in different countries' was jointly written with contributions from our colleagues around the world.

Frank Banks
David Barlex
August 2020

Acknowledgments

We could not have written this book without the considerable help of the following colleagues to whom we are most grateful.

Mike Watts for his contribution to Chapters 1 and 12. Clare Lee who provided the DEPTH mathematics diagram in Chapter 2. Torben Steeg and Celia Hoyles for giving interviews concerning the relationship between design & technology and science and mathematics in Chapter 4. Charlie Stripp who commented in detail on Chapter 5. Moshe Barak for the extensive use of his ideas on 'problem solving' in Chapter 6. Torben Steeg and also the staff of Ysgol Uwchradd Caergybi in Wales, in particular Adam Williams, Charlie Wilson and Wyn Owen, for their case studies in Chapter 9. David Ellis, John Williams, Didier Van de Velde, Vitor Mann, Yang Chunling, Ke Shan, Osnat Dagan, Sergey Gorinskiy, Kuen-Yi Lin, Yu-Jen Sie and John Wells for their major contributions to Chapter 11, and Matthew James of Lewis Girls School and also Paul Gittins for their help with Chapter 12.

We are particularly grateful to Annamarie Kino at Routledge for her invitation to write this second edition and her invaluable comments on drafts of chapters, and to Molly Selby, Natalie Hamil and their colleagues for the unfailing support throughout the production process.

Finally, once again, we wish to acknowledge the Pullman Hotel on the Euston Road in London. The lobby of the hotel continued to provide the authors with the essential comfort and coffee when we met to discuss the new STEM developments and necessary chapter revisions of this book.

1

What is STEM?

Introduction

The other day I repeated something I had done five years ago. I asked some upper-primary school and lower secondary school pupils to draw a picture of a 'Scientist' and a picture of an 'Engineer'. Of course, not many of them had ever met a scientist and so, just as five years ago, some drew the cliché often seen in films – white, male, middle aged, balding or 'mad'-haired and white-coated – a bit like Doc in *Back to the Future* – with Dr Frankenstein wild eyes, and a bubbling conical flask in their hand as a modern-day Dr Jekyll. But this time, there were some significant differences. Some pupils, both boys and girls, drew their scientist as female, dressed more as an 'explorer' rather than wearing a white coat, and with a sunhat, magnifying glass, notebook and pencil. And the engineer? Well, like before, all male, with a hard hat and carrying a larger-than-life spanner. While accepting that the very act of asking for pictures to be drawn might have led them to offer me a caricature of how scientist and engineers are commonly represented in the media, I was intrigued that although it seems the stereotype of a scientist is changing, engineering is generally still seen as 'male' despite the impetus over the years to broaden the appeal of both engineering and the physical sciences.

The STEM subjects – Science, Technology & Engineering, and Mathematics – are separate in most national curriculum documents around the world but with common links at a range of levels, and with at least a nod to relevance in the 'real world' and to vocational usefulness. These links are structural too. For example, I looked up what is said about the UK Parliament's Science and Technology Committee yesterday wondering what it was and what it did. I found out that:

> The Science and Technology Committee exists to ensure that Government policy and decision-making are based on good scientific and engineering advice and evidence. [It] scrutinises the *Government Office for Science* (GO-Science), which is a 'semi-autonomous organisation' based within the Department for Business, Energy and Industrial Strategy. GO-Science 'supports the Government

> Chief Scientific Adviser and works to ensure that Government policy and decision-making is underpinned by robust scientific evidence'.

Notice the words 'scientific evidence' and 'Business, Energy and Industrial Strategy', Guessing that this was not unique, I wondered about thinking in the USA, which has an Office of Science and Technology:

> In 1976, Congress established the White House Office of Science and Technology Policy (OSTP) to provide the President and others within the Executive Office of the President with advice on the scientific, engineering, and technological aspects of the economy, national security, homeland security, health, foreign relations, the environment, and the technological recovery and use of resources, among other topics.

Again, notice 'scientific, engineering, and technological aspects of the economy'. Finally, I looked up what happens in Australia. There, I discovered that as part of the work of the Department of Industry, Innovation and Science there was a specific policy concerning STEM:

> Increasing science, technology, engineering and mathematics (STEM) capability is at the core of the government's science agenda […] The global economy is changing which means new industries are emerging and new skills are required for workers at all levels.

Action on STEM is critical to:

- Australia's ability to compete in international markets
- creating new opportunities for industries
- supporting high living standards

Ensuring all Australians can be engaged with STEM is a key priority.

Although they might not draw the same pictures of the scientist and engineers as the youngsters, it is clear that politicians, too, have some stereotypical views and often refer to 'science and technology' as an epistemological unit, more-or-less the same thing, a single activity inseparably linked, which is the principal driver of the modern economy.

The aims and processes of science, however, are fundamentally different from those of technology and the links between them are not as formal as many people think. Maybe the confusion is because science is seen, erroneously, as necessarily always underpinning technology – providing the foundation to develop 'useful knowledge'. Disappointingly, the confusion is also present in the school curriculum where, in perhaps rather crude and simplistic terms, science is often seen as 'theory', i.e. 'know why', and technology as practical, i.e. 'know how', and that in some way technology is dependent on science. Before we consider curriculum links across STEM subjects, which we will do in Chapter 2, we must first clarify our understanding of why STEM has gained such interest in recent years and, in particular discuss 'science', 'technology' and maths, and how science knowledge and mathematical ability is 'exploited' in technology and vice versa. This chapter considers:

- the birth of STEM; when did we start thinking of this area of knowledge in linked capital letters?
- some milestones in the development of STEM subjects in schools;
- the difference between science knowledge and technology knowledge;
- technology before science? What does history tell us?
- common ground between science and technology learning;
- the contribution of M in STEM;
- what else do the STEM subjects contribute? Affective knowledge and personal values, problem solving, and systems thinking;
- why should all pupils learn STEM?

The birth of STEM

When did we start thinking of this area of knowledge in linked capital letters? In 1944, US President Franklin D. Roosevelt wrote a letter to the Director of The Office of Scientific Research and Development. He made the point that, under a great secrecy, extraordinary developments had been made for the war effort and it was time to consider how similar progress could be promoted in peacetime. He wrote:

> What can be done, consistent with military security, and with the prior approval of the military authorities, to make known to the world as soon as possible the contributions which have been made during our war effort to scientific knowledge? The diffusion of such knowledge should help us stimulate new enterprises, provide jobs for our returning servicemen and other workers, and make possible great strides for the improvement of the national well-being [...]
>
> New frontiers of the mind are before us, and if they are pioneered with the same vision, boldness, and drive with which we have waged this war we can create a fuller and more fruitful employment and a fuller and more fruitful life.
>
> FRANKLIN D. ROOSEVELT
> THE WHITE HOUSE
> Washington, D.C.
> November 17, 1944

The post-war period was one where the STEM subjects were indeed to the fore as the US economy boomed with consumption of new cars and domestic white goods raising the standard of living to a level that few had experienced before. A slower post-war revival in Europe also promoted and encouraged an interest in STEM as means to follow the US and 'stimulate new enterprises, provide jobs for our returning servicemen and other workers'. In Britain, the first commercial jet airliner and the first nuclear power station were held up as examples of British competence in 'science and engineering', again building on remarkable advances that had taken place during the war years. But in 1957, the capitalist west was to be shaken to the core by the launch of Sputnik. The shock was profound as it implied that communism, so despised in post-war America, was ahead of the capitalist West in 'science and technology'. There was a sudden realisation that STEM education – considered so important for developing the industrial base and providing jobs – seemed to be lagging behind the 'Russians'. The Space Race had begun and the STEM education starting pistol had been fired.

The race was initially a sprint as President Kennedy proposed that the USA would land a man on the moon 'within the decade' (the 1960s). Kennedy's powerful rhetoric stirred the nation and put the STEM subjects stage centre in the endeavour as he said,

> We choose to go to the moon in this decade and do the other things, not because they are easy, but because they are hard, because that goal will serve to organise and measure the best of our energies and skills.

But as is indicated in the table below, after the initial sprint, the development of STEM education in schools turned into a steady-paced marathon.

Some milestones in the STEM subjects in schools

TABLE 1.1 Some STEM education milestones

1956	Physical Science Study Committee (PSSC) formed	Concern about the quality of high school physics teaching led Massachusetts Institute of Technology (MIT) physicist Jerrold Zacharias to assemble the PSSC to push for an updated curriculum.
1957	Launch by USSR of Sputnik – the first artificial satellite	This was the starting pistol for the Space Race between the USSR and the USA. It caused shock in the Western hemisphere as 'Russia' went into the lead. What should be done about our lagging science and technology education? In the USA, the concerns of PSSC were picked up and $1 billion was put into the implementation of the National Defense Education Act to promote science, mathematics and foreign language education.
1962	School Mathematic Project (SMP)	Although moves to change the mathematics taught to secondary (high) school students has its roots before the Second World War, the change to a discovery approach to learning mathematics accelerated when new school texts books were published. Introducing ideas such as set theory and using number bases other than ten, this approach exposed all pupils to a wider appreciation of the wonder of mathematics. It was criticised by many as being too abstract and not a good grounding for science and engineering. There was a 'back-to-basics' backlash – including a need for more arithmetic, for example – a decade later.
1962–1972	Harvard Project Physics (HPP)	In the USA, HPP by using aspects of the history of science was an alternative to the technical physics of PSSC. Taking a humanistic approach, it sought to widen the appeal of physics by considering the people behind the discoveries.
1966	Nuffield Science Teaching Project	Pupil and teacher guides were produced in the UK that encouraged an experiential approach to teaching of science through a range of new practical ideas and pupil experiments. This, coupled with an assessment regime that encouraged application of scientific ideas rather than simple recall of facts, was a revolution in child-focused learning.

(Continued)

TABLE 1.1 (Continued)

1969	1st moon landing	Space race that initiated so much STEM funding comes to a climax. Next decade sees education funding cut as the rise in oil prices causes economic inflation across the West. Computers start to appear in schools – the computer on the moon lander(s) had a small fraction of the memory of a mobile phone in 2020, and far less processing power than even a modern washing machine, although it was one of the very first computers to use integrated circuits and so minute by the standards of the time. It was also to automatically re-start.
1980	Assessment of Performance Unit (APU)	Series of tests of 11, 13 and 15-year-olds on their scientific understanding of topics such as electricity and the chemistry of metals – and their practical manipulation of apparatus to investigate their scientific thinking – helped to inform changes to the curriculum. This led to a Secondary Science Curriculum Review (1980–1989).
1980–1989	Children's Learning in Science Project (CLISP)	Directed by Ros Driver at Leeds University, CLISP was very influential in promoting a 'Constructivist' view of learning in science. In a nutshell, pupils construct their understanding of the world around them and teachers should appreciate that: • What is already in the learner's mind matters • Individuals construct their own meanings • The construction of meaning is a continuous and active process • Learning may involve conceptual change • The construction of meaning does not always lead to belief • Learners have final responsibility for their learning • Some constructed meanings are shared, but pupils' original untutored constructions are resistant to change.
1982	Singapore Math	In Singapore, a new country-specific mathematics program with a focus on problem solving and on heuristic model drawing was introduced. Trends in International Mathematics and Science Study (TIMSS) in 2003 showed Singapore at the top of the world in 4th and 8th grade mathematics.
1983	Technical and Vocational Educational Initiative (TVEI)	Funded by the UK Department of Industry rather than the Department of Education. By the time of its eventual demise in 1997 almost £1 billion was spent on this initiative. There were two broad aims of TVEI; first to align the school curriculum more closely to the 'needs' of industry and commerce and second to rectify some of the knowledge, skill and particularly the 'attitude deficits' of school leavers. Through the funding, new topics like Microelectronics, Pneumatics and system approaches were introduced across science and technology.
1985	The Department of Education's 1985 'Science 5–16: A Statement of Policy'	'The essential characteristic of education in science is that it introduces pupils to the methods of science.' Also, the findings from the Assessment of Performance Unit (APU), about children's understanding of the concepts of science, led to the view that 'science should be an active process whereby learners construct and make sense of the world by constructing meaning for themselves'. The project also followed on from the publication of 'Insight to Science' by Inner London Education Authority (ILEA) in 1978/1979. The 'Science in Process' materials were developed by a team of ILEA teachers and were trialled in schools.

(Continued)

TABLE 1.1 (Continued)

1988	The Great Educational Reform Act – Introduction of a prescribed National Curriculum in Science and Mathematics from age 5–16 in England, Northern Ireland and Wales	Core subjects were established for science and mathematics and technology (which included design & technology and also information technology) was designated as a foundation subject. The difference between Core and Foundation subjects was never clear. The specification for science and maths was published in 1988 and technology in 1990.
1990–1999	The Science Processes and Concepts Exploration (SPACE) project	The SPACE research was conducted at the University of Liverpool and King's College, London, with Wynne Harlen and Paul Black as joint directors. It investigated the science 'misconceptions' of primary (elementary) school pupils aged 5–11 in topics such as Light, Sound, Forces and the Earth in Space.
1990–1999	Nuffield Design & Technology Project	Launched as technology became part of the national curriculum, Nuffield design & technology was very influential. It recommended 'Resource Tasks' to address specific skills and knowledge to be used in larger 'Capability Tasks' and these were adopted into a revised curriculum structure (under different names).
1992	Publication of 'Technology in the National Curriculum –Getting It Right'	Commissioned by the Engineering Council and written by Alan Smithers and Pamela Robinson, this was a blistering critique of Technology in the National Curriculum – suggesting it was 'a mess' – and led to a series of consultation and changes to the attainment targets programme of study, finally settling on design & technology and IT as separate subjects.
2000	Young Foresight – an example of school – industry links for STEM	Young Foresight is a curriculum initiative giving pupils aged 14 the opportunity to work co-operatively to conceive products and services for the future in consultation with mentors from industry.
2001	No Child Left Behind Act (NCLB) in USA	All USA school districts must demonstrate annual yearly progress (AYP) in student standardised test scores. The first years of the law required achievement gains only in mathematics and reading. Testing in science started in 2007.
2002	Changes to the National Curriculum for England, Wales and Northern Ireland	Science and mathematics (and ICT in England, not Northern Ireland and Wales) still a compulsory subject to age 16. Design & technology, however, only compulsory to age 14 – but it must be offered as a subject in all schools.
2013–2017	Implementation of a revised National Curriculum in England and the 'English Baccalaureate'	As a means of monitoring schools in England, a cluster of subjects known as the English Baccalaureate (EBacc) is devised. Not a qualification, but it privileges certain subjects including maths, science (all sciences or computer science) but design & technology is not included.
2022–2026	Implementation of a revised curriculum in Wales	Wales will move away from specific subjects and introduce six 'Areas of Learning and Experience': <ul><li>Expressive Arts</li><li>Health and Well-being</li><li>Humanities</li><li>Languages, Literacy and Communication</li><li>Mathematics and Numeracy</li><li>Science and Technology</li></ul>

What is the difference between science knowledge and technology/engineering knowledge?

I think most would agree that young people want to know *why* something is the way it is or *how* something works; they seem to want answers to two sorts of questions. One type of question seeks knowledge of the 'knowing how' variety – how a thing works, how it is used, how it is possible to improve the function of something or the way something is done, or how to create something which has a new purpose. This can be thought of as technological knowledge. It is the practical knowledge of application, i.e. 'know-how'. The other type of question seeks knowledge of the 'knowing why' variety – why the world is the way it is, first to help us understand the rules that confirm generally accepted agreement about what we know, and second to help us rationalise the experience of our senses. This type of knowledge can be thought of as scientific knowledge. Therefore, in a nutshell, as Per Nostrom would summarise it: 'Technology is about creating artefacts and solving problems, while science is primarily about describing and explaining phenomena in the world' (Per Nostrom, 2011).

The philosophy of science and of technology are two well developed disciplines and I have, perhaps, been rather simplistic and rode rough-shod over them in the way that I have set out starkly the two types of knowledge, so let us look more carefully at the subtleties.

The press cliché is that we live in a 'technological age'. Some would say that all should have an understanding of the workings of what we use, yet most of us lead perfectly satisfactory lives on the basis of knowing how rather than knowing why. One can know *how* to drive a car without having much idea of *why* the engine and all its control systems do the job they do. Similarly, a motor mechanic (or a TV repairer and other similar people) can mend engines without any knowledge of gas laws, combustion principles, materials properties, quantum mechanics of semiconduction or other scientific knowledge of the 'knowing why' variety. The level of 'knowing why' needs to be appropriately matched to the needs for the 'knowing how' for them together be 'useful knowledge' for creating appropriate products.

Technology before science: What does history tell us?

Science and mathematics have been in the school curriculum for a long time, yet the subject of technology is a relative newcomer and engineering is rarely taught as a separate subject at secondary (high) school level. In many countries, technology in the curriculum fights for its survival as curriculum designers have perhaps tended to cling to the belief that science education provides a more appropriate preparation for pupils intending to follow careers in industry and that without a thorough understanding of scientific principles there can be little progress in the various fields of application. Engineering, too, has had a niche in some secondary schools but again is fighting for its place in an increasingly 'academically' defined curriculum as it has been associated with vocational preparation. The role of the E in STEM will be considered fully in Chapter 7.

The assumption that science knowledge always precedes technology knowledge can be challenged through some wide-ranging examples. For example, how to refine copper has been known since ancient times, millennia before the concepts of oxidation and reduction were understood. Around 1795, the Paris confectioner Appert devised a method of preserving food by heating it (to kill bacteria) and, without delay, sealing it in a container. The idea caught on, and a cannery using 'tins' was already functioning in Bermondsey in 1814 when Louis Pasteur proposed a 'Theory of Bacterial Action'. England became the 'steam workshop' of the world following the invention of the first commercial steam engines by Thomas Savery and Thomas Newcomen in the late seventeenth and early eighteenth centuries. Their knowledge of how to design steam engines spread as 'know how' across Europe and to North America. Yet the concept that heat was a form of energy able to do work came later. Later still, Sadi Carnot, an officer in the French Army, became preoccupied with the concept of heat engines but it was years before his findings influenced steam engine design. The science of thermodynamics followed from the intellectual challenge to understand the operation of better steam engines.

With the exception of the ancient techniques to refine copper, the examples of machines and process that I have given begin in the early eighteenth century. This period is often called the 'Age of Enlightenment' or the 'Age of Reason' and it marked a significant change in thinking of the world and how we can manipulate it. The modern relationship between science and technology has not always existed. Yuval Harari (2011) points out that it was not until the Enlightenment that humans realised that there was much new knowledge to be discovered and exploited. He explains that this is where the relationship between science and technology developed apace. Discovering new knowledge and then exploiting it, however, is expensive. It requires significant funding, and this is where Harari suggests a third party enters the relationship – capitalism. Those with funds to invest paid for scientific discovery and its exploitation, with the expectation that they would make a significant profit, some of which they could re-invest in finding and exploiting new knowledge. Hence there has been an alliance between science, technology and capitalism (through both private and government investment) that has enabled science and technology to develop a significant synergy that has led to the world we have today. In communist countries like the post-war Soviet Union and in modern China, the government investment and enabling of technological enterprise has lifted significant numbers of their citizens out of poverty. The principal point, however, is that technology is more than the application of fully understood scientific knowledge; a point acknowledged by the economist Nathan Rosenberg (1982: 143):

> It is knowledge of techniques, methods, and designs that work, and that work in certain ways and with certain consequences, even when we cannot explain exactly why. It is […] a form of knowledge which has generated a certain rate of economic progress for thousands of years. Indeed, if the human race had been confined to technologies that were understood in a scientific sense, it would have passed from the scene long ago.

Technologists today use a host of ideas and 'rules-of-thumb' that are helpful but not scientifically sound. Examples include the idea of a centrifugal force, heat flow (like a fluid) and the notion that a vacuum 'sucks'. For example, heat flow in science is often conceptualised using the kinetic theory of molecular motion. This is of limited value in technology where the idea of heat flow, related to conductivity or perhaps 'U values' and temperature difference, is usually much more useful in solving problems in everyday practical situations. In order to use a particular idea for practical action, it is sometimes the case that a full scientific explanation is unnecessary and too abstract to be useful:

> [Reconstruction of knowledge] involves creating or inventing new 'concepts' which are more appropriate than the scientific ones to the practical task being worked upon. … Science frequently advances by the simplification of complex real-life situations; its beams in elementary physics are perfectly rigid; its levers rarely bend; balls rolling down inclined planes are truly spherical and unhampered by air resistance and friction. Decontextualisation, the separation of general knowledge from particular experience, is one of its most successful strategies. Solving technological problems necessitates building back into the situation all the complications of 'real life', reversing the process of reductionism by recontextualising knowledge. What results may be applicable in a particular context or set of circumstances only.
>
> (Layton, 1993: 59)

In technology, if the knowledge is 'useful' then it continues to be exploited until it is no longer of use. In science, a concept that is not 'correct', in that it does not match experimental results or other related theory, is discarded. However, the rejection of certain established scientific ideas such as phlogiston and the caloric theory of heat, and acceptance of energy as quanta took many years!

It is obviously true, however, that many new technologies have arisen from scientific discoveries. Microelectronics is founded on the 'blue skies' fundamental science of semiconductors and similar fundamental research has led to:

- improved knowledge of the intrinsic properties of materials such as lightweight alloys, plastic, carbon fibres, fullerene and graphene;
- to the development of new types of superconductor, the laser and another electronic devices;
- high yielding, disease-resistant crops through an improved understanding of the scientific basis of genetics.

There is a link between scientific discoveries and new or improved technologies and, in turn, technology *can* stimulate new directions for science too. Space research is an example of this. Technological developments, for example rockets that can launch space telescopes, explorer satellites and Mars landers – extraordinary technological achievements in their own right – can promote new challenges for science by revealing new and unexpected features of the universe. In 2019 the first picture of a black hole was a combination of the technology of world-wide connected radio telescopes linked to very sophisticated computer programming.

Common ground between science and technology learning

As we have seen, science does not need to precede technology, but technology can be stimulated by the findings of science. Indeed, the above extracts from government science and technology committees around the world illustrate that in response to today's economic demands there are policy pressures to structure scientific research with the specific purpose of stimulating technology, so creating new products or existing ones more efficiently. Of course, the 'laws of nature' as formulated by science set particular constraints within which all technological activity has to take place. For example, the Second Law of Thermodynamics suggests that the building of a perpetual motion machine is futile despite inventors' persistent efforts to 'break' the law! Other constraints may be economic, affected by cultural influences, resource availability and so on. Furthermore, scientific discoveries can suggest new products such as lasers and nuclear magnetic resonance imaging in medicine. Conversely, as illustrated above, technology does make a contribution to science in several ways. Examples include providing the stimulus for science to explain why things work in the way they do. The contribution of technology is especially evident in the way scientific concepts are deployed in technological activities.

It is useful to make a distinction between concepts that are directly related to 'knowing how' (i.e. technological concepts as defined above) and concepts related to 'knowing why' (i.e. scientific concepts). It is very difficult to make hard and fast distinctions between these two types of concepts but consider the following examples. An electron is a concept, a fundamental atomic particle; science is able to describe its mass, charge and other properties. In these terms, the concept of an electron has no obvious practical application and is an example of a 'knowing why' concept. On the other hand, a light switch is a technological concept for it has been designed for the particular purpose of switching on and off a flow of electrons. It is a 'knowing how' concept.

To see how the concepts are deployed in teaching science and technology, take the concept of *insulation* (a technological concept), which has relevance to understanding *conduction* (a scientific concept) of electricity and of heat. In the context of a science lesson, a teacher might involve children in exploring *electrical conduction* through simple experiments. For example, by using an ohmmeter to compare the resistance of a variety of materials, or using a simple circuit and noting the effect on the brightness of a lamp when different materials are placed in series with the lamp. In a study of *heat conduction*, pupils might be encouraged to plot temperature/time graphs that compare the rate of cooling of a beaker of hot water wrapped around with different materials. Very often, such a science activity would be placed within an 'everyday' context (see Figure 1.1). The aim, in a scientific sense, is to find out the property of the material. This would lead on to the idea that if there is a lot of trapped air, then that material is a good insulator as it stops conduction (as gases are poor conductors). However, as Patricia Murphy (Murphy, 2007) notes, some pupils are distracted by this pseudo-realistic 'technological' context. As emphasised in Layton's quote that in science 'Decontextualisation, the separation of general knowledge from particular experience, is one of its most successful strategies' (Layton, 1993: 59) and indeed the important first step in this science lesson is to strip away the 'mountaineer jacket' context to just set up a comparison experiment between beakers lagged with different materials;

Imagine you are stranded on a mountainside in cold, dry, windy weather. You can choose a jacket made from one of the fabrics in front of you.

This is what you have to find out:

> Which fabric would keep you warmer?

You can use any of the things in front of you. Choose whatever you need to answer the question.

You can use:

a can instead of a person
put water inside to make it more life-like
make it a 'jacket' from the fabric
use a hairdryer to make an imitation on wind (without the heater switched on, of course!)

Make a clear record of your results so that other people can understand exactly why you have decided which fabric would be best.

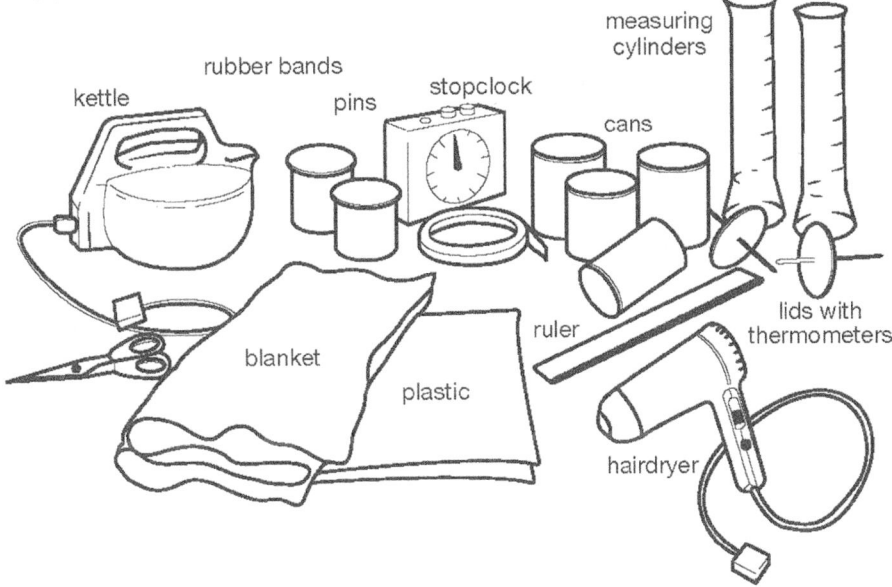

FIGURE 1.1 Investigating the 'best' material for a mountaineer's jacket

yet some pupils will wish to stick with the real-life problem presented and instead make a little 'jacket'. After all, that is what was asked for! – find out the material that would best make a mountaineer's jacket, not focussing on some abstract experimental method. Rather than making the science lesson 'real', the context has provided a serious distraction. The science lesson has different learning objectives from a textile technology lesson. Some pupils see that distinction and will 'play the school lesson game', but others do not.

This is an example of how knowledge that is important for a science lesson is not the totality of what is useful for a technology lesson. In technology, such an understanding of material properties would be an important factor to consider, but it would not be the *only* criterion. In addition, the pupils would need to consider non-scientific factors such as cost and availability, water resistance and toxicity, strength and flame-proofness, colour and density of the insulating materials that might be used. So, whereas scientific knowledge of heat conduction would contribute to the design process, a range of other factors could also influence the choice of insulating material. Further, suppose scientific experiments in a country with few 'advanced' material resources show that the stripped and powdered bark of a local tree, or the cotton-like seed heads of a local plant would make a suitable low-cost heat insulating material. Why then, should the technologists in that country use a hard-to-obtain and costly imported insulating material when the collection and preparation of this indigenous material also provides local employment? These wider considerations that are grounded in 'know-how' and the value systems of the people using the technology are an important aspect of technological design activities. We will be considering design in more detail in Chapter 4.

In summary, science often has a contribution to make to enhancing design & technology projects. However, teachers need to be clear about what that contribution may be, and plan to teach it to pupils. It is also important to realise that in designing and making, scientific understanding is but one contributory factor among many competing concerns. Although scientific ideas *can* enhance projects, it is possible, in fact usual, for a pupil to conduct complex technological activity without first exploring and understanding all aspects of the science involved.

The contribution of M in STEM

In this chapter on 'What is STEM', I have so far focused on the way each of the STE subjects interact one with another. As we have discussed already, this is particularly important as so many people talk about 'science-and-technology' as if it was one area of knowledge, or at least technology and engineering as always a user of science – the 'appliance of science' view. But what about the 'appliance of mathematics'?

In my early secondary school years, I studied 'new math', which included such topics as the use of numbers systems other than ten, probability and statistics, set theory and manipulation of matrices. In many ways, it was learning mathematics for the love of mathematics. It was part of the developments in the learning of mathematics that was gaining ground around the 'Anglo-Saxon' world, particularly the USA (see Table 1.1). It was argued that in traditional mathematics, numbers and equations were difficult and many lacked the necessary curiosity that is at the heart of maths, so they were replaced with what was more up to date and that all could understand. So, what was 'new' – and could you learn the 'new' without the basics of the 'old'? When challenged about why a topic such as set theory was important for all children to understand, the argument was that it taught logic and, for example, Venn diagrams were the graphic representation of such logical thought. But some of the 'old maths' such as geometry was excluded in the new curriculum and with it was rejected the logical build-up from simple axioms that had served young mathematicians for thousands of years.

The criticism of the 'new math' approach was mainly twofold. First, it was suggested that there was little coherence in the topics that were covered – it was a little bit of this and a little bit of that – which led to a certain learning of the abstract ideas by rote rather than gaining a full grasp of the underlying principles. Second, teachers have themselves to understand it and some of the new maths was just that – new – and many teachers were working at the limits of their own understanding. As is usual in many aspects of curriculum reform, there is a pendulum effect and the swing away from the old to the new has to happen for the curriculum to settle to a more middle line. However, we must first recognise that mathematics is a wonderful domain of learning in its own right and 'new math' brought that to prominence in school mathematics.

It hardly needs saying that an elegant solution to a problem or the construction of curves and geometrical shapes are things of beauty, both metaphorical and physical. But as we shall explore much more in later chapters, and in particular Chapter 5, using mathematics enhances understanding in science and facilitates designing and making in technology and engineering.

I confess that at the time I was learning 'new math', I wondered 'what is the point of all this?' and was much happier when I studied applied mathematics in the upper school. However, the 'new math' topics eventually did prove very useful. For example, in programming computers I have used both binary and hexadecimal numbering systems, I have applied statistics to explain molecular movement in gases, and I used matrices to help understand (some of!) Dirac's formulation of quantum mechanics. What seemed remote and abstract when I learnt it at the age of 14 was later practically useful.

This balance between an appreciation of the exploration of mathematics as a subject of intrinsic value and its usefulness as a tool to tackle problems and represent data in science, technology and engineering (and indeed across all subject domains) is central to any consideration of the learning of the subject. There are many examples where mathematics serves, in a utilitarian way, STE subjects and here are a few:

- The mathematics of error-correcting codes is applied to CD players, ATM machines, and cleaning up pictures from radio telescopes and from space probes such as *Curiosity* and *Voyager*.
- Statistics is essential in medicine for analysing data on epidemiology and on the safety of new drugs.
- Maths and logic are at the heart of computer software design.
- The physical sciences in particular (chemistry, physics, oceanography, astronomy) require mathematics for the development of their theories.
- In biological and ecological systems, mathematics is used when studying the laws of population change, for example what might happen if badgers are culled in an attempt reduce bovine tuberculosis.

At the school level, mathematics related to laws of motion can be algebraic or graphical. The use of performance characteristic graphs in technology can be used to make decisions – which electric motor and what batteries are best for my purpose?

What is the strength and stiffness of the component parts of an artifact? Simple calculations about whether 'will it break', or 'how much will it bend/stretch if it is 'only this thick' will provide invaluable information. Is not a set of common principles regarding introducing equations and graphs across the school STEM subjects, linked to these clear practical uses inside and outside school more likely to lead to a positive attitude to mathematics?

What else do the STEM subjects contribute?

Affective knowledge and values

The STEM subjects cannot be divorced from other dimensions of human thinking and behaviour since the beliefs and values of individuals and communities are influenced by, and exert pressure on, both science and technology themselves. In technological activities it is just as important to involve pupils in making value judgements about the *human*, or rather *humane*, dimensions of technology as it is to focus solely on technical details about the functioning of the technological product. In science, experiments involving animals and humans have ethical dimensions that are paramount. Given that the *purpose* of technology is to respond to certain sorts of need, pupils should be expected to find answers to questions such as:

- Whose needs are to be met?
- Who has identified the needs?
- Are proposals for a particular technological development acceptable to the individuals and communities who are to use or be influenced by the development?

In science, despite an assumption that it leads to answers that are 'right or wrong', there is also a strong *values* dimension, especially in the design of experiments that affect living creatures or have an impact on the environment.

- How should a particular experiment be constructed – and what does it tell us about 'the nature of science'?
- What is the impact of the findings? How might they affect people, animals and the environment generally?
- How are the scientific ideas communicated to others?

And, in mathematics, how statistics are gathered manipulated and displayed have a moral dimension too. The 'lies, damn lies and statistics' epithet has a grain of truth when information in newspapers or online talks about percentage falls and increases without a clear reference to the value of the base figure; or when graphs are shown not starting from zero or with misleading scale divisions. The very nature of mathematics as a subject domain of clarity and truth with 'just one answer' means one should be on guard to how data is analysed and presented.

Decisions about various scientific and technological *processes* are affected by a range of criteria, each of which depends on different kinds of values. For example, materials used may be in short supply or come from environmentally sensitive regions of the globe or be extracted using child labour; new construction projects may disturb or destroy wildlife and so on. Evaluation of the *products* of technological activities is subject to decisions about fitness for purpose, the calculation of cost effectiveness and possible health hazards. People's values affect every stage of the technological process from decisions taken about whether to embark on a particular innovation, through the process of development, to the acceptability of the subsequent product. The clarification of values is a responsibility of all engaged in scientific and technological activities and it has a central role to play in the affective dimension of a pupil's education.

The different *social* meanings attached to science and technology is nowhere more evident than in the use of the terms like *big science, high tech* and *intermediate technology*. The former is used to describe large-scale, capital-intensive projects such as atom-smashing machines or manufacturing technologies such as for microelectronics that use a highly skilled workforce; and the latter is used to describe small-scale, labour-intensive technologies advocated for small communities that capitalise on local skills and resources that are at the community's disposal. It is, of course, quite possible that relatively high-technology electronics may be *appropriate* in small communities (e.g. those in remote areas) but this leads to issues about the control of technology and economic power. It is these influences that make the projects and applications considered in science and in design & technology so rich in educational terms. The interpretation of what is needed, how it is to be done, how outcomes are measured, calculated and analysed, who is to benefit, and who might be disadvantaged – who wins, who loses – should be made explicit and debated in order to question the value judgements that underlie any assumptions about a course of action.

Increasingly, the values pupils hold in relation to climate change and the deterio-rating state of the planet that they will inherit are generally out of step with those of many adults including politicians and teachers. In 2019, the then 16-year-old Swedish student Greta Thunberg became an international celebrity: speaking to the United Nations, leading anti-climate-change rallies around the world and stimulating through social media world-wide protests by young people furious at the slow progress by politicians in addressing climate change (see Thunberg, 2019b). That technology is both the problem and a means towards a possible solution can be tackled head on. The largest contributor to global warming is carbon dioxide (CO_2), emitted when fossil fuels are burnt, but there are also other 'greenhouse gases' such as methane (CH_4) resulting from industrial processes and agriculture.

Attitudes and values raised by considering the social impact of STEM activities are vitally important to all, and of profound interest to young people. What needs to be done and what can be addressed in school? There are ways to reduce greenhouse gas emissions globally as for each country:

- By reducing its own emissions, a country is supporting wider international efforts.
- In a future world where greenhouse gases are restricted, the cost of emitting those gases (i.e. carbon price) will be high. Early action to reduce emissions – here and elsewhere – can help reduce future costs.
- Investment in and development of low-carbon technologies will contribute to a new and expanding global markets.

Examples of using STEM to address global concerns will be considered in later chapters.

Developing problem solving

As you can see from the STEM milestones (see Table 1.1), problem solving is a key activity in all STEM subjects and we will consider this in more detail in Chapter 6. 'Doing problems' is what many think of when they recall mathematics lessons.

However, a worksheet that requires a pupil to practise a particular algorithm repeatedly is better described as mathematical exercises. Francisco and Maher (2005: 362) put is as follows:

> Our perspective of problem solving recognises the power of children's construction of their own personal knowledge under research conditions that emphasise minimal interventions in the students' mathematical activity and an invitation to students to explore patterns, make conjectures, test hypotheses, reflect on extensions and applications of learnt concepts, explain, and justify their reasoning and work collaboratively. Such a view regards mathematical learning and reasoning as integral parts of the process of problem solving.

Scott Chamberlin (2008) worked with a number of mathematics educators to set out what were the processes that children engaged in when problem solving in mathematics. The following were agreed as being present in true maths problem solving. Pupil:

- engage in cognition (they learn from the process);
- seek a solution to a mathematical situation for which they have no immediately accessible/obvious process or method;
- communicate ideas to peers;
- engage in iterative cycles;
- create a written record of their thinking;
- 'mathematise' a situation to solve it (it requires more than just common sense);
- create assumptions and consider those assumptions in relation to the final solution;
- revise current knowledge to solve a problem;
- create new techniques to solve a problem;
- create mathematical models;
- define a mathematical goal or situation;
- seek a goal.

And the characteristics of a mathematics problem were agreed as:

- have realistic contexts;
- can be solved with more than one tool;
- can be solved with more than one approach;
- can be used to assess level of understanding;
- require the implementation of multiple algorithms for a successful solution;
- DO NOT lend themselves to automatic responses;
- promote flexibility in thinking;
- may be purely contrived mathematical problems;
- can be puzzles;
- can be games of logic.

(Chamberlin, 2008)

The scientific process has been a key part of the science curriculum, too, for many years and the UK government policy document *Science 5–13*, which pre-dated any prescribed curriculum, did not merely define what should be taught in terms of *content* such as electricity or plants, rather it emphasised the importance of a *process* approach. Science curriculum innovation in the middle to late 1980s saw a large number of new courses such as 'Warwick Process Science' and 'Science in Process' for secondary schools, which focused not on science concepts but rather on science methodology such as observation, interpretation and classification. This mood was picked up in the developing primary science curriculum, too. In the 1980s, the teaching profession generally welcomed a move away from what was considered as often merely the memorising of poorly understood facts, to a curriculum that might be more accessible to *all* pupils and which emphasised problem-solving approaches and skills applicable to other areas of life both in and outside school. The attention to 'doing' science – raising questions that could be answered by an investigation – became the cornerstone of the developing investigation driven, problem-solving approach especially for primary science. For example, the question 'What is the best carrier bag?' would be turned into an investigable question such as 'Which carrier bag carries the greatest weight?' To answer such a question, so-called 'dependent and independent' variables were identified. At the time, primary teachers were very concerned about the introduction of science into their day to day work, and the rhetoric from those advocating that science should indeed be part of the primary curriculum was that the teachers could 'learn with the pupils' as only the *process* was important, not the science facts or concepts that the teacher knew or did not know. Now, those intending to become primary teachers in all nations of the United Kingdom are required to hold a basic qualification in science as well as maths and English as a pre-requisite for their teacher-training course.

Process was all important and science content relegated to a side issue. In an almost content-free science curriculum 'good' pedagogy was that which promoted a questioning attitude amongst pupils and the means of answering such questions. What was important was knowing how to conduct practical work and in particular 'fair tests' to find things out. The doing of the practical work was the most important not the 'right' answer as such. The process is more important than the answer.

As was the case with 'new math', in time the science pendulum swing away from content to process came back into a more central balanced position. Murphy and Scanlon (1994: 105) summarised it as follows:

> there emerged a consensus that scientific inquiry was not about following a set of rules or a hierarchy of processes but 'the practice of a craft – in deciding what to observe, in selecting which observations to pay attention to, in interpreting and discussing inferences and in drawing conclusions from and in drawing conclusions from experimental data' (from Millar, in Woolnough and Toh, 1990). There was also considerable agreement evident in the various published discussions about the nature of scientific observation.

The 1980s not only saw the introduction of primary science but a new emphasis in the initial and in-service education of teachers of a view of learning that recognised that pupils construct meaning by interacting with the environment around them. Rather than being 'empty vessels' into which new knowledge and understanding could be poured, teachers came to recognise that, for a fuller understanding, pupils themselves had to make sense of the world around them by seeing how their new experiences, along with the views of others, matched their own preconceived ideas and notions. Teachers failed to take sufficient notice of what was involved when pupils attempted to construct new understandings and integrate these with their existing knowledge of the world. Ros Driver (1983: ii) pointed out some problems with 'discovery' pedagogy for science:

> Discovery methods in science teaching put pupils in the role of investigator, giving them the opportunities to perform experiments and test ideas for themselves. What actually happens in classrooms when this approach is used? Although, of course, pupils' ideas are less sophisticated than those of practising scientists, some interesting parallels can be drawn. The work of Thomas Kuhn indicates that, once a scientific theory or paradigm becomes established, scientists as a community are slow to change their thinking. Pupils, like scientists, view the world through the spectacles of their own preconceptions, and many have difficulty in making the journey from their own intuitions to the ideas presented in science lessons.

A focus on the investigative problem-solving process rather than content might have been considered 'good practice' as suggested above, but questions for investigation have to link to some real content when they are answered. A primary science question such as 'Can you make your plant grow sideways?' or 'What happens if you pinch the leaves off a young growing plant?' might be more concerned with the practical activity itself but they lead, for that particular group of pupils, to some understanding of tropism in plants. Before there was a national curriculum, secondary school teachers could not easily cope with the variety of primary school science experiences and so chose to ignore that the pupils had received any scientific experience at all in the primary school. Science teachers at the secondary school would 'start again'. Alternatively, secondary teachers would complain that primary teachers had stolen the 'best bits' of the theatre of lower secondary science such as the 'collapsing can' demonstration of air pressure, so from their point of view spoiling some of the excitement and spectacle of lower secondary science lessons. Some 35 years after the publication of *Science 5–13*, in-service work with secondary teachers still tries to tackle the lack of progress by pupils in the first few years of secondary (high) school caused by a failure to fully recognise the now quite extensive and structured science understanding gained by pupils in the primary (elementary) school.

Indeed, discussions about science and technology as vehicles for the teaching of problem-solving sometimes become emotionally charged. Over the years, those proposing different technology curricula, the emphasis of processes in science and the dominance of STEM in general have used this argument as a principal way of advocating that STEM should have an enhanced status in the school curriculum because a general ability to solve problems is central to satisfying human needs.

For some time, researchers such as Perkins and Salomon (1992) and McCormick (2006) have pointed out that learning is heavily influenced by the *context* in which it occurs. Pupils do not easily transfer their ability in a particular activity from one learning 'domain' to another. Technology teachers have assumed that if pupils are taught to investigate the factors influencing the design decisions for making one product, for example a moisture sensor, then they will be able to transfer those techniques to consider the different design decisions for, say, developing a food product that meets certain dietary requirements. The evidence is that pupils do not easily transfer their understanding across these different contexts and require considerable support from their teacher to help them do so. Also, pupils may know what they want to do but not be able to realise their solution because they do not have the required knowledge or skills. More critically, when planning their work pupils may not consider certain approaches to a problem because they are ignorant of the existence of specific equipment or a particular technique which might help them. For these pupils 'problem solving' is doing little more than applying their common sense.

There is a close association in a particular context between the conceptual knowledge associated with the particular problem and an understanding of what action needs to be done to tackle that problem (procedural knowledge). People think within the context in which they find themselves – 'situated-cognition' – and when pupils are presented with problems in unfamiliar contexts they tend to use 'common sense' intuitive understanding as opposed to science concepts to tackle them.

So, what is the best approach in technology and engineering? Should pupils learn knowledge and skills in isolation that might prove useful later but for which they perceive little immediate value? Should pupils learn skills 'as needed' within projects when they appreciate the usefulness of what they are learning but without a coherent structure and without realising that there *was* something new that they should know, to transfer to future work? The best approach is probably to steer a middle line. A carefully planned selection of shorter projects or 'focused tasks' emphasises particular, skills and techniques, together with the longer, more open task or 'project' which allows pupils to develop their capability by drawing on their accumulated experiences.

Systems thinking: Black boxes

As has been emphasised a number of times, it is necessary for teachers to think carefully about the purpose of teaching a particular scientific concept for use in technology lessons and this will be considered in detail in Chapters 3 and 4. 'Systems thinking' is important in both biology and technology. Some examples of organs working together to perform a certain task include the digestive system, blood circulation system and the nervous system. Such human systems are, of course, present in other animals but in all cases they can be considered as a functional block that does a job – but with component parts. For the blood example, components are the heart, blood and blood vessels; for the nervous system the brain, spinal cord and peripheral nerves. The approach to first aid is also systemic as is triage, the process of determining the priority of patients' treatments based on the severity of their condition, dealing with breathing and bleeding problems before taking action on broken bones. The design and use of systems is an example of the value of using 'know-how' rather than 'know-why' and these can be extended to activities in the classroom.

Let's take the example of electronic systems in technology. In technological activities pupils are expected to have a clear idea of what they want the electronics systems to do; it is a goal-oriented approach. Rather than focusing on any scientific understanding of the way in which the devices and circuits work, the emphasis is on the *functional* aspects of the electronic devices and circuits that the pupils are to use – considering each unit as a functional block. Pupils should be expected to ask questions such as:

■   What do I want my electronics system to do?

■   What operating conditions, e.g. power supply requirements, does it need to work?

■   Will the device stand up to rigours of use in its intended environment?

■   How much will it cost to make and run?

■   What characteristics of this device are better for this design than other similar devices?

■   Will it be safe and easy to use?

■   Can the components needed be obtained easily?

■   Will it be acceptable, culturally and economically, to the people in the community in which it is to be used?

To a technologist, meeting these functional and contextual criteria is as important, if not more important a consideration as knowing why the electronic devices used work in the way they do. The emphasis on *function* and *context* rather than *theory* and *fundamentals* may be misleading, seeming to lack opportunities for rigorous thought. However, the design and assembly of circuits and systems for specific purposes requires knowledge and understanding at the operational level. These operating precepts are just as demanding intellectually as the operating aetiology used by science to explain concepts such as electrical conductivity and potential. An example or two will make these points clearer.

An *electronics system* can be represented by three linked building blocks.

It is an assembly of functional electronic *building blocks* that are connected together to achieve a *particular purpose*, e.g. sounding an alarm when smoke is in the air. Examples of *input* building blocks include switches, e.g. mechanical and semiconductor types, microphones and light-dependent resistors. *Processor* building blocks include amplifiers, comparators, oscillators and counters. *Output* building blocks include light-emitting diodes, seven segment displays and loudspeakers and meters. Thus, the *input* building block of a smoke detector would be a smoke sensor. Its *processor* building block might comprise a comparator to switch on an audio frequency oscillator when the smoke level detected by the sensor has reached a pre-set danger point followed, perhaps, by an amplifier. The detector's *output* building block would be a small loudspeaker or piezoelectric device to generate an audio frequency sound when signals are received from the oscillator. Pupils quickly learn to associate a circuit board with a particular 'job'. For example, a 14-year-old pupil would easily solve the problem of making a 'rain alarm' by linking a moisture detector (*input*) to a buzzer (*output*) by using a transistor switch (*process*).

Such 'black boxes' can also be used to make more complex devices, and explain more complex systems, too, such as the biological examples considered above. Design decisions are based on how the product is to be used and pupils are constrained by their specification criteria, not by a lack of understanding of why the circuit functions. A *detailed* knowledge at the component level is unnecessary. Let's assume that a pupil is aiming to design and make an anti-theft warning device to clip onto a bicycle and provide an ear-piercing sound if the bicycle is about to be stolen, i.e. it is a portable device to be used by an individual. First and foremost, there needs to be a clear specification of what the system is to do. Second, a consideration of the environment it is to be used in, not just the physical environment (e.g. wet, dusty, hot, cold or dry) but the human environment, too:

- who is to use it;
- what is it to look like – its shape, colour, size and so on;
- how it is to be used, e.g. whether fixed to the wheels, handlebars or forks;
- how much it is to cost to make and to sell;
- whether the user needs to have any technical skills to use it.

Only after these criteria are established through appropriate research is it possible for the pupil to select the functional building blocks that will enable a prototype system to be made which meets the criteria. There are several concepts that arise in this analysis of need. For example, in terms of *energy* there is a consideration of the power supply requirements. In terms of the *process*, a pupil will need to consider how the device can control the sound long and loud enough to alert attention. Is it to have an automatic cut-out? What is to be the operating principle of the sensor that first detects the movement of the bicycle?

In terms of *materials*, cost, ruggedness, waterproofness and design of the casing for the unit and similar considerations for the components need to be tackled.

When it comes to the manufacture of the anti-theft bicycle alarm, however, the technical factors to be considered are more than simply selecting appropriate input, process and output devices, plugging them together and expecting the system to work. What is most often missed in designing electronic systems is the need to consider the requirements that enable each building block to respond to the signal it receives and send an appropriate signal to the building block that follows it. The concept being highlighted here is called *matching*. This is more complex, but at a basic level, pupils are able to use computer software that takes matching into account and will give the design for a printed circuit board combining the contributory functional blocks. So, considering electronic devices in terms of input, process and output blocks can simplify the learning of electronics. A technologist does not need to know about the detailed working of an integrated circuit, or even a transistor, in terms of the physics involved, just how to *use* it in a range of circumstances.

Before leaving systems thinking in technology, it is also worth thinking about using systems in a wider context and this often involves a consideration about where to put the 'system boundary'. A few years ago, I was talking about an examination entry by a 16-year-old pupil with his teacher. The pupil had designed and made a 'panic alarm' in case he was attacked late at night. In a technical sense it was very well done indeed with proper consideration of the alarm's weight, power supply, loudness, ease of action and so on. If anyone had attacked that boy, everyone would have heard about it! I asked his teacher, who was very proud about what his pupil had achieved, whether the pupil had considered the issue of *why* such an alarm was needed in his neighbourhood. The teacher looked puzzled by the question as he obviously thought it irrelevant; why such a panic alarm was needed (in terms of the wider values exhibited by those in the pupil's locality) was not part of the examination marking scheme and so not important as it did not 'gain marks'. However, I wondered if by drawing the system boundary narrowly around the alarm itself this was the best solution to the problem he faced. By not considering *why* he was afraid at night due to few late-night buses or limited and poor street lighting, his solution was, in some senses, restricted. Maybe the 16-year-old could not do much himself about the wider context of supplying maybe free buses or better street-lighting. However, the narrow system boundary around the well-crafted and technically sound panic alarm provided only a partial solution to the youth's problem, nor a consideration of alternative approaches to crime prevention. A wider system view could consider not just burglar and panic alarms looking at the result of crime, but also the engage with the possibility of changing the behaviour of the thieves – soft system thinking as well as hard system thinking (see Hallstrom & Klasander, 2020).

Why should all pupils learn STEM?

You might think this is a question that I should have asked at the start of this chapter rather than the end. However, in illustrating the links between the subjects of STEM, I hope you might have already formed your own answer(s). One answer can readily be found by returning to the government statements from the UK, USA and Australia, which emphasise the links between STEM learning and economic growth. After the Second World War, Roosevelt was clear that STEM success was important to stimulate the economy and provide jobs for the returning troops. Similarly, following the 2008 economic crash that just preceded his first term in office, President Barack Obama eventually looked to STEM education as a route out of the prolonged depression. In February 2013 gave his 'State of the Union' speech and said:

> Tonight, I'm announcing a new challenge, to redesign America's high schools so they better equip graduates for the demands of a high-tech economy. And we'll reward schools that develop new partnerships with colleges and employers, and create classes that focus on science, technology, engineering and math, the skills today's employers are looking for to fill the jobs that are there right now and will be there in the future.

Although I accept that STEM is useful in many work-related contexts, I think that more important is STEM learning for *all* pupils as its outcomes are based on the use of *evidence*. All citizens need the processes, ideas and thinking tools of STEM to be informed citizens. I asked Mike Watts of Brunel University London why he thought all pupils should learn STEM and he gave this cogent response:

John Falk (2005) has calculated that people spend on average just 15 per cent of their 'learning lives' in formal educational systems (schools, colleges, universities); the major portion (85%) of their learning lives is spent outside of formal institutions.

Let me begin at the formal beginning. Early years and pre-school education in the England is governed by something called the Early Years Foundation Stage (EYFS, Department for Education, 2012). One question here is: a foundation for what, exactly? Well, in general, it is a foundation for further schooling as soon as the child reaches age four-plus or five. But it is also a foundation for learning in general, as in learning-for-life. So, should there be STEM in this early-years foundation curriculum? It is there, of course, in the form of early number work, the M of STEM. Moreover, as we point out in Silby and Watts (2018) and Bilton and Watts (2019), there is outdoor play in natural settings, block play, fabric boxes, cooking utensils, Lego and Meccano construction kits, mud kitchens, nature tables, water play and sand pits – to list just a few of the exploratory provisions that good pre-school specialists provide. If I include visits to local parks, city farms, riverside walks, study centres and the like, I can make a reasonable case that these also cater for some of the S, T and E in early STEM – even if nursery teachers wouldn't recognise this as such.

I could make a similar case for what happens in primary and secondary schools, and for colleges and universities. But, back to John Falk, such formal education covers but a small fraction of life. There has been considerable attention paid to entities such as 'scientific literacy', 'science-for-life' and 'citizen science' and the enormous importance of science and technology beyond the school years. The intention behind these ideas is that everyday people in everyday life should know and understand some aspects of science for a variety of good reasons – not least of which is that possessing some working knowledge of science is far more welcome than is wholesale ignorance. Arguably, a foundation curriculum in early years is part of the foundation for scientific literacy in later life. Less attention, though, has been given to, say, foundations for STEM literacy, STEM-for-life or to what I might call 'Citizen STEM'.

So, if Citizen STEM operates in the 85% – that measure of the iceberg that lies deep beneath the surface – then just what is it made of? How does it happen? There are two broad answers. First, taking each initial in term, it is possible to argue for how and when people use the 'M' numbers and calculations, for example, and how they acquire and develop such skills and competencies in their employment, their leisure activities, their sporting lives, travel arrangements, domestic contexts and so on. However, second, the emphasis of STEM education is less a juxtaposition of the four separate initials, each one treated as a discrete component, as an integrative approach that draws on the capabilities and 'affordances' of each – of the four in combination. Such integrative STEM education – both formal and informal – tends to encompass real-world problems that include social, economic and environmental issues alongside many others. STEM education has a strong hands-on group-minded team-based problem-solving flavour, which generates multiple possible answers and reframes moments of failure as a necessary part of learning.

So, from where does Citizen STEM derive his or her informal STEM education? One answer is in the early years nursery 'writ large': in decorating the house, DIY, landscaping the garden, exploring the beach, trekking the woodlands, working out in the gym, delving into the sewing box, foraging the fridge and larder, navigating computer apps and software. There's YouTube and 'How stuff works'. These are the Wikipedia for finding out how things happen and how to fix them when they fail. I'm sure I join an enormous audience of doers and fixers in YouTube-ing (if there is such a verb) at the point when I need precise STEM information, illustration, advice and guidance.

I live in an area of the UK that is earmarked for the development of fracking. 'Fracking' is shorthand for hydraulic fracturing and refers to ways that rock is fractured apart to release oil and gas inside. The practice entails drilling down or, more commonly, horizontally into rock layers before a high-pressure 'slickwater' – a mix of water, sand and chemicals – is injected into the rock. The process allows drilling firms to access otherwise difficult-to-reach resources, creating new pathways to force oil and gas out through the head of the well. But what does this have to do with STEM education? For anyone living in my part of the UK – or anywhere in the world – where fracking is envisaged, then acceptance of – or resistance to – the development of fracking requires considerable informal STEM learning. Citizen STEM (either 'pro' or 'anti' fracking) needs to develop an understanding, gain an appreciable grasp of chemistry, earth science, seismology, engineering, scale and magnitude, economic impact, environmental footprints, sustainability, social upheaval – and much more – in order to form a clear and reasoned opinion.

What, then, might being Citizen STEM entail? Well, a sense of relevance and purposive curiosity within everyday experiences are powerfully effective STEM-engagement tools (Watts, 2015). Grandpa is having a hip-replacement operation – what is entailed in that exactly? Aunt Edith has suffered a detached retina – why would the eye surgeon use a laser to fix it? Brother David sleep-walks at night. What can we do to keep him safe? When Dad is driving, he always turns down the car radio the moment he gets lost. Is there a way we could do that automatically every time he's stressed? Mum bought some sea salt in Anglesey, called Halen Mon, and swears Jamie Oliver says it's more healthy than supermarket salt. But salt is just salt, isn't it? Sister Emma wants to eat gluten-free. What on earth is a gluten and what does it do to us? Why does some music give me goosebumps and how exactly can I get more music to do that? I want to design a bendy mobile phone, I keep dropping and breaking mine. None of these 'personal interest problems' would be found on the school STEM curriculum, they belong outside the system in the 85%. The question here, though, is how to teach for Citizen STEM? What is the *School* (and university) Years 'Foundation Stage' (the SYFS and UYFS) that would prepare and enable life-long learning in STEM beyond school?

To begin, STEMers need a sense of visualisation, to grasp the size of the problem, what 'problem quantum' is manageable, the 'bite-size' of that is do-able. They need strong purpose and perseverance to set off and to see projects through – there is little to be gained from failing wholly disheartened at the first fence. They need clear purposive curiosity sustained over long periods: curiosity as a way of life. They need a mischievous urge to fiddle, break things to fix things, to meddle, model and make. They need access to resources, data, information, ideas, perspectives, opinions, supportive advice, direction. They need the skills of 'critical synthesis' in order to bring together, make sense of, to 'see through' the plethora of information (physical, verbal, aural, visual) they gather. They need access to tools and materials, not necessarily their own but as part of networks of like-minded makers and doers. They need good interpersonal and team-building skills to draw on the wit and wisdom of others, as they collaborate with colleagues to solve problems. They need emotion management to cater for the affective ups-and-downs in working on a project.

As Steve Alsop (2005: 48) says:

> Some emotions such as joy, excitement, interest, enthusiasm, curiosity and hope can act to enhance cognition, while others (such as fear, anxiety, boredom) might serve to deaden curiosity and insight.

As Mike Watts argues, every citizen needs tools from science, mathematics and technology to address everyday issues such as, for example, the grave reality of global warming especially in the face of those who would deny it for political or commercial advantage; the community benefit and safety of vaccines; the interpretation of statistics for government spending on public services; or the social and environmental impact of cheap fashion clothing. Values are important in creating appropriate solutions to technological problems and it is interesting that, in 2019, New Zealand government ministers were first tasked with finding collaborative ways to meet five wellbeing goals. They are to support mental health, particularly among young people; reduce child poverty; increase support for Maori and Pacific Islander peoples; transform the economy for a low-carbon future; and boost productivity and digital innovation. But in this and in every society, consideration of the impact on individuals and the environment is dependent on assessing the available evidence. In those societies where rhetoric and political expediency trumps critical engagement, an appreciation of the methods and concepts learned in STEM provides a means for all citizens to evaluate what is best for themselves, their families and the wider society and enables them to engage critically with their government's policies.

Concluding remarks

STEM subjects have a number of common threads such as problem solving, discovery approaches and direct applicability to questions affecting everyday life. As we will see in later chapters, teachers can benefit their pupils if they 'look sideways' to take advantage of teaching and learning across related STEM subjects. But there are clear differences too. If technology is merely seen as 'applied science', then technology educators miss the point about the subject for which they are responsible. Technology is founded on the addressing of human needs, science in understanding the whys and wherefores of the world around us and Mathematics is a service to both and an exciting and intriguing aspect of human endeavour in its own right. Put very simply, the 'know-why' of science is a fundamentally different goal from the 'know-how' of technology. Science and mathematics knowledge and understanding will often contribute to project work in schools, but it is necessary to keep in mind the sometimes-limited extent of such knowledge which is actually required. The contribution of science needs to be set against the other dominant factors such as *sustainability*, *aesthetics* and *appropriateness*.

But as Plant (1994: 29) notes:

> it is also important to recognise the different STEM subjects have a part to play in stimulating technological activities. First, by revealing new frontiers to spur technological inventiveness. Second, by using the vocabulary of science for providing convincing explanations of the behaviour of technological devices. Third, in the provision of tools to develop convincing explanations of the behaviour of technological devices. Lastly, in the provision of resources for the constraints on technological processes.

History has shown that no one contributory subject is more important than any other in STEM; sometimes science follows technology and mathematics is often the key to help improve our understanding of both. The outcomes of STEM subjects are steeped in the culture and social values of the society which uses them and provide evidence to help in decision making. Indeed, the response of governments across the world to the 2020 Covid-19 pandemic provided starkly the need to 'follow the science' in order to save lives. It is this evidence outcomes of science and mathematics along with the value-laden aspects of teaching and learning across all STEM subjects that highlight the distinctive role for engineering and technology in enhancing human life and behaviour.

Recommended reading

The birth of STEM and technology before science? what does history tell us?

Bronowski, J. (1973) *The ascent of man.* London: BBC Publications.
Bunch, B. & Hellemans, A. (2004) *The history of science and technology.* New York: Scientific Publishing Inc.

The difference between science knowledge and technology knowledge

de Vries, M. (2019) Technology education in the context of STEM Education. In A. Koch, S. Kruse, & P. Labudde (eds), *Zur Bedeutung der Technischen Bildung in Fächerverbünden.* Wiesbaden: Springer Spektrum.

The contribution of M in STEM

Eastaway, R. (2019) *Maths on the back of an envelope: Clever ways to (roughly) calculate anything.* London: Harper Collins.

What else do the STEM subjects contribute? affective knowledge and personal values, problem solving

Barak, M. (2007) Problem-solving in technology education: The role of strategies, schemes & heuristics. In D. Barlex (ed.), *Design and technology for the next generation.* Whitchurch, UK: Cliffe & Company.
Lave, J. (1988) *Cognition in practice.* Cambridge, MA: Cambridge University Press.

Hennessy, S. & McCormick, R. (2002) The general problem-solving process in technology education. Myth or reality. In G. Owen-Jackson (ed.), *Teaching design and technology in secondary schools*. London: Routledge/Falmer.

Thunberg, G. (2019a) *No one is too small to make a difference*. London: Penguin.

References

Alsop, S. J. (Ed.) (2005) *Beyond Cartesian dualism: Encountering affect in the teaching and learning of science*. Dordrecht, Netherlands: Springer.

Bilton, H. & Watts, D. M. (2019) Early STEM education: Policies and perspectives. Editors of a special edition of *Early Child Development and Care*, 190(1), 1–2.

Chamberlin, S. A. (2008) What is problem solving in the mathematics classroom? *Philosophy of Mathematics Education Journal*, 23, 1–25.

Department for Education (2012) *The early years foundation stage*. www.education.gov.uk/publications (accessed 4 June 2020).

Driver, R. (1983) *The pupil as scientist?* Milton Keynes: Open University Press

Falk, J. H. (2005) Free-choice environmental learning: framing the discussion, *Environmental Education Research*, 11(3), 265–280.

Francisco, J. F. & Maher, C. A. (2005) Conditions for promoting reasoning in problem solving: Insights from a longitudinal study. *Journal of Mathematical Behavior*, 24, 361–372.

Hallstrom, J. & Klasander, C. (2020) Making the invisible visible: Pedagogies related to learning about technological systems. In J. Williams & D. Barlex (eds.), *Pedagogy for technology education in secondary schools – Research informed perspectives for classroom teachers*. Dordrecht, Netherlands: Springer.

Harari, Y. (2011) *Sapiens: A brief history of humankind*. London: Harvill Secker.

Layton, D. (1993) *Technology's challenge to science education*. Buckingham, UK: Open University Press.

McCormick, R. (2006) Technology and knowledge. In J. Dakers (ed.), *Defining technological literacy*. New York: Palgrave.

Murphy, P. (2007) Gender & pedagogy. In D. Barlex (ed.), *Design and technology for the next generation*. Whitchurch, UK: Cliffe & Company.

Murphy, P. & Scanlon, E. (1994) Perceptions of process and content in the science curriculum. In J. Bourne (ed.), *Thinking through primary practice*. London: Routledge.

Nostrôm, P. (2011) 'Engineers' non-scientific models in technology education. *Intentional Journal of Technology & Design Education* (on-line first). DOI 10.1007/s10798-011-9184-2

Perkins, D. N. & Salomon, G. (1992) Transfer of learning. *International encyclopaedia of education* (2nd ed.) Oxford, UK: Pergamon Press.

Plant, M. (1994) *How is science useful to technology? Course Booklet for E886*, Milton Keynes, UK: The Open University.

Rosenberg, N. (1982) *Inside the black box: Technology and economics*. Cambridge: Cambridge University Press.

Schumacher, E. E. (1973) *Small is beautiful*. New York: Harper and Row.

Silby, A. & Watts, D. M. (2018) *Early years science education: A contemporary look*. London: Routledge.

Thunberg, G. (2019b) *No one is too small to make a difference*. London: Penguin.

Watts, D. M. (2015) Public understanding of plant biology: Voices from the bottom of the garden. *International Journal of Science Education B*, 1–18, DOI: 10.1080/21548455.2015.1004380

Woolnough, B. E. & Toh, K. A. (1990) Alternative approaches to the assessment of practical work in science. *School Science Review*, 71(256), 127–131.

A curriculum for STEM

'Looking sideways'

Introduction

We nearly called this book *Looking Sideways* as we believe that pupil learning can be significantly enhanced if specialist teachers across the curriculum look sideways at what their colleagues are teaching and build on their pupils' experience in other classes. For many years, however, teachers have been working just in their own 'curriculum silo'. As long ago as 1996 Karen Zuga said 'Communities of technology and science educators have been passing as two ships pass silently in the night without speaking to each other about their relationships' (Zuga in Yager, 1996: 227). As we have already discussed in Chapter 1, most people would agree that there is a significant and symbiotic relationship between science, technology, engineering and mathematics but how is it possible for us to best exploit that relationship in our day-to-day teaching for the benefit of our pupils?

There was never a better time to consider new ways of constructing a relevant curriculum and the associated assessment regime. Around the world, particularly in developing countries of Africa and South Asia, it is recognised that building enough schools of the right quality for the vast numbers of young people is a necessary first step, but if the curriculum offered is considered irrelevant by them or their parents, then school attendance with be patchy and student drop-out will be huge. In resource-rich countries, schools are pressed for ever improved outcomes, and here there is a realisation that a tightly regulated academic curriculum at state or national level is also not serving the needs of all pupils; and an external over-prescription of what should be taught can stifle creativity in teaching. Those who desire to make teachers 'accountable' for their classroom work and to improve standards now believe that giving schools more freedom will encourage new ways of levering improvement rather than simply imposing greater and greater external control by government.

In Scandinavia, as in the UK, the detail of national curriculum documents has diminished over the years, but in England, although what is prescribed for schools maintained by local government is still very specific, over 70 per cent of state secondary (high) schools are now directly funded by central government and designated 'academies' or 'free schools'. Counter-intuitively, it is possible for these national

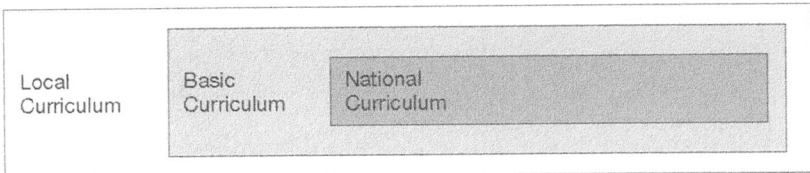

FIGURE 2.1 The school curriculum

government funded schools to create their own curriculum rather than adopting the national curriculum that is prescribed by national government for the local-government funded schools! Some people in the media are quite shocked by this new lack of external regulation on what is taught, and yet it has only been since 1988 that such a national 'one-size-fits-all' view of what should be taught in English and Welsh schools, articulated as a set of different subjects each with their own individual specification of what pupils should attain, has been detailed.

In many countries, then, there is a developing view that any state specification of the curriculum should be restricted, both in the number of subjects that are required to be taught and in the extent of prescription of those subjects.

Figure 2.1 sets out a curriculum model that many countries adopt either explicitly through detailed local or national legislation or implicitly in the way that teachers are provided with government sponsored guidebooks and pupils are provided with specific resources such as a national textbook. First there is a 'national curriculum', which is specified in some detail by external people at national or state level outside the school. It is what society at a broad level requires that all its citizens should 'know, understand and be able to do'.

Most schools also adopt a local curriculum that parents would like to see added to what their child studies at school. Usually negotiated with the school board of governors or other local representatives, these school-based curricula are very often a means to promote or perpetuate local cultural identities such as aspects of the art, dance or poetry of a particular tradition, or lessons in the community language.

In many countries there is also the idea of a basic curriculum; one that is required should be offered but not specified in detail by external prescription and, perhaps, not required to be studied by all pupils. For example, this might be an agreed approach to ethics or religious education, a requirement to offer careers education or experience of learning in the workplace.

A cubic curriculum

Some years ago, Ted Wragg a professor of Education at the University of Exeter in the UK proposed what he called a 'Cubic Curriculum'. What he really said was, 'Actually, it isn't a cube. It's a multi-dimensional hyperspace, but *The Multi-dimensional Hyperspace Curriculum* does not exactly have a ring to it'. It is based on a vision of a future curriculum wider than subjects alone, important though they are, and founded on some linked propositions:

- Education must incorporate a vision of the future.

- There are escalating demands on citizens due to what has been called 'spiralling credentialism' – examination grades are needed for more and more jobs.
- Children's learning must be inspired by several influences. *How* something is learned is as important as *what* is learned.
- It is essential to see the curriculum as much more than a mere collection of subjects and syllabuses.

In other words, a pupil's experience in school – their 'experienced curriculum', that part that is *designed* to be experienced rather than other aspects of the so called 'hidden curriculum' – embraces subjects, cross-curricular themes and issues that affect development (like language and thought) and different forms of learning such as 'telling' and 'observing'.

Most important here is a need to plan such a curriculum. Staff at all levels from senior management to the day-by-day planning by the newly qualified teacher need to embrace what a curriculum means for them, and the implications it has for improving pupil learning.

So, around the world, and particularly in the STEM subjects, there is an opportunity and an urgency to think of the school curriculum in new ways: to develop new relevant content, and to explore new organisation patterns.

What approaches to a STEM curriculum might suit your school?

As a way to undertake an exploration of curriculum links between science, technology, engineering and mathematics, I am going to set out what I call the 'specified curriculum', the 'enacted curriculum' and the 'experienced curriculum'. This is similar to the approach taken by school inspectors in England who look at the *intent* of a school's curriculum, its *implementation* across a sequence of lessons, and consequently the *impact* on pupils by talking to them and inspecting their books (OfSTED, 2019). I am using the term curriculum widely to comprise most of what children learn in school, including under 'experienced curriculum' the values and behaviours that schools hope to inculcate such as respect for others, or the acceptance of authority. As you read what follows, consider how the three dimensions of subjects, cross-curricular processes and different pupil learning experiences are being addressed.

In brief, by specified, enacted and experienced curriculum, I mean:

- *Specified curriculum*: The curriculum content (as found in official documents and local agreements. The case study example for England is shown in Table 2.1).
- *Enacted curriculum*: This is what teachers teach and that is highly dependent on 'teacher knowledge' (The types of professional knowledge teachers need to bring to bear to plan and implement their teaching).
- *Experienced curriculum*: This the understanding gained – the pupil learning (how both of the above are interpreted and made sense of by pupils).

The specified STEM curriculum

The following statements are from the National Curriculum in England published in July 2013 and I will use these merely as a case study to consider aspects of the STEM

TABLE 2.1 Extracts from the National Curriculum for 11–14-year-old pupils in England

Science	Mathematics	Computing	Design & technology
Measurement: • use and derive simple equations and carry out appropriate calculations • undertake basic data analysis including simple statistical techniques. **Current electricity:** • potential difference, measured in volts, battery and bulb ratings; resistance measured in ohms as the ratio of potential difference (p.d.) to current	**Algebra:** • substitute numerical values into formulae and expressions, including scientific formulae • understand and use standard mathematical formulae; rearrange formulae to change the subject • model situations or procedures by translating them into algebraic expressions or formulae and by using graphs • use algebraic methods to solve linear equations in one variable (including all forms that require rearrangement) • work with coordinates in all four quadrants • recognise, sketch and produce graphs of linear and quadratic functions of one variable with appropriate scaling, using equations in x and y and the Cartesian plane • interpret mathematical relationships both algebraically and graphically	• design, use and evaluate computational abstractions that model the state and behaviour of real-world problems and physical systems • understand how instructions are stored and executed within a computer system; understand how data of various types (including text, sounds and pictures) can be represented and manipulated digitally, in the form of binary digits • understand the hardware and software components that make up computer systems, and how they communicate with one another and with other systems • undertake creative projects that involve selecting, using, and combining multiple applications, preferably across a range of devices, to achieve challenging goals, including collecting and analysing data and meeting the needs of known users • create, re-use, revise and re-purpose digital artefacts for a given audience, with attention to trustworthiness, design and usability • understand simple Boolean logic (for example AND, OR and NOT) and some of its uses in circuits and programming; understand how numbers can be represented in binary, and be able to carry out simple operations on binary numbers [for example, binary addition, and conversion between binary and decimal]	**Design** • use research and exploration, such as the study of different cultures, to identify and understand user needs • identify and solve their own design problems and understand how to reformulate problems given to them **Technical knowledge** • understand how more advanced electrical and electronic systems can be powered and used in their products [for example, circuits with heat, light, sound and movement as inputs and outputs] • apply computing and use electronics to embed intelligence in products that respond to inputs [for example, sensors], and control outputs [for example, actuators], using programmable components [for example, microcontrollers].

curriculum internationally and how they might interact with each other in a 'Cubic Curriculum' sense.

The purpose of science

A high-quality science education provides the foundations for understanding the world through the specific disciplines of biology, chemistry and physics. Science has changed our lives and is vital to the world's future prosperity, and all pupils should be taught essential aspects of the knowledge, methods, processes and uses of science. Through building up a body of key foundational knowledge and concepts, pupils should be encouraged to recognise the power of rational explanation and develop a sense of excitement and curiosity about natural phenomena. They should be encouraged to understand how science can be used to explain what is occurring, predict how things will behave, and analyse causes.

The purpose of design & technology

Design & technology is an inspiring, rigorous and practical subject. Using creativity and imagination, pupils design and make products that solve real and relevant problems within a variety of contexts, considering their own and others' needs, wants and values. They acquire a broad range of subject knowledge and draw on disciplines such as mathematics, science, engineering, computing and art. Pupils learn how to take risks, becoming resourceful, innovative, enterprising and capable citizens. Through the evaluation of past and present design & technology, they develop a critical understanding of its impact on daily life and the wider world. High-quality design & technology education makes an essential contribution to the creativity, culture, wealth and well-being of the nation.

The purpose of computing

A high-quality computing education equips pupils to use computational thinking and creativity to understand and change the world. Computing has deep links with mathematics, science, and design & technology, and provides insights into both natural and artificial systems. The core of computing is computer science, in which pupils are taught the principles of information and computation, how digital systems work, and how to put this knowledge to use through programming. Building on this knowledge and understanding, pupils are equipped to use information technology to create programs, systems and a range of content. Computing also ensures that pupils become digitally literate – able to use, and express themselves and develop their ideas through, information and communication technology – at a level suitable for the future workplace and as active participants in a digital world

The purpose of mathematics

Mathematics is a creative and highly inter-connected discipline that has been developed over centuries, providing the solution to some of history's most intriguing problems. It is essential to everyday life, critical to science, technology and engineering,

TABLE 2.2 Topics that could be taught to mutual advantage

Science	Mathematics	Computing	Design & technology
• use and derive simple equations • resistance measured in ohms as the ratio of potential difference (p.d.) to current	• use formulae by substitution to calculate the value of a variable, including for scientific formulae • begin to model simple contextual and subject-based problems algebraically	• explain how data of various types can be represented and manipulated in the form of binary digits including numbers, text, sounds and pictures, and be able to carry out some such manipulations by hand • use logical reasoning to evaluate the performance trade-offs of using alternative algorithms to solve the same problem	• apply computing and use electronics to embed intelligence in products that respond to inputs [for example, sensors], and control outputs [for example, actuators], using programmable components [for example, microcontrollers]

and necessary for financial literacy and most forms of employment. A high-quality mathematics education therefore provides a foundation for understanding the world, the ability to reason mathematically, an appreciation of the beauty and power of mathematics, and a sense of enjoyment and curiosity about the subject.

These four statements lay out the rationale for the designation of these separate subjects as areas of study during the ages of 5 to 14 years. In Chapter 1 we considered some of the differences between the STEM subjects and also some of the common themes such as Problem Solving and Systems thinking. Let us now look across the STEM subjects in terms of common requirements. In Table 2.1 some brief extracts from the different STEM subjects are set out to show where some common themes exist. It is particularly noticeable in the mathematics column how it refers across to exemplification through science. Similarly, in design & technology there are explicit links to the use of computing; and computing suggests examples to 'monitor and control physical systems' which has importance for both science and design & technology. The specified curriculum is emphasising the possible curriculum links.

The enacted curriculum

How can we enable the pupils to see the links between the specified STEM curricula and how can we help teachers to consider the curriculum that they enact in practice?

To explore the nature of the 'enacted' curriculum, we draw on work we have done with teachers across a number of countries. A key lesson to be learned by the rapid revisions of the specified curriculum of science, technology and mathematics over the last 30 years is that it is very difficult to impose a curriculum onto teachers be it from central government or from within a school management structure. A top-down method of seeking to describe the curriculum in close detail without working with teachers, and those involved in pre-service and in-service teacher education, to develop a common understanding of purpose of the changes, leads to a mismatch between a teacher's own view about their subject and why they teach the way they do, and what is specified by others to be taught. Teachers have a personal view about the purpose of their subject is about and, although they wish their pupils to do well in externally set examinations, when the specified curriculum moves independently of these deeply held views teachers feel obliged to just 'teach to the tests'. In doing so, they lose some of the fire and passion for their subject. It is therefore imperative that the teachers are involved in curriculum development, and that the test or examinations accurately reflects the intensions of the curriculum designers. How can those involved in teacher professional development, and teachers working with colleagues in school, find out the rational for the way they 'enact' the curriculum?

Sharing teacher professional knowledge across STEM

In our observation of teaching it is evident that success or failure of lessons organised by teachers was often linked, not only to their college-based subject knowledge and their choice of teaching strategies, but also to their appreciation of how their subject is transformed into a school subject. Figure 2.2 is a useful blank diagram for STEM teachers to fill in, and then compare with that of others as it draws out what they know and what they feel is important in teaching. To be clear, the figure is *not* a Venn diagram. Rather, it is intended to convey the close interplay between the following aspects of teacher professional knowledge.

Subject content knowledge

Teachers' subject matter knowledge influences the way in which they teach, and teachers who know more about a subject will be more interesting and adventurous in their methods and, consequently, more effective. Teachers with only a limited knowledge of a subject may avoid teaching difficult or complex aspects of it and teach in a manner that avoids pupil participation and questioning and which fails to draw upon children's experience.

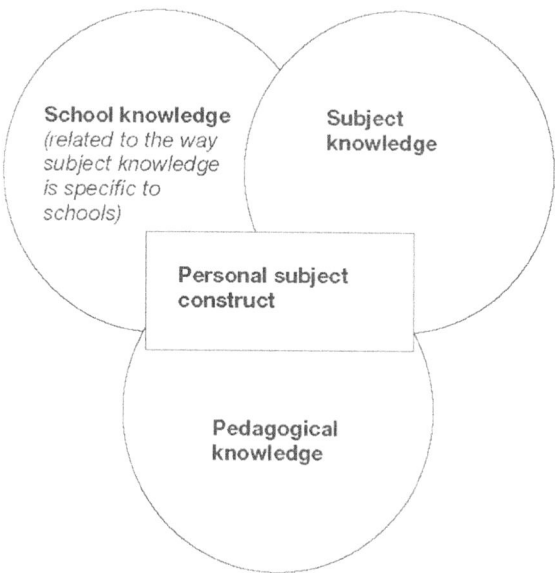

FIGURE 2.2 Framework of teacher professional knowledge

Pedagogical knowledge

At the heart of teaching is the notion of forms of representation and, to a significant degree, teaching entails knowing about and understanding ways of representing and formulating subject matter so that children may engage with it and develop knowledge, skills and understanding. This includes, pedagogical content knowledge (PCK), the *materia medica or pharmacopoeia*, as Shulman puts it, from which teachers draw their 'equipment' that present or exemplify particular content. However, PCK implies a rather restricted pedagogy focused on 'explaining'. We prefer the term 'Pedagogical Knowledge' requiring a wider creative approach to teaching and learning; and an appreciation of learning through experience supported by an understanding of how children learn.

School knowledge

To these types of teacher knowledge, we would wish to add 'school knowledge'. All schools are professional communities with their own distinctive approach to practice and it is important to understand what each school is like in deploying subject knowledge and pedagogic knowledge. We call this 'school knowledge'. Drawing on work by Chevellard (1991), *La transposition didactique* is a process of change, alteration and restructuring that the subject matter must undergo if it is to become teachable and accessible to children. By altering a subject to make it accessible to learners, a distinctive type of knowledge is formulated in its own right – 'school mathematics' or 'school technology'. In the same way that school science has differences from science conducted outside the school laboratory, so school design & technology and engineering has some differences from technology as practised in the world outside the school, but it is important to try to keep the tasks done through STEM activities as authentic as possible.

One might initially see 'school knowledge' as being intermediary between subject knowledge (knowledge of technology as practised by different types of technologists, for example) and pedagogical knowledge as used by teachers ('the most powerful analogies, illustrations, examples, explanations and demonstrations'). This would be to underplay the dynamic relationship between the categories of knowledge implied. For example, a teacher's subject knowledge is enhanced by his or her own pedagogy in practice and by the resources that form part of their local school knowledge. Which teacher has not confessed to only really understanding a topic when they were required to teach it to others! It is the active intersection of subject knowledge, school knowledge and pedagogical knowledge that brings teacher professional knowledge into being.

I think you would agree that lying at the heart of this dynamic process are the 'personal constructs' of teacher and pupils; a complex amalgam of past knowledge, experiences of learning, a personal view of what constitutes 'good' teaching and belief in the purposes of the subject. This all underpins a teacher's professional knowledge. This is as true for any teacher. A student teacher has to question his or her personal beliefs about their subject as they work out a rationale for their classroom behaviours.

Figure 2.2 has been discussed with a number of education professionals in the UK, Canada, Spain, the Netherlands, Sweden, Finland, New Zealand, Australia and

South Africa. These professionals have been different groups of school teachers of science, mathematics and technology, and of English too, and including student teachers, teacher educators, and researchers. The reaction to the model across this spectrum of teacher professional expertise has been remarkably similar:

- The different aspects of teacher knowledge are recognised by all these groups as being meaningful. Teachers, in particular, are excited by the categories and value the model as a way of easily articulating what they know and are able to do.
- The model can be interpreted at different levels. Some see it as a tool for categorising personal understanding. Others see it as being useful for planning in-service development for a group of teachers.

But a tool such as this is more than just a means to an end. The means is, perhaps, more important than the end as it enables STEM teachers to engage in what they do or 'enact' as they work with their STEM curriculum. In practice, the discussion of what is appropriate for different parts of the diagram and the relationship between the circles helps with reflection on practice more than any completed picture ever could. The process of thinking about what is important in a subject, initiated by the diagram, is more important than the diagram itself. Similarly, a completed diagram such as Figure 2.3, done by Clare Lee a mathematics teacher, can engender considerable debate and further reflection on practice in explicit terms.

Looking sideways

The first step in looking sideways is to spend some time with others in your department filling in your *own* version of Figure 2.3 and comparing what you think of your subject to what they have highlighted. Next, compare your diagram with that of your colleagues across the school in the other STEM areas of the curriculum. In some of the following chapters we will look in detail about how the different STEM subjects can support each other if each teacher spends a little time looking sideways at what is being taught in the other subjects – and when! However, what then happens is down to the approach that the departments wish to take in moving to working with each other for the support and benefit of the pupils.

I will discuss three possibilities of working across the STEM subjects in turn through a *coordinated* approach, a *collaborative* approach and an *integrated* approach.

A coordinated approach

What science teacher, especially someone teaching physics, has not asked the question, 'Have you done this in maths yet?' to a class scratching their collective heads trying to manipulate an equation. The silo nature of the traditional subjects has militated against proper coordination of the subjects for mutual benefit. In a properly coordinated approach, teachers in each subject become familiar with the work carried out in the others and plan their curricula so that the *timing* of topics within each subject is sensitive to each other's needs; and taught in a way that supports the pupils' developing understanding rather than one that causes confusion. For example, proficiency with

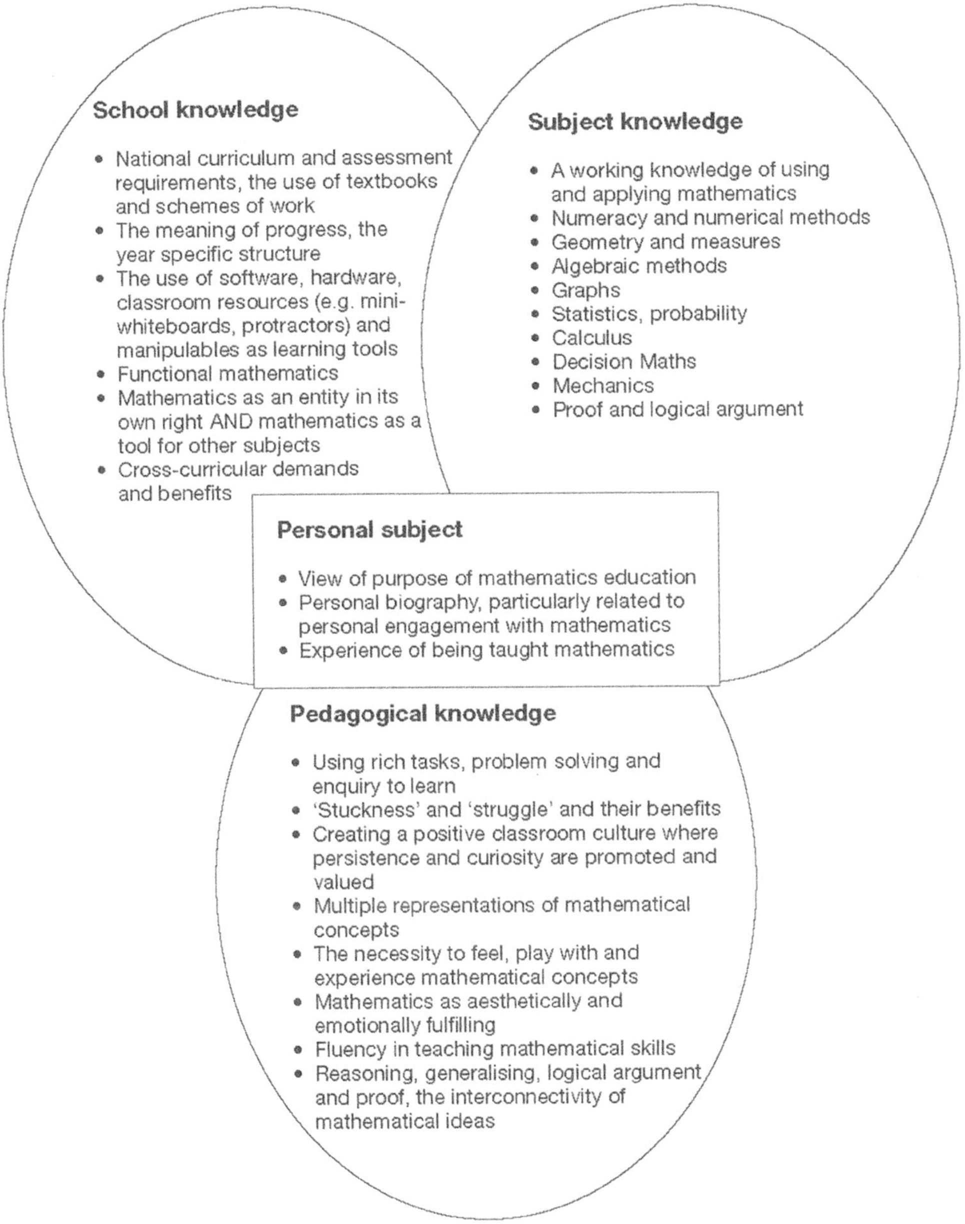

School knowledge

- National curriculum and assessment requirements, the use of textbooks and schemes of work
- The meaning of progress, the year specific structure
- The use of software, hardware, classroom resources (e.g. mini-whiteboards, protractors) and manipulables as learning tools
- Functional mathematics
- Mathematics as an entity in its own right AND mathematics as a tool for other subjects
- Cross-curricular demands and benefits

Subject knowledge

- A working knowledge of using and applying mathematics
- Numeracy and numerical methods
- Geometry and measures
- Algebraic methods
- Graphs
- Statistics, probability
- Calculus
- Decision Maths
- Mechanics
- Proof and logical argument

Personal subject

- View of purpose of mathematics education
- Personal biography, particularly related to personal engagement with mathematics
- Experience of being taught mathematics

Pedagogical knowledge

- Using rich tasks, problem solving and enquiry to learn
- 'Stuckness' and 'struggle' and their benefits
- Creating a positive classroom culture where persistence and curiosity are promoted and valued
- Multiple representations of mathematical concepts
- The necessity to feel, play with and experience mathematical concepts
- Mathematics as aesthetically and emotionally fulfilling
- Fluency in teaching mathematical skills
- Reasoning, generalising, logical argument and proof, the interconnectivity of mathematical ideas

FIGURE 2.3 A completed framework of teacher professional knowledge from a mathematics teacher

the use of measuring in millimetres and collating data from respondents if covered in lower school mathematics would benefit technology, and if electricity is explained in technology using similar analogies and terminology to science, the pupils' developing ideas are reinforced.

Building on what is in Table 2.1, Table 2.2 shows some common links between some topics taught to 11–14-year-olds that can be coordinated with minimum disruption.

A collaborative approach

In a collaborative approach, teachers in each subject plan their curricula so that some activities within each subject are designed and planned together to establish an effective relationship.

In developing teaching resources for the curriculum in Scotland, Education Scotland created as part of a STEM initiative an interdisciplinary unit of work concerned with renewable energy (ES, 2018). The study of renewable energy is introduced by a short video in which a prominent populariser of science and technology interviews young professional engineers who are working in the renewable energy industry in Scotland. Pupils undergo four 'learning journeys'. The first, 'From fossil fuels to wind', meets some of the science requirements of the curriculum with possible links to social science and technology. The second, 'Wind, wave and tidal', meets some of the technology requirements of the curriculum with some links to science. The third, 'Calculating the wind', meets some of the mathematics requirements of the curriculum with some links across to science and to technology.

In the fourth learning journey, 'This island is going renewable', pupils are challenged with making the case for the use of renewable energy on a small island community. In this challenge the pupils will need to use their learning from the first three learning journeys, and also develop skills in using maps and geographical information systems to gather, interpret and present data relating to location of renewable technologies. This large challenge is divided into three smaller challenges.

Challenge 1

An important part of any energy plan for a community would include consideration of energy consumption and ways to reduce this. Advise one of the following user groups about the use of energy to support their lifestyle/business:

■   an elderly couple who are retired and live in a small cottage;

■   a family consisting of a mother, father and two teenage children, living in a three-bedroom detached house. The father works at the local school, the mother works at the slate mine and the children attend the local school;

■   a family consisting of a mother and father and a baby aged six months, the mother is a full-time mum, and the father works in the timber mill;

■   the local post office/community shop;

■   the head teacher of the island school, which has 250 pupils.

Challenge 2

Based on your findings from Challenge 1 on individual user groups, work out an approximate energy usage for the whole island.

■   Could all the energy needs of the island be provided by wind, tides or waves?

■   Decide as a team the kind of information you will need to know about renewable technologies to help you answer this question.

- How will you analyse this information?
- What criteria will you use for comparing the different possible renewable technologies?
- Which other factors will you need to consider?

Challenge 3

Create an exhibition stand displaying the findings of your investigations into the feasibility of using renewable energy on the island to help inform the islanders about the issues around energy such as:

- energy usage and consumption;
- options for generating energy from renewable sources;
- best locations for particular technologies;
- a scaled model of the island to demonstrate the potential impact that the technologies could have on the landscape.

You could include examples or photographs of the working models you have been making in class, charts, diagrams, written explanations, PowerPoint presentations, leaflets, annotated maps and so forth.

The above approach to collaborative interdisciplinary work in Scotland is not dissimilar to that of the Nuffield Key Stage 3 (pupils aged 11 to 14 years) STEM Futures project (Nuffield, 2010) but there are significant differences. The Nuffield challenge is set by the teacher rather than being negotiated with the class and the pupils' response to the challenge is clearly structured.

In April 2019, a curriculum was launched in Wales built around 'areas of learning and experience'. Known as a Curriculum for Wales 2022, the need for a coordinated approach to STEM is built into the curriculum design. For example, one area of learning is 'Mathematics and Numeracy' and another 'Science and Technology' and the curriculum stretches across the age range 3 to 16. Looking at the 'Science and Technology' area of learning, links are made to Literacy, Numeracy and Digital Competence, and the document itself stresses the need for 'looking sideways':

FUTURES CASE STUDY

The STEM Futures resource is composed of a series of 'pods'. A pod is a series of lessons organised around a particular sustainability theme. Typically, a pod contains an overview, teacher notes, pupil tasks, video clips, animations and a pupil presentation. The activities in each pod are ideally conducted in order, to scaffold the concept development.

Pod 1: Introduction

Pupils are introduced to the idea that many current human problems relate to food, energy and materials. They look at a brief history of civilisation, to emphasise that humanity's quest for resources is nothing new. Advances in technology have increased the depletion rate of fossil fuels and other materials.

Pupils engage with the idea that our linear take → make → dump culture is not sustainable. We need to learn some 'closed loop' lessons from nature where all waste is recycled through natural

Pod 2: Waste

Pupils start by classifying debris on a beach according to whether it will decay or not. Pupils analyse product life cycles and generate questions about natural closed loop systems. They consider how cradle to cradle design could help provide closed loop systems for human activities.

Pod 3: Cars

Pupils consider conventional car engine design and review new green alternatives. They collect evidence for pollution in their local area and analyse the data. Pupils interpret graphs showing past and predicted oil consumption. They use reports and data to assess the impact of legislation on traffic pollution. Pupils produce and present suggestions for alternative closed loop approaches to local transport.

Pod 4: Climate change

Pupils investigate the key components of the carbon cycle. They analyse evidence relating CO_2 to climate change. Pupils compare the carbon footprint of different activities and different societies. They use closed loop thinking to consider new ways of reducing CO_2 in the atmosphere. Finally, they present the case for the construction of a local wind farm.

Pods 5: Pupil project

Pupils use the learning skills they have acquired in earlier pods to carry out a piece of project work. Pupils identify a problem or question relating to sustainability, and use STEM knowledge and understanding to present a closed loop solution. Their project involves research, analysing, evaluating and synthesising information, and communicating possible solutions creatively through a variety of media.

Here are some of the main topics covered in Futures.

Science

Carbon, nitrogen and water cycle
Photosynthesis and respiration
Energy from combustion renewable energy
Global warming
Pollution
Properties of materials

Design & technology

Materials
Product life cycle
Car design
Sustainable products
Sustainable systems
Renewable energy

Making the most of a multidisciplinary approach to curriculum development provides learners with a more coherent learning experience. Deep understanding can develop through planning across all six statements of what matters in science and technology learning, and the connection and application of these in a range of contexts. The learning undertaken in one aspect can reinforce and support work across different disciplines in a timely manner. To achieve this, cohesive and coherent curriculum planning across traditional disciplinary boundaries is crucial. However, practitioners will also wish to consider the need for more discrete disciplinary learning and teaching; this becomes increasingly important as learners progress.

The integration of STEM subjects

There are two ways of considering the integration of the STEM subjects. One is getting synchronous inputs from a range of staff for an off-timetable event or project. Here all the staff support the activities through team-teaching and pupils turn to a member of staff for advice and support when they are available. Around the world, pupils of all ages take part in competitions and challenges, or just attend workshops in science museums and higher-education institutions. In Taiwan, for example, robot building from using Lego through to full combat 'robot wars' models is a common out-of-school activity. In Japan, 'STEMinars' occur early in the school year where pupils are encouraged to attend a University for an intensive one-week 'deep dive' into an STEM area of interest (see Chapter 8).

The second is a full integration of the STEM subjects in school so that the one teacher follows a themed project across a number of lessons, as is often the case in primary schools. If this is followed at the secondary level this assumes a lot of expertise is available in the one teacher, or resources are needed for a team-teaching approach. As we discussed in Chapter 1, science and design & technology, for example, are significantly different from one another and it is difficult to ensure that it is a true integration of subjects as equals and one of the subjects does not dominate and subsume the other. Integration has been successful in Belgium and in Israel (see Chapter 11).

The Israeli approach to the relationship between science and technology is based on science and technology teachers working together and focusing on problem-solving in a social context. The curriculum developers in ORT established a didactic model between the disciplines, known as the STSS (Society-Technology-Science-Society). This is shown in Figure 2.4. This model serves as both a conceptual and a

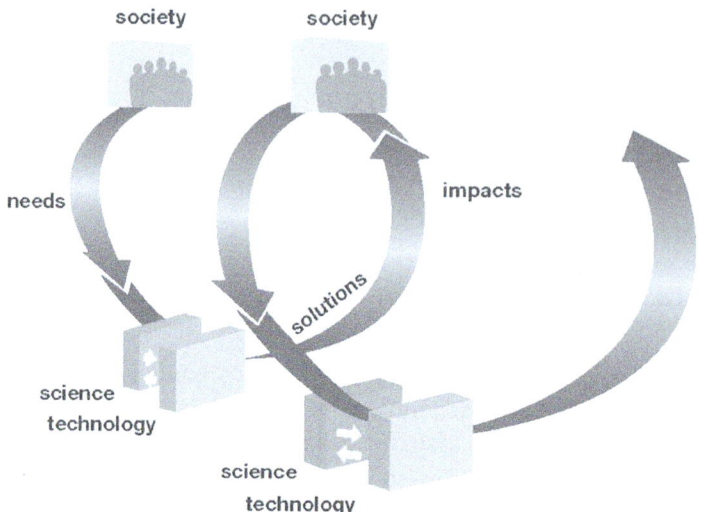

society   society

needs

impacts

solutions

science
technology

science
technology

FIGURE 2.4 Science-Technology-Science-Society model

curriculum framework for dealing with social and environmental issues (e.g. 'The Noise Around us'). The STSS model is underpinned by four elements:

- problem solving;
- the use of social, scientific and technological knowledge for problem solving and decision-making;
- the view that science and technology are two distinct but interacting disciplines;
- the gap between the needs of society and reality; this gap has the role of a 'driving force' for development in both science and technology.

Although there are considerable advantages to linking science and technology in the way described in the STSS model, there were a number of difficulties in implementing the model in schools. The two major impediments were:

- the lack of appropriate curriculum materials;
- this model is predicated upon appropriate curriculum coverage between teachers from both subjects, and this is not easy to achieve.

At the implementation level, collaboration in the STSS model is currently more wishful thinking than reality. The reasons for this are the differences between science and technology teachers (including status, academic background), the lack of a collaboration component within the teacher training programmes, and timetabling and other organisational difficulties within schools.

The experienced curriculum

On initial teacher education courses, it is common to ask the student–teacher to do a 'pupil trail' and follow one or maybe two pupils as they move from class to class and are taught by different teachers (as is common in many countries). The novice teacher often expresses surprise at the way the same pupil reacts very differently to the different teachers and to the different environments such as a workshop, laboratory or classroom. Another way of looking at this – and I recommend everyone to do this once a year whatever the stage of their career as a teacher – is seeing the curriculum actually experienced by the pupil.

From our research, and that of Bob McCormick and other colleagues at The Open University in the UK, there is considerable evidence that problem solving – a key aspect of all STEM subjects – is often conducted in a sort of 'ritual' way in school classrooms (see Banks, 2009; McCormick & Davidson, 1996; McCormick, Murphy, & Davidson, 1994). As it had the potential to bring together aspects of science and mathematics through design & technology, let us follow a small case study of a teacher with 12–13-year-old students working on an electronic badge project based on a 'face' with LEDs for eyes. (These cases are taken from Banks & McCormick, 2006 and based on classroom research undertaken by McCormick.)

McCormick noted that the teacher deliberately did not emphasise the underlying processes; it was not one of his main aims, and he seemed to view designing as a logical approach rather than as a process that involved sub-processes to be taught and learnt. He said:

> although I'd like them to understand and use the design process and I think it's quite a nice framework for them to fit things on to, I don't think there's a great need to be dogmatic about it and say you must learn it … the nature of projects leads them through the design process despite the teacher's bit, going through it with them in front of the class.

The particular view that a teacher takes of the process being taught affects the way tasks are structured, the kinds of interventions that are made by the teacher, and the assessment of students' work. Not all of these approaches will be consistent either with each other, or with the view espoused by a teacher, but collectively they will have a profound effect on the students' perceptions and activities. But, whatever view is taken of designing, there is a tendency to see it as an algorithm to be applied in a variety of situations. The teacher involved in the *electronic badge project* began it with the 'Situation' being presented:

> A theme park has opened in [place] and it wants to advertise itself. It plans to sell cheap lapel badges based on cartoon characters in the park. To make these badges more interesting, a basic electronic circuit will make something happen on the badge.

This was set within the general title of 'Festivals', but the links to the 'Situation' were not discussed, and from then on no further reference was made to festivals. The teacher continued in the session by asking the students to draw up a spider diagram of 'Considerations' (a specification), a task that all the students seemed familiar with. He did not, however, elaborate on the 'Situation' or the 'Design brief', nor invite students to discuss them in the context of the planned project. Three students were followed (we'll call them Bill, Tanvir and Rose) who produced different design briefs that illustrated how the 'Situation' was interpreted by them. Bill and Tanvir interpreted it as a 'button is pressed to light up the eyes', whereas Rose makes no such inference as she wants 'to design and make a clock badge'. Their initial ideas of their personal 'briefs' lingered and influenced future tasks; for example, Rose continued to talk about a 'clock face' for several lessons and abandoned the idea only when she realised that the electronics would not be like that of a watch. She also imagines that the battery would resemble that in a watch and was almost incredulous when the teacher showed a comparatively large conventional dry 9-volt battery that she (rightly) considered too heavy for a lapel badge. The teacher's discussion with Rose about this issue indicated that, unlike Rose, he had not entered into the 'Situation' and 'Design brief' in a meaningful way, but only ritualistically – his ultimate answer to the problem was to 'have a strong pin for the badge', a response Rose felt very dissatisfied with!

Next the teacher gave several tasks relating to drawing the faces for the badge, which implicitly reflected the sub-processes of 'generating ideas', 'developing a chosen idea' and 'planning the making'. However, this was again done in a 'ritualistic' way as the following sequence of events indicates. At the end of the first session students were asked, for homework, to create four cartoon faces as potential designs for the badge. No parameters were given other than that all four should fit into the design sheet and that students should be 'creative'. As with the 'Situation', 'Design brief' and 'Considerations', this step of producing four designs appeared to be a standard one and, again, was accepted without question by the students. However, in the next session students were asked to re-draw the faces so that they touch the sides of a fixed drawn square (70 × 70 mm). The reason for this was not made clear until a later session. Evidence from the students' folders indicates that students had to modify their designs in order to fit these new demands. For example, Rose had originally drawn a thin 'carrot' character, which she had to distort to make it fat enough for it to touch the sides of the square. The fact that the creation of several designs is sometimes perceived by students to be merely a ritual is seen in Rose's comments to the teacher implying she had in fact already made a final choice while she is still completing the four 'possible outcomes' drawings.

In looking at STEM the pupil experience of the curriculum in the classroom we discovered some of the strategies that students *actually* adopted in response to the various ways the teachers viewed and enacted the problem-solving process. These pupil strategies certainly do not resemble the 'algorithms' or 'ways of problem solving' that are so often taught. The first strategy is what we characterised as *problem solving as dealing with classroom culture*. This occurs when students try to 'work out' the rules the teacher sets in the classroom, and play to those rules – we discussed this in Chapter 1 when considering the 'real' purpose of investigating suitable materials for making a mountaineers jacket as teaching a 'cooling curve' experiment. Another example of pupils seeking out this classroom culture is provided by the contrasted experience of two girls (Kathy and Alice) in producing a mobile. Alice wanted to do something that clinks when the wind blows, and so had an idea of using metal. So, given a restricted choice of material, she chose to cut thick mild steel in the form of disks about 5 cm in diameter. Because, through experience, she played the rules of the classroom, Alice ended up with very sore hands, and took a long time; her endeavour resulted in a very inappropriate way of creating the effect she wanted. (But she did learn quite a lot about mild steel, as it turned out!) Kathy had designed a moon and planets going around it, and wanted some kind of glinting material. When presented with the choice of material, Kathy, in contrast to Alice, looked elsewhere and saw some aluminium (not available to the class) and asked to use this. The teacher agreed, and she cut this easily with tin snips. Kathy took this approach many times throughout the project. She broke the rules of the classroom, knowing what she could and couldn't get away with. She experienced different kinds of issues and problems from Alice, but she was avoiding many technological problems that Alice faced.

The second strategy is *problem solving as giving and finding a solution*, illustrated in a case-study project involving the construction of a 'moisture sensor'. This had to be appropriate to the situation of detecting moisture (or lack of it) to indicate when, for example, a plant needed to be watered. The teacher in this example defined the task in terms of making a box in which to put the electronics (the transistor circuit, the bulb or the little speaker, switch, etc.). He taught them to cut the material (styrene) in straight lines with a steel ruler and a knife because when he said 'box', he had in mind a rectangular box. He also gave them a jig so that they could put the two edges together at right angles and run the solvent along to stick the two together. But some pupils wanted curved shaped boxes, which gave some of them at least three emergent problems. First they had to cut a curved shape, and students asked each other and the teacher how to cut the shape as the steel ruler method wouldn't work (the solution was to cut it slowly). Second, a curved profile on one part of the box required one side to bend to follow the profile, but the styrene they were given was too thick. One pupil complained to the teacher who simply gave her a thinner gauge of styrene, without any discussion. Third, the pupil did not know how to support or hold the thinner styrene in place to apply the solvent, and so again asked the teacher. This time the teacher had to think and was obviously solving the problem himself, but again he gave the *results* of his thinking as a ready-made solution to the pupil and did not involve her in his problem-solving process. All she received was the solution without being involved in the problem solving. This continually being 'given solutions' becomes a pupil experience of the curriculum at the expense of a 'problem-solving' culture.

The above case study indicates just how pivotal the teacher is in enabling pupils to engage in genuine problem solving. The behaviour of the teacher in the study is a litany of what NOT to do if the intention is to give pupils the possibility of generating and developing their own ideas and dealing with the problems that emerge as they pursue their intentions.

The final strategy is the *student collaboration model*. In both science and technology, pupils are often set individual projects, so they may be working alongside each other on a table or a bench, and they can co-operate because they are doing similar things; they are not identical, but similar enough to help each other and share tasks. The second form of collaboration involved students in dividing up the task: 'You do this bit, I'll do that bit. You're good at that and I'm good at this'. Some of the learning is lost in this approach, but at least it is a way of collaborating, because they have to put the two bits together at some stage, and that has an element of good collaborative problem solving. The final form of collaboration occurs when pupils have a shared task, and they can talk about it. This means the design of the task must *require* the students to collaborate. Designed correctly, tasks should require solutions to a problem to be considered by all pupils through discussion and decision making.

Conclusion

I have covered many issues in passing so let's draw together what I consider are the crucial points. Classrooms are social environments and the specified curriculum leads to what is enacted by teachers and, ultimately, what is experienced by pupils.

The specified curriculum

- It is very difficult to control the intended learning of pupils by an elaborate spec-ification in law of what pupils should know.

In most parts of the world the specified curriculum as a legal document is being down-played and schools are freer to construct their own curriculum. However, if teachers themselves are not part of the discussion on what STEM in school should be, they will merely 'teach to the test' leading to teaching strategies that have, for example, elements of 'ritual'. There will be a clash between their personal view of their subject and that specified by others and classroom practice will go through a period of extremes until some commonly shared beliefs of what constitutes 'good' teaching emerge.

The enacted curriculum

- In an effort to direct the learning outcomes for all pupils and make the tasks manage-able in the classroom, teachers sometimes tend to closely direct the activity of pupils.

By teachers 'looking sideways', pupil learning can be enhanced. Technology teach-ers have much to teach science teachers on the handling of processes and the sci-ence teachers much to teach technology teachers about the problems associated with acquiring conceptual knowledge. Mathematics teachers can help both with data han-dling and can learn about making their subject relevant to all.

The experienced curriculum

Through constraints of time and resources, teachers transfer their subject into 'school knowledge' and pupils play the game of discovering what that is. Some pupils never quite work out the rules of the game and the relevance of the subject becomes lost to them; others pick up incidental aspects either because teachers have not made clear what is salient, or their classroom culture produces effects at odds with their rhetoric.

- The way that students engage in problem solving in design & technology and in science and in mathematics depends on the view of designing and of investigating held by the teacher.

An overwhelming conclusion, however, would be that good practice in STEM classrooms is not shared well across schools and between schools. As new equipment produces yet more teaching opportunities, we need to find out about their impact on the curriculum experi-enced by pupils. STEM offers some very exciting opportunities for schools and these will be enhanced if teachers keep 'looking sideways' at what their colleagues are doing.

Recommended reading

Subject knowledge

Al-Khalli, J. (2020) *The world according to physics*. Oxford: Princeton University Press.
Attenborough, D. (2018) *Life on earth*. London: William Collins.

Houston, K. (2009) *How to think like a mathematician*. Cambridge: Cambridge University Press.

Lowe, D. B. (2016) *The chemistry book: From gunpowder to graphene*. New York: Stirling.

Norman, D. A. (2013) *The design of everyday things*. New York: Basic Books.

Pedagogical knowledge

Barton, C. (2018) *How I wish I'd taught maths: Lessons learned from research, conversations with experts, and 12 years of mistakes*. Woodbridge,UK: John Catt Education Ltd.

Johnston-Wilder, S., Lee, C., & Pimm, D. (eds) (2016) *Learning to teach mathematics in the secondary school*. Abingdon: Routledge.

Owen-Jackson, G. (ed) (2015) *Learning to Teach Design and Technology in the secondary school*. Abingdon: Routledge.

Toplis, R. (ed) (2015) *Learning to teach science in the secondary school*. Abingdon: Routledge.

School knowledge

Dintersmith, T. (2018) *What school could be*. Princeton:Princeton University Press.

Drew, D. E. (2011) *STEM the tide*. Baltimore: Johns Hopkins University Press.

References

Banks, F. (2009) Research on teaching and learning in technology education. In A. Jones and M. J. de Vries (eds), *International handbook of research and development in technology education*. Rotterdam, Netherlands: Sense Publishers

Banks, F. & McCormick, R. (2006) A case study of the relationship between science and technology: England 1984–2004. In M. J. de Vries & I. Mottier (eds), *International handbook of technology education: Reviewing the past twenty years*. Rotterdam, Netherlands: Sense Publishers.

Chevellard, Y. (1991) *La transposition didactique: Du savoir savant au savoir enseigné*. Paris: La Pensee Sauvage.

ES (2018) Education Scotland National Improvement Hub – STEM Central. https://education.gov.scot/improvement/learning-resources/STEM%20Central (accessed 4 June 2020).

McCormick, R., Murphy, P., & Davidson, M. (1994). Design and technology as revelation and ritual. In J. S. Smith (ed.), *IDATER 94 – International Conference on Design and technology Educational Research and Curriculum Development* (pp. 38–42). Loughborough, UK: University of Loughborough.

McCormick, R. & Davidson, M. (1996). Problem solving and the tyranny of product outcomes. *Journal of Design and Technology Education*, 1(3), 230–241.

Nuffield (2010) Nuffield Key Stage 3 STEM Futures project www.nuffieldfoundation.org/stem/nuffield-stem-futures (accessed 4 June 2020).

OfSTED (2019) *The education inspection framework*. Manchester: Crown.

Zuga, K. F. (1996) STS promotes the rejoining of technology and science. In R. E. Yager (ed.), *Science/technology/society as reform in science education*. Albany, NY: State University of New York Press.

Teaching science in the light of STEM

The nature and purpose of science education

In 2010, Wynne Harlen and a group of distinguished colleagues described ten principles of science education and within these identified three aims for science education:

- understanding of a set of big ideas in science, which include ideas *of* science and ideas *about* science and its role in society;
- scientific capabilities concerned with gathering and using evidence;
- scientific attitudes.

The big ideas are shown in Table 3.1. It is worth noting that one of the ideas *about* science acknowledges that the knowledge produced by science is used in some technologies to create products to serve human ends. This knowledge is, of course, to be found in the ideas *of* science and immediately provides some justification for developing a curriculum relationship between science and design & technology as we considered in earlier chapters.

In his presidential address to the Association for Science Education in January 2012, Robin Millar made a compelling case for 'science for all'. Parts of his address were relevant to the idea of teaching science in the light of pupils' learning in design & technology. He quoted Jon Ogborn (2004: 70) who signals that the economic argument, educating the next generation of scientists, has little worth in justifying science for all.

> A central fact about science is that it is actually done by a very small fraction of the population. The total of all scientists and engineers with graduate level qualifications is only a few percent of the whole population of an industrialized country. Thus the primary goal of a good science education cannot be to train this minority who will actually do science.

However, the voice from government, in the report of the Science and Learning Expert Group (2010) indicates that this minority will play a crucial role in the future of the UK:

TABLE 3.1 Fourteen big ideas in science (taken from *Principles and Big Ideas of Science Education*, edited by Wynne Harlen)

Ideas *of* science

1. All material in the universe is made of very small particles.

2. Objects can affect other objects at a distance.

3. Changing the movement of an object requires a net force to be acting on it.

4. The total amount of energy in the universe is always the same but energy can be transformed when things change or are made to happen.

5. The composition of the Earth and its atmosphere and the processes occurring within them shape the Earth's surface and its climate.

6. The solar system is a very small part of one of millions of galaxies in the universe.

7. Organisms are organised on a cellular basis.

8. Organisms require a supply of energy and materials for which they are often dependent on or in competition with other organisms.

9. Genetic information is passed down from one generation of organisms to another.

10. The diversity of organisms, living and extinct, is the result of evolution.

Ideas *about* science

11. Science assumes that for every effect there is one or more causes.

12. Scientific explanations, theories and models are those that best fit the facts known at a particular time.

13. The knowledge produced by science is used in some technologies to create products to serve human ends.

14. Applications of science often have ethical, social, economic and political implications.

Global development means that the competition and market for the products of science, engineering and technology are greater than ever before. It is a truism to state that the future of the UK depends critically on the education of future generations. Science, technology, engineering and mathematics (STEM) must be at the forefront of education in order for the UK to address some of the most important challenges facing society.

Those concerned about the place of STEM in the curriculum have voiced concern about how the science curriculum appeals to pupils. The review by Sir Gareth Roberts (2002) 'SET for success' noted:

widespread concern that science is taught in a way that does not appeal to many pupils and that the curriculum places too much emphasis on rote learning rather than relating theory to situations relevant to the pupil.

Recently, the Australian government has identified STEM education as 'critically important for our current and future productivity as well as for informed personal decision making and effective community, national and global citizenship' as part of a National Innovation and Science Agenda (Pyne, 2015).

This provides more than a hint that relating science learning to the wider world in which it is applied might pay dividends.

However, there is a problem with much science understanding. It is counterintuitive. A common-sense approach will almost certainly lead to ideas that aren't scientific. Learners have to contend with what Lewis Wolpert (1992) has called 'the unnatural nature of science'. This is compounded by the way in which pupils might respond personally to the nature of science termed the 'affective challenge' by James Donnelly (2003: 19).

> Scientific knowledge offers a materialistic worldview which, in its substance, is devoid of humane reference, whatever might be said of its practices and its implications. Science is profoundly successful, on its own terms, and scientific knowledge profoundly authoritative. In consequence, creating scope for the individuality of pupils to come into play is difficult … these characteristics of science challenge pupils affectively and cognitively. … It might even be said that they are somewhat at odds with the tenor of modern cultural life.

Robin Millar finds a telling quote from a student that supports this position. In science, 'there's no room to put anything of you into it'.

Again, there is the suggestion that enabling pupils to relate science learning to aspects of learning 'outside science' might pay dividends.

In England there has been some resistance to the idea of accommodating the apparently intrinsically unappealing nature of science by curriculum reform from Michael Young and David Lambert (2014). They argue that forging cross-curricular links or developing learning programmes based on integrating separate subjects with a view to developing particular skills sets are neither worthwhile nor intellectually defensible. They make the case for the curriculum being based on 'powerful' knowledge, i.e. knowledge that has been created by wider disciplinary communities and has, according to Michael Young (2008), the following features:

- It is abstract and theoretical (conceptual) – it is concerned with the general not the particular.
- The concepts associated with it as interrelated – they are part of a system.
- It is reliable but open to challenge.
- It is often counterintuitive to experience.
- It has a reality that is independent of direct experience of the teacher and the learner.

Intriguingly, this echoes, to some extent, the words of Neil Postman (1993) over 20 years earlier who argued that a curriculum should not be child-centred, nor skill-centred, nor training-centred, not even problem-centred but idea-centred and coherence-centred and otherworldly in as much as it does not assume that what one learns in school must be directly and urgently related to a problem of today.

Proponents of 'powerful knowledge' as a basis for the curriculum argue it is the right of all young people and by acquiring such knowledge through the study of subjects based on such knowledge they will extend considerably their understanding of the world around them. They will become engaged with the best that has been thought, said, written and done. It will of necessity be a difficult journey for all young people, especially those who are late in developing and alienated from schooling and also those who are neither late in developing nor alienated but come to school from backgrounds where 'social capital' from home doesn't support this approach to knowledge. But this does not deny them their right to powerful knowledge, and it is the role of schools and teachers to develop pedagogy that succeeds in engaging them in a curriculum based on powerful knowledge. Being able to look at the world as a place that is 'understandable' is very empowering hence this approach is likely to promote social justice. A curriculum in which attempts are made to consider interaction between the STEM subjects might be seen as diluting those subjects and as such would seem anathema to those who support a curriculum based on powerful knowledge derived from established disciplinary bases. Their view of the curriculum would seem to justify and solidify subject silos. Yet this might not be the case given the way august bodies such as the Institute of Physics, the Royal Society of Chemistry and the Royal Society of Biology are developing guidelines for curriculum development in particular science subjects.

In contrast to Neil Postman Charles Tracey, Head of Education at the Institute of Physics, (2018) in providing guidelines for future physics curricula identifies three dimensions of learning to be considered:

1   A representative set of practices and ways of thinking and reasoning
    In the study of:

2   Ideas and explanations based on a set of constituents and the way that they behave
    Illustrated by and designing:

3   Solutions to practical problems outside the laboratory.

Tracey formulates these in terms of Big Ideas:

1   Big Ideas about physics and its practices.

2   Big ideas of physics and its explanations.

3   Big Ideas from physics in applications.

The use of Big Ideas has resonance with Young and Lambert's view of 'powerful knowledge' but it is interesting to note that number three in this list considers how the Big Ideas from physics might be exploited in solving practical problems, which are invariably 'messy' and require more than physics for their solution, particularly, as Tracey notes, responding to society's big questions and big challenges and the design activities of engineers.

Daniel Gibney, Programme Manager, Curriculum, Qualifications and Assessment in Education Policy Team at the Royal Society of Chemistry (2018b) in considering the contents of an ideal chemistry curriculum also adopts a threefold approach but frames this in terms of Big Questions as follows:

1   **Chemistry as a science**
    How do we think about chemistry?
    How do we do chemistry?

2   **Chemical concepts**
    What are things made of?
    How do we find out what things are made of?
    How do we explain how substances behave?
    How can substances be made and changed?

3   **Chemistry and the world**
    What is the impact of chemistry?
    These Big Questions are considered in terms of the powerful knowledge that is
    required to provide answers. Interestingly Gibney develops the impact dimension
    as follows

    ■   investigating the world;

    ■   making things and developing processes;

    ■   making decisions about chemistry.

This mirrors the application dimension identified by Tracey in physics and considers
the ethical, moral, economic, political and environmental considerations that feature
in decision making about chemistry.

   Lauren McLeod, Head of Education Policy at the Royal Society of Biology
(2018b) in developing a framework for the biology curriculum takes Wynne Harlen's
Big Ideas of science that concern biology and develops them into five Big Questions
of biology as follows:

1   What are organisms and what are they made of?

2   How do organisms grow and reproduce?

3   How do organisms stay healthy?

4   How do organisms live together?

5   Why are organisms so different?

At first sight it seems that the idea of application given significance in the curriculum guidance being provided by the Institute of Physics and the Royal Society of
Chemistry is absent from the guidance provided by the Royal Society of Biology.
However, perusal of the detail reveals *Inheritance and the genome* in How do organisms grow and reproduce? *Biodiversity and human impact* in How do organisms live
together? and *Treating disease* in How do organisms stay healthy? Within this detail
is powerful knowledge highly relevant to current future health care systems and the
relationship between humans and other living creatures on Planet Earth. Such considerations move learners to consider the wider implications of their biology knowledge.

Science and design & technology

David Layton, an acknowledged expert of in both science and design & technology education, played a key role in the conception of design & technology in the
National Curriculum in England. He acknowledged the difficulty faced by science

education. Writing as long ago as 1975: 'At the school level … the acquisition of scientific knowledge is inescapably tinged with dogmatism'. But almost 20 years later he used the following metaphorical question to explore the relationship between science and technology, 'Should science be seen as a cathedral, a quarry or a company store?' (1993). This has significant implications for the curriculum relationship between science and design & technology. In the cathedral of science, the purpose of the endeavour is to explore and explain natural phenomena without much in the way of considering possible application or exploitation. The goal is understanding – 'worshiping science for its own sake'. This is a purist position and in reality, of course, there is a dynamic relationship between science and technology in which there is a spectrum of response –from pure/fundamental science: driven by curiosity and speculation about the natural world without the thought of possible applications; through strategic science: yielding a reservoir of knowledge, out of which the as yet unidentified winning products and processes will occur; to applied science: related to a specific project and tied closely to a timetable with a practical outcome often specified by a client. Technologists, Layton argues, can rarely specify in advance what items in the cathedral will be most useful and so they treat it more as a 'quarry' to be visited and revisited less to marvel at the beauty of the creations there than to search out for items that might be of use. Note that in the middle ground Layton suggests the idea of the company store – spaces where strategic investigations predominate. We would identify such spaces as research and development centres where, to quote David Layton, 'the products of the cathedral are reorganised and remodelled to make them more accessible to practical users rather than worshippers'. So perhaps the science teacher wishing to teach in the light of pupils' learning in design & technology will need to view the knowledge, skill and understanding she teaches not only as a place of wonder and awe but also as a region into which her pupils can make forays of exploration for a variety of design & technological purposes – a space to be raided for that most precious of commodities – ideas that work. Appreciating the utility of subject knowledge outside the subject might well enhance learners' engagement. And it is ideas that work that are given some prominence in the writings of Tracey, Gibney and McLeod. As one might expect, they protect the subject's identity and integrity yet they each make the case for a view wider than a purist and inevitably narrow consideration of the individual subjects themselves. From the curriculum guidance issued by the Institute of Physics, the Royal Society of Chemistry and the Royal Society of Biology, it appears that there is now an acknowledgement of the view that it is important to value the knowledge of science both for its intrinsic worth and also for it 'applicability'. In addition, a publication for the Institution of Mechanical Engineers (2017: 7) argues that it is important to develop 'curricula that better reflect the importance of the made world to modern society, and make explicit reference to the engineering applications of science, mathematics, and design and technology'. It would appear that there is no contradiction between powerful subject knowledge informing school science subjects and an approach to STEM in the secondary school in which there is an interaction between these subjects and design & technology; quite the reverse, in fact. And with regard to interaction between the subjects, it is becoming increasingly clear that the problems now facing the world will need robust disciplinary knowledge used by interdisciplinary teams for their solution hence an interaction at school level might be a useful precursor.

Science and mathematics

So far, this discussion has concentrated on the possible significance of a curriculum relationship between science and design & technology. We must now consider the relationship between mathematics and science. There is general agreement that mathematical thinking provides the ability to identify and describe patterns in a wide range of phenomena. Clearly, this ability will prove useful in science, particularly in the move from qualitative to quantitative thinking. Consideration of, for example, speed, velocity and acceleration, only becomes worthwhile and potentially useful once such phenomena can be described algebraically and the 'describing' formula can be used to justify, for example, speed limits with regard to road safety. Here again, we have an example that takes the 'dogmatism' of science required for particular and precise definition and can be linked to the wider world and the personal interests and welfare of pupils. Such work could be extended in a variety of ways that make use of mathematics, including the derivation and interpretation of graphs to describe the motion of driven vehicles and the collation and interpretation of road safety statistics. However, it is worth noting that there are some mathematics teachers who argue for a purity in mathematics, unsullied by context, which confuses learners treating mathematics as a 'cathedral', which like science can be worshipped for its own sake (Onion, 2010).

We have seen that the views of significant members of the science education community concerning the implementation of one science curriculum 'for all' have revealed how the very nature of science can cause tension. We have noted that there is the possibility of resolving this tension to some extent by relating science learning to its application in the wider world through developing curriculum relationships with mathematics and design & technology. However, it is important to consider the status of those engaged in this relationship. We discuss this issue in the following section.

A relationship among equals

There is little doubt that science and mathematics are privileged subjects in terms of their curriculum status. Mathematics has long been regarded as an essential element in the education of all pupils. It has significant cultural status having been developed over centuries and providing the solution to some of history's most intractable problems. Jonathan Osborne (2012), in his address at the 50-year celebration of the Nuffield Foundation, reminded the audience that this was not always the case for science education. At the end of the nineteenth century and beginning of the twentieth century, some thought that the 'proper' education for an elite was deeply rooted in the Classics and humanities. This view was that science and technology were a necessary evil and that they did not offer proper training for the mind and that science had nothing to say about the human condition. It was believed that 'ordinary' people need only be educated in the three Rs. A particularly well-known rebuttal of this anti-science education position came in 1959 from C. P. Snow, who argued that anyone who did not know and understand the implications of the Second Law of Thermodynamics could hardly be considered educated. And now, some 60 years later, science along with mathematics has an apparently unassailable opposition in the curriculum. But what of design & technology – a relative newcomer to the curriculum existing in England as a defined entity within the national curriculum only since 1990? Both Jonathan Osborne (2012) and Robin Millar are clear that science education has definite instructional goals. There are singular answers to particular questions. Millar (2012: 23) sums this up well when he writes

> There is no merit in helping a learner to construct an idiosyncratic personal theory of matter or of motion (to take two examples) – it is the kinetic particle model and Newton's Laws that we want them to understand and be able to use.

Similarly, in mathematics teaching the educative goal is to some extent to impart agreed mathematical truths and procedures to enable mathematical thinking. This indicates a very real difference from design & technology where the values of both the designer and end user are integral to the process. Of course, there are clearly identified matters to be taught and learned in design & technology: properties of materials, ways to manipulate and join materials, ways to enable control and systems thinking for which there is an agreed understanding – but this is only half the story. Pupils then use this learning to develop products and systems to meet needs, wants and opportunities and it is perfectly possible, and indeed desirable, that the outcomes of this development produced by different pupils vary widely from one another. The extent to which particular developments meet needs, wants and opportunities is a matter of judgment and it is possible for quite different responses to be worthwhile. Hence design & technology does not suffer from the 'dogmatism' identified by David Layton. John Holman and Michael Reiss were very clear in their report to the Royal Society, *S-T-E-M Working Together for Schools and Colleges*, that it was important that any form of curriculum collaboration between science, mathematics and design & technology respected the legitimate differences between the subjects as well capitalising on areas of common interest. A difference of particular importance is the legitimacy of individual interpretations and responses in design & technology compared with the almost exact opposite in mathematics and science. But it is both interesting and significant that the curriculum guidance now being offered by the Institute of Physics, The Royal Society of Chemistry and the Royal Society of Biology acknowledge the role that their subject knowledge can play in developing such individual interpretations and responses.

In the previous section we considered how difficulties in pursuing science for all caused by the nature of science itself might be resolved to some extent by forging curriculum relationships with design & technology and mathematics. In this section we have noted that within these relationships it is essential that legitimate differences between the subjects are both recognised and valued. In the following section we will provide examples of how science activities can be related to pupil learning in design & technology and mathematics, taking these issues into account.

Examples of teaching science in the light of STEM

It is important to illustrate that all areas of science can benefit from teaching in the light of learning in design & technology and mathematics. Hence the following examples cover the breadth of science. They take into account the importance of recognising and valuing legitimate differences in the subjects involved.

Example 1: The magnetic effects of electric currents

TEACHING THE MAGNETIC EFFECTS OF ELECTRIC CURRENTS

Imagine a sequence of lessons concerned with teaching the effects of an electric current. You, as the teacher, could use iron filings and button compasses to show that as a direct electric current flows through a straight wire it generates a circular magnetic field around that wire. You could then challenge the pupils to explore what sorts of magnetic fields are formed when a coil of wire is used. With some guidance the pupils should be able to find out that the circular fields combine to give a field like that of a bar magnet. You could then challenge pupils to investigate the reverse possibilities with the question 'What effect does a magnetic field have on the electricity in a wire?' With some scaffolding of their investigations you could show that a moving magnetic field causes an electric current to flow in a wire. You could place this learning in the 'cathedral' of science indicating that these breakthroughs were made by the great Danish scientist Hans Oersted and the great English scientist Michael Faraday in the first half of the nineteenth century.

In the early days of scientific investigation, the prize sought was understanding with little thought of application. What we would now call 'blue sky research'. Yet the results of this understanding led to the development of solenoids, electric motors and dynamos. But it took over 50 years before these discoveries about the relationship between electricity and magnetism led to a powerful and useful electric motor. This is a clear example of Wynne Harlen's ideas of science – objects can affect other objects at a distance – and ideas about science – for every effect there is one or more causes and the knowledge produced by science is used in some technologies to create products to serve human ends. At this stage it would be worth reminding the class of the electric motors that they use in their design & technology lessons – small direct current motors containing permanent magnets. It would also be worth showing the class the internal structure of such a motor and giving them the opportunity to construct a simple electric motor for themselves (see instructions for assembling electric motors from everyday items 2020). Through these activities the pupils will begin to appreciate how the 'cathedral' can become the 'quarry' that can be mined for ideas of practical application; in this case, the understanding of the phenomena of electromagnetism being exploited to develop the electric motor.

Now it is worth the pupils considering how they might use their understanding of magnetism and electromagnetism in their design & technology lessons. The Design & Technology Association (2019) have developed visual materials that allow teachers and pupils to explore open starting points for their designing and making. These enable a class to explore a range of possible options without starting with a pre-defined product. Six starting points have been identified – playtime, keeping in touch, keeping secure, staying safe, thinking machines and other worlds. For example, let's say that a design & technology teacher is exploring 'playtime' with the class. She could suggest that whatever is designed uses electromagnetism and or magnetism. This would provide a technical focus for the activity without overly limiting the variety of toys that the pupils might choose to develop. The simple electric motor the pupils have already constructed could be a starting point for some pupils. It is not powerful but does spin very quickly and could be made the basis for a wide range of amusing and intriguing visual effects. Here, we have an example that illustrates the moves from the 'non-negotiable precision' required by science to the flexible interpretation necessary in design & technology. A science teacher could deal with the teaching of electromagnetism through demonstration only, communicating explanations to be learned. By requiring pupils to carry out investigations and simple making, the activity moves to a place where pupils have a greater responsibility to construct their own understanding. Using the understanding and the artefact, both 'constructed' by the pupils, in a designing and making task moves the pupils into situations where speculation is crucial – 'What if I do this? 'Can I do this?' 'Will this work' 'What about this?' The pupils are treating their science knowledge and understanding as a resource to be exploited, pushing it to the limits in their quest to produce an engaging toy. This almost playful pushing to the limits will in some cases require pupils to reformulate and increase their understanding.

If the construction of a simple electric motor is seen as too complex and too time consuming then a simpler item to consider might be a doorbell circuit that works by means of an electromagnet operating a make and break circuit causing the clapper, which is part of the circuit, to vibrate and ring the bell (Explain that stuff, 2019). Pupils can be given a homework task to find out how door bells have changed since this simple device was first introduced, at the beginning of the twentieth century and can now be part of the Internet of Things enabling a homeowner with a smartphone to see who is ringing the doorbell even when they are not at home.

With all electrical items powered by batteries there are opportunities to consider battery life. This requires the use of calculations that use data about the current that flows when the item is used. In this case of simple electric motor-driven toys, some pupils will be able to measure current consumption and use this to calculate how long particular batteries or arrangements of batteries might last. This is not a trivial task and for pupils aged 11–14 this would almost certainly be seen as extension work for those who had shown an aptitude to using mathematics. Conversations between all the subject specialist teachers are important here to ensure that the relevant science concepts are used appropriately, that the mathematical manipulations are sound and that the circuits under consideration are appropriate.

Example 2: Floating and sinking

The work of Archimedes in the second century BC is perhaps one of the first historical examples of scientific activity that can be seen to occupy David Layton's 'company store'. Archimedes was set the task of determining whether a crown made for King Hiero had been made from pure gold as supplied by the king or whether silver had been added by the goldsmith. Using the principle of buoyancy – the loss of weight when an object is immersed in water – Archimedes was able to show that the goldsmith was dishonest and had adulterated the pure gold with silver.

TEACHING BUOYANCY

Young pupils are introduced to the idea of buoyancy in the primary school by means of clas-sifying materials as floaters or sinkers. This leads to the idea that some materials are more dense than others. It is not until they become older that they are asked to explain the mecha-nism that causes some materials to float and others sink. Here, they are required to consider the forces acting on the material and conduct experiments in which they weigh materials in air and immersed in water and compare the apparent loss of weight on immersion with the weight of the water displaced by the material. Ultimately, they come to a statement of Archimedes Principle: any object, wholly or partially immersed in a fluid, is buoyed up by a force equal to the weight of the fluid displaced by the object. This is a clear example of Wynne Harlen's ideas about science – for every effect there is one or more causes and scientific explanation best fitting the facts known at a particular time.

The science teacher can help pupils relate this principle to their everyday experi-ences of floating in the bath, the swimming pool or the sea. So, given a table of the density of materials it becomes relatively easy to spot the pattern that if a material has a density greater than that of water (1g per cubic centimetre), then it will sink. It doesn't matter how much of the material is present – a single gram of lead will sink just as surely as a kilogram. The calculation of the density of different materials, involving weighing and the measurement or calculation of volume, would be an appropriate mathematics activity and there is no reason why some of this investi-gation into why some materials sink and others float should not be carried out as a part of mathematics lessons as well as science lessons. Mathematics teachers appre-ciate the difficulty pupils experience in understanding compound measures and exploring density would provide a useful activity to support learning in this area.

Both the mathematics teacher and the science teacher can challenge pupils' under-standing with the question, 'If iron and steel sink then how come ships made of iron and steel float?' Discussion can lead to the idea of shaping materials so that the shape can displace more than the volume of the material. This is easy to demonstrate in the design & technology workshop as follows. A disc of thin aluminium sheet when placed in water sinks. The disc can then be formed into a bowl by beating with a pear-shaped mallet in a dishing block and will then float when placed in water. This can be the starting point for a designing and making activity in design & technology in which pupils make water toys. These can include bath toys and small-scale replicas of yachts and powerboats. This links to Wynne Harlen's ideas about science – its use to create products. In all cases the way the toy floats will be dependent on both the material of the hull and the form of the hull. A hull made of solid wood will float very low in the water – it will be almost submerged. So, in order to float in a realistic way, the wood must not only be shaped to resemble a hull but also hollowed out to some extent. A hull made by vacuum forming thin sheet plastic over a hull-shaped solid block floats very high in the water, almost skimming along the surface. In this case additional weight needs to be added to reduce the buoyancy and achieve realistic floating. If the hull is to be made from sheet metal then it has to be formed from flat 'tin plate' into a hull shape. This is difficult to achieve without cutting the sheet into a net that is then folded up into a hull shape and soldered at the joins to prevent water leaking into the hull. The need to meet these making challenges in design & technology can be seen in the context of the learning about buoyancy that takes place in science lessons.

Example 3: Clean, accessible drinking water for all

It is not surprising that in *Chemistry for Tomorrow's World: A Roadmap for the Chemical Sciences* the Royal Society of Chemistry identified the availability of clean drinking water as one of the ten challenges in which chemistry has a key role. The UN Millennium Development Goal 6 target is to **ensure availability and sustainable management of water and sanitation for all.** Current population forecasts suggest that an additional 784 million people worldwide will need to gain access to improved drinking water sources to meet the targets for 2030, which are as follows:

- **Target 6.1:** By 2030, achieve universal and equitable access to safe and affordable drinking water for all.
- **Target 6.2**: By 2030, achieve access to adequate and equitable sanitation for all and end open defecation paying special attention to the needs of girls and women and those in vulnerable situations..
- **Target 6.3:** By 2030, improve water quality by reducing pollution, eliminating dumping and minimising release of hazardous chemicals and materials, halving the proportion of untreated wastewater and substantially increasing recycling and safe reuse globally.
- **Target 6.4:** By 2030, substantially increase water-use efficiency across all sectors and ensure sustainable withdrawals and supply of freshwater to address water scarcity and substantially reduce the number of people suffering from water scarcity.
- **Target 6.5:** By 2030, implement integrated water resources management at all levels, including through transboundary cooperation as appropriate.
- **Target 6.6:** By 2020, protect and restore water-related ecosystems, including mountains, forests, wetlands, rivers, aquifers and lakes.
- **Target 6.a:** By 2030, expand international cooperation and capacity-building support to developing countries in water- and sanitation-related activities and programmes, including water harvesting, desalination, water efficiency, wastewater treatment, recycling and reuse technologies.
- **Target 6.b:** Support and strengthen the participation of local communities in improving water and sanitation management.

The World Health Organization estimates that safe water could prevent 1.4 million child deaths from diarrhoea each year. Technology breakthroughs required include: energy efficient desalination processes; energy-efficient point of use purification, for example, disinfection processes and novel membrane technologies; developing low-cost portable technologies for analysing and treating contaminated groundwater that are effective and appropriate for use by local populations in the developing world, such as for testing arsenic-contaminated groundwater. There are strong links here with Wynne Harlen's ideas about science – the knowledge produced by science is used in some technologies to create products to serve human ends.

TEACHING PURIFICATION

For the science teacher, teaching about the simple purification techniques such as filtration and distillation is the starting point to developing the knowledge and understanding required to tackle this global problem. Situating the teaching of this elementary science in the context of this problem would indicate clearly the importance of science knowledge and understanding in tackling the large problems faced by the world.

There is a growing movement to encourage designers to tackle the problems encountered in the developing world. Books such as *Design for the Other 90 Per Cent* (Smith, 2008), *Design Like You Give a Damn* (Architecture for Humanity, 2006) and *Design Revolution* (Pilloton & Chochinov, 2009) are written to inform the general public about the way design can be a powerful means to improving the situation and to provoke a response from the design community. There is no reason why this approach cannot be extended to pupils in schools through the design & technology curriculum and build on the work carried out in science. Pupils could be challenged to develop simple filtration devices that 'clean' cloudy water and simple distillation devices to desalinate salt water. Such devices need only be developed into preliminary working prototypes that indicate their effectiveness. The pupils could then compare their designs with the devices developed by professional designers to identify differences and similarities, note where the basic science separation techniques were used, and enhance their appreciation of user-centred design and the importance of designs that empower people to improve their situation.

This learning of science in the light of learning in design & technology can also be extended to include mathematics quite simply. Asking pupils in mathematics lessons to estimate the amount of water they and their families use each day and compare this to that available to those living in developing countries would be a valuable learning activity. It could also lead to pupils considering the water shortages that might take place in the UK and the way such shortages are dealt with. This would inevitably lead to the question 'Where does our water come from?' which would take the learning science in the light of STEM full circle back to the science curriculum and the water cycle.

Example 4: The properties and applications of metals

TEACHING METALS

Consider the topic of 'metals'. The classification of some elements as metals or non-metals was an important step towards understanding the behaviour of these elements in chemical reactions. Most chemistry teachers will teach about metals: what they are like (their properties); where they come from (natural resources); how we get them (reduction of metal ores); what they are used for and why (properties related to use).

The design & technology curriculum is concerned with teaching what metals are like (properties); what they can be used for (properties related to use); how we can manipulate them (making skills) to design and make products that people need and want. Clearly, there is an overlap in teaching intention here and there is the possibility of capitalising on this through a collaborative scheme of work in which the science teaching about metals is tackled in the light of the learning that is taking place in design & technology. It will, of course, be important to engage with pupils in their daily lives and a simple 'homework' investigation of where there is metal at home and what is it used for will reveal its ubiquity – door handles, coat hooks, door knockers, casings for white goods, interiors of washing machines, filaments of light bulbs, electrical wiring, cooking utensils, cutlery, plumbing pipe work, mobile phone cases, etc. Pupils' attention can also be directed to the underlying structural framework of most modern buildings plus the chassis, body shell and moving parts of most cars and railway lines and the trains that run on them. Such a homework exercise can be divided between science and design & technology. Of course, the bigger picture must also be considered (i.e. the amount of production and its impact, the use and disposal of metal and the multifarious ways that metal is utilised in the made world). To find and use data about this requires some understanding of statistics and the ability to question what such data means. Hence there is an interesting role for the mathematics teacher in this collaboration.

Chemistry teachers will teach about the reduction of metal ores through practical activities that are reliable and intriguing. In design & technology lessons, the designing and making of simple body adornments is a relatively standard yet highly enjoyable exercise – copper rings, brooches, bangles and bracelets are all possible. It could be possible to link the production of copper from its ores learned in a science lesson to the use made of copper in the manufacture of body adornment. One possible way is to start with the copper ore in a science 'lesson' and from this produce enough copper to be used in making copper rings and bangles. This is perhaps not amenable to the everyday timetable but, as a STEM club activity, the production of copper from a small sack of green ore (malachite) by means of a home-made blast furnace would provide insight into industrial processes that produce the materials pupils use in their daily lives difficult to achieve in any other way.

Chemistry teachers may also teach about the electrolytic purification of crude copper that has been obtained by the reduction of copper ores. The electrolytic deposition of copper from a solution containing copper ions can also be used in design & technology lessons to decorate brass with designs made from copper. It is a relatively simple activity involving using the brass as the anode onto which copper will be deposited and masking areas of the brass so that electrolytic deposition takes place in the form of the desired decoration. Only a thin film of copper need be deposited to give a good effect. Once pupils have been taught about electrolysis in science it would be an interesting assessment of their learning to challenge them with decorating brass with copper as part of a sequence of design & technology lessons concerned with producing body adornment.

Example 5: Nutrition

TEACHING DIGESTION AND RESPIRATION

Many science courses for pupils aged 11–14 years teach digestion and its relationship with respiration. The 'key idea' here is that in many foodstuffs the molecules are large, in the form of polymers, and that for these materials to become useful to the body the large molecules need to be broken down in order to be absorbed through the gut wall (small intestine) and enter the bloodstream. Once in the bloodstream they can be transported throughout the body. Often, but not always, the treatment of digestion is limited to the digestion of starch, which breaks down to form glucose and that the glucose is then available as a source of energy to drive the chemical reactions on which the body depends and the use of muscles.

The formation of glucose from starch is often demonstrated through an experiment involving visking tubing, a semi-permeable membrane representing the gut wall. A starch solution is placed in some visking tubing sealed at one end and saliva is added. The tubing is lowered into water at body temperature. The pupils test for the presence of starch and glucose at regular intervals in small samples taken from within the visking tubing and the surrounding water. Care must be taken to avoid contamination between the solution inside the visking tubing and the surrounding water. Ideally, the test results should show the disappearance of starch from within the visking tubing and the appearance of glucose in the surrounding water. Students are given information about the size of starch molecules, glucose molecules and the perforations (holes) in the visking tubing and have to deduce the activity of the enzymes in the saliva in breaking down the large molecules of starch to much smaller molecules of glucose that can pass through the holes in the visking tubing, which are too small to allow starch molecules to pass. This explanation is linked strongly to Wynne Harlen's idea of science – all material in the universe is made of very small particles – but extended to include the idea that these very small particles can be of different sizes. It can be related to another of Wynne Harlen's ideas of science – organisms require a supply of energy and materials for which they are often dependent on or in competition with other organisms – by discussing with pupils where we might acquire starch in our diet in order to produce glucose.

It is at this point when the science teacher might consider reminding pupils of their work in food technology. Through discussion with stakeholders, Marion Rutland (2009) identified the following conceptual framework as being essential for a modern food technology course:

1   Designing and making food products.

2   Underpinned by an understanding of the science of food and cooking and nutrition.

3   Incorporating an exploration of both existing, new and emerging food technologies.

4   In the context of the sustainable development of food supplies locally, nationally and globally.

5   Incorporating an appreciation of the roles of the consumer, the food industry and
      government agencies in influencing, monitoring, regulating and developing the
      food we eat.

Learning about digestion as described above links strongly to the second concept mentioned above (understanding of nutrition in particular) and also the fourth concept
(concerning food supplies). So, it should be relatively easy for the science teacher and
the food technology teacher to collaborate around the teaching of digestion, where
the food technology teacher continues the work of the science teacher but deals with
sources of starch in various food stuffs, naturally occurring, processed and synthetic,
and relates this to the labelling of food stuffs that now indicate their calorific value.

   Calorific values are a cause of considerable conceptual confusion for pupils.
Calorific values are derived by burning foods and measuring the heat produced. For
many pupils, the relation of this to the energy released during respiration of small
molecules derived from such foods in our bodies is so counterintuitive that it seems
to be completely mysterious. Pupils can, of course, calculate the heat released when
potato crisps or breakfast cereals are burned but this is not a trivial task. This might
take place in either the science or food technology classroom – perhaps even both.
First, the material to be burned must be weighed. Then, as it is burned, the heat from
the burning must be transferred to a measured amount of water. The temperature
rise caused by the burning must be accurately measured. This temperature rise must
be converted into the amount of calories that caused the temperature rise. This, plus
the weight of material burned, must be used to calculate the heat released per gram.
This involves a lot of calculations relying on sound arithmetic (or competent use of a
calculator) and a strong understanding of ratio and proportion. So, a conversation with
the mathematics teacher would not be amiss here. Indeed, the mathematics teacher
might welcome the opportunity for pupils to be engaged in using mathematics for
such a 'real world' purpose. The results obtained will probably not mirror those produced by professional food technologists and listed on food packaging. If the order of
magnitude is similar then this is of great credit to the pupils. There are several sources
of error over which they have little control given the equipment available in junior
high schools. For example, some of the heat released will be absorbed by the atmosphere and some will be transferred to the container holding the water; measuring the
temperature will be accurate only to one or two degrees; weighing the material to be
burned will be accurate to only 0.1 gram. These inadequacies in experimental design
can, of course, be discussed with the pupils in terms of Wynne Harlen's ideas about
science – scientific explanations, theories and models are those that best fit the facts
known at a particular time. Here, the facts are being derived from experimental data
and are used to support a scientific explanation of respiration of foodstuffs providing
the body with energy. To achieve genuine understanding, it is necessary to make the
link between the measurement of calorific value of food by heat transfer and the
energy made available to cells throughout the body through respiration of small molecules derived from the food by digestion that is expressed on food packaging. It is
difficult to see how else this can be achieved other than by collaboration between the
science, food technology and mathematics teachers. In food technology lessons pupils
sometimes develop products for those with special dietary requirements and controlling calorie intake is often essential. Demystifying the energy content of various

foodstuffs by enabling pupils to understand how such values are obtained gives an important scientific dimension to such activities.

Example 6: Genetic modification

TEACHING GENETICS

Most school biology courses now describe and explain genetic modification (GM) and discuss possible costs and benefits of applying this knowledge to the development of genetically modified organisms, particularly with regard to GM crops. Here we have examples of Wynne Harlen's ideas of science – genetic information is passed down from one generation of organisms to another and ideas about science – the knowledge produced by science is used in some technologies to create products to serve human ends and applications of science often have ethical, social, economic and political implications.

The production and use of GM crops is a topic that raises strong emotions. In 1999, the UK reaction against GM crops was so strong that supermarkets removed products containing or associated with GM crops from their shelves. The science writer Bernard Dixon (New Scientist, 2012) provides an interesting account of the influences of many different stakeholders that led to this rejection citing, in particular, the circulation wars in the popular press leading to sensationalism which coupled with public ignorance led to an irrational fear of the new and emerging technology. Thee treatment in most science courses is now much more balanced giving voice to concerns by those who are apprehensive towards or against the use of GM crops, providing counter arguments and engaging pupils in discussion of the issues from different stakeholder perspectives. Such an approach is exemplified by the twenty-first century science biology course for pupils aged 14–16 years (Nuffield, York Twenty First Century Science, 2011). Hopefully, in the future such courses will lead to a general public who are better informed. However, as far as the UK is concerned the damage has been done and GM products are not available. Yet, the debate still continues so it is important for the science teacher to keep up to date with developments. For example, at the time of writing, the state of California is about to vote on Proposition 37, which would require the labelling of all foods containing GM ingredients and prohibit such foods from being marketed as 'natural'.

There appears to be little evidence that GM ingredients are intrinsically harmful to humans. GM can be used to develop pesticide resistance in crops, allowing farmers to use pesticides on such GM crops without harming the crops while preventing weeds from competing with the crops. But there is some evidence that this can lead to pests evolving resistance more rapidly. The situation is complex, and proponents of GM argue that rather than preventing the use of GM, it should be extended to include a wider range of GM crops, which could reduce the likelihood of resistance emerging by allowing farmers to switch the chemicals they use before pests evolve resistance. There are some (e.g. Greenpeace) who are philosophically opposed to the use of GM on the grounds that it is not natural and that natural plant-breeding methods can and should be used. Greenpeace have cited a developed strain of sweet potato that has four to six times the beta-carotene of an average sweet potato without recourse to genetic modification. Beta-carotene is a precursor to vitamin A in the body and important in preventing blindness in young children. A two-year project in Uganda involving 110,000 households demonstrated that eating the improved variety almost doubled the number of children who escaped vitamin A deficiency. This approach is in direct contrast to the Golden Rice project, which has developed GM golden rice (golden in colour as opposed to the white) that significant amounts of beta-carotene. Ordinary rice contains no beta- carotene. The latest trials of golden rice using isotopic labelling indicated that just 100 to 150 grams of the rice – about half the children's daily intake – provided 60 per cent of the recommended daily intake of vitamin A. Greenpeace's position is in some cases more nuanced than an outright rejection of GM per se. They argue that the herbicide tolerance deliberately engineered into GM crops encourages the use of glyphosphate-based herbicides that kill all other vegetation leaving the GM crop free to grow without competition but there is mounting evidence that questions the safety of glyphosphate. This leads them to argue for a ban on developing GM crops that are glyphosphate tolerant. An editorial in the *New Scientist* magazine in 2012 called for multiple solutions to be adopted to combat preventable blindness; not just natural breeding, not just GM but both.

There are those who argue that natural breeding methods will be totally inadequate, and it is only through significant investment and deployment of GM that the world food problem can be addressed. Mark Lynas, once a committed activist against GM, has now completely changed his position in response to consideration of planetary boundaries with particular regard to the nitrogen boundary (Lynas, 2011). The natural mechanisms of the nitrogen cycle do not provide enough nitrogen in forms that can be used as fertiliser to support growing the food the world needs. The synthesis of ammonia and nitrates from nitrogen in the atmosphere has enabled the world to produce significant amounts of fertiliser that are used worldwide but at significant environmental cost. He argues that it will be essential to develop GM crops that are more efficient at utilising nitrogen thus reducing and even eliminating our dependency on synthetic fertilisers.

Where does all this leave the food technology teacher? Marion Rutland (2011) has argued that a modern food technology programme should involve a consideration of new and emerging food technologies and GM would seem an important example. Often in design & technology classes, pupils are challenged with the question 'What would you use this or that technology for?' Clearly, if the technology in question is GM it is important that the food technology teacher is *au fait* with the learning that has taken place in science. Similarly, it is important that the science teacher who is teaching about GM knows that there is the possibility that pupils will be asked to consider its applications as a new and emerging technology in their food technology lessons. In considering GM in food technology, especially when speculating about possible uses, it is important that the speculation deals with what is feasible and that it is underpinned by current scientific understanding. Marion has also argued that in a modern food technology programme pupils should consider the sustainable development of food supplies locally, nationally and globally. Considering the role of GM crops would certainly enable pupils to engage with the global picture. And it is here that the use of mathematics will be significant in helping pupils understand the sheer scale of the GM problem: the projected growth in world population, the different foods needed to feed the increasing population, the extent of malnutrition in the world and resultant dietary related disease, and the amount of fertiliser and pesticides that are needed to maintain the food production. The statistics used to track these issues and their interpretation to inform global food policy will provide a rich context for learning and using mathematics.

Example 7: Building your own laboratory equipment

TEACHING EXPERIMENTAL EQUIPMENT

In attempts to modernise science curricula, many teachers use data logging equipment. But such equipment is expensive, and when finances are limited schools can only afford to purchase one or two of such items then their use is confined to demonstrations. Joshua Pearce, an associate professor at Michigan Tech in the US, has found a solution to this problem, one he applies to his own work in higher education and one that he feels is also appropriate for secondary (high) schools. Joshua advocates the use of a 3D printer to produce structural parts and the open source Arduino microcontroller to drive the 3D printer and provide data capture functionality. For schools in the UK, the PICAXE microcontroller could be used to achieve the same result.

Joshua Pearce (2012) waxes lyrical about the benefits of using 3D printers in educational settings:

> the open-source microcontroller is key. The beauty of this tool is that it's very easy to learn. It makes it so simple to automate processes. Here's how it works. The Arduino chip – which retails for about $35 at RadioShack – can run any number of scientific instruments, among them a Geiger counter, an oscilloscope and a DNA sequencer. But it really shines when it operates 3D printers like the open-source RepRap. This microwave-sized contraption starts at about $500 and can actually make parts for itself. Once you have one RepRap, you can make an entire flock. My lab has five 3D printers make stuff by laying down sub-millimetre-thick layers of plastic one after another in a specific pattern. This allows users to make devices to their own specifications, so they don't have to make do with what's available off the shelf. The Arduino controls the process, telling the printer to make anything from toy trains to a lab jack. Lab jacks raise and lower optical equipment and aren't radically different from the jacks that raise and lower your car, except that they are more precise. I received a quote for a $1,000 version, which inspired me to design my own. Using a RepRap, inexpensive plastic filament and a few nuts and bolts, my students and I made one for under a buck. Then we posted the OpenSCAD code used to make the lab jack on Thingiverse, a web repository of designs where members of the 'maker community' can submit their designs for all kinds of objects and receive feedback. Immediately someone I'd never met said, 'This isn't going to work quite right, you need to do this'. We made a simple change, and now I have a lab jack that's superior to our original design. The Thingiverse community already has a whole line of open-source designs for over 30,000 'things', and everyday it's only getting better. Using open-source hardware has easily saved our research group thousands of dollars, and we are only getting warmed up. This will change the way things are done.

3D printers are now available to secondary schools in England, too. The Department for Education initiated a small pilot study to explore their use in the design & technology curriculum. The use of microcontrollers, usually PICAXE, has become well established in some design & technology departments due to in-service training programmes that focus on the use of digital technologies. Hence, for a science teacher who wants her pupils to develop their own experiments, there is the possibility of students using the learning in design & technology to devise and manufacture the necessary equipment. This would not be a trivial task. The science and design & technology teachers would need to discuss at some length the sorts of experiment that the pupils might wish to carry out, the equipment required and the programming of the microcontrollers necessary to enable sensors to collect the required data. In all probability, the design & technology department would have computer-assisted design (CAD) software that students could use to design structural parts and produce files that could be loaded directly into the 3D printer. This would eliminate the need to programme a microcontroller system to drive the 3D printer. In developing the structural parts for their experiment there is scope for considerable mathematical thinking that uses pupils' knowledge and understanding of both measurement and geometry. Achieving the correct size, shape and form for laboratory equipment such that the result can contain the electronics and required battery, allows access for battery replacement and microcontroller programming, connection of sensors is no mean feat. A three-way conversation between the science, design & technology and mathematics teachers would reveal the extent to which mathematics learning could be used to enhance and facilitate this task.

The National STEM Programme for England was published in 2008 and the Action Programme 10 was devoted to improving the quality of practical work in science. The body responsible for this aspect of the programme was SCORE (Science Community Representing Education). SCORE is a partnership between the Association for Science Education, the Biosciences Federation, the Institute of Biology, the Institute of Physics, the Royal Society, the Royal Society of Chemistry and the Science Council. SCORE acts under the auspices of the Royal Society. SCORE developed a framework for practical work in science and an accompanying professional development programme enabling teachers to discuss the framework and build it into their teaching. The framework identified a range of features present in high quality practical work including the following:

- Self-directed enquiry by individuals, or more commonly by groups, which promotes 'pupil ownership' of science and can be motivating and enjoyable.
- Investigations to encourage teamwork with members being given particular roles in the planning, implementing, interpreting and communication of the work.
- Extended enquiry or projects, which encourage pupil autonomy and opportunities for decision making.
- Use of ICT for handing and presenting data and contemporary technical equipment to relate science techniques in school to modern practice.

The emphasis on the importance of practical work in science lessons has not changed as indicated by these quotes from the curriculum guidance given by the Royal Society of Biology, The Royal Society of Chemistry and The Institute of Physics:

From the Royal Society of Biology:

- Carry out practical work and manage associated risks associated with practical work.
- Use technology and scientific instruments to facilitate and support practical work.

From the Royal Society of Chemistry:

- Development of investigative skills.
- Development of basic laboratory skills such as careful measurement.
- Use of those skills in specific processes.

From the Institute of Physics:

- Carry out practical investigations, performing practical tasks to develop laboratory techniques and aspects of procedural knowledge, including isolating phenomena, controlling variables, making observations and measurements, analysing and interpreting data, testing plausibility of results, developing and refining explanations.

Developing your own laboratory equipment in the way advocated by Joshua Pearce would meet the features initially identified by SCORE and later endorsed by the separate institutions in high quality practical work and provides an inspirational example of teaching science in the light of STEM.

Maintaining the integrity of learning in the interacting subjects

Each of the examples of teaching science in the light of STEM can be justified in terms of meeting an aspect of Wynne Harlen's ideas of science and about science. Hence, although the science teaching has been undertaken in the light of learning that has already or might take place in mathematics and design & technology, it has not been compromised. The design & technology activities that were described in this chapter, as in all of the examples, link science learning to either design & making activities or to designing activities to which there are multifarious, as opposed to single correct answers. Hence, the particularly significant difference between science and design & technology has been respected and preserved. And in each example, science understanding has informed designing and making such that the utility of the science is exemplified without the science teaching becoming distorted. The nature of the mathematics activities is varied and includes simple measurement and estimation, the use of both arithmetic and algebraic calculations, the use of nets, understanding compound measures, using ratio and proportion and the interpretation of statistical data concerned with real world activity. All of these activities relate to topics that are prominent in most mathematics courses for pupils aged 11–16 years today. The mathematical perspective provided through making these links informs both science and design & technology and in some cases provides a mathematical window onto significant global problems. Hence, as with science, the utility of the mathematics is exemplified without mathematics teaching becoming distorted.

Conclusion

This chapter raises many questions. Does the nature of science really make it that difficult to teach because it is essentially 'outside students' individuality'? Is there room for a personal response to science within science for young people at school? One approach is to rely on creative pedagogy. The use of a blast furnace to create copper may appear intrinsically boring but if the response to learning about this and showing understanding is through creative writing then maybe it could actually be much more interesting. Asked to respond in such a manner we know of one student who wrote a short parody of a James Bond movie. 'My name is Bon, Car Bon. When things get hot I don't sweat, I just get stronger. That devil iron is no match for me when it comes to a contest over oxygen!' Of course, many teachers do use creative pedagogy but the thrust of this chapter is that there is another very powerful weapon in teachers' armoury that can be used to combat some students' disillusion with science and to enhance the science curriculum for all. That is, science teaching should be carried out in the light of STEM where links with design & technology enable students to 'raid' science for useful ideas and links with mathematics can reveal the elegance of relationships within phenomena that can be described in no other way. We hope that the examples of teaching science in the light of STEM we developed here have shown what is possible. You might find some implausible or inappropriate for your particular situation; that is almost inevitable given the different circumstances in which schools and teachers find themselves.

Our hope is that you will be able to look at your own curriculum and see where you might teach science in the light of STEM to the advantage of your students. We firmly believe that professional development is an essential means for you to explore how you might teach science in the light of STEM. The experience of exploring practical activities and tricky concepts with a view to teaching science in the light of STEM is an important first step. Good ideas will come from the conversations you have with your colleagues.

Finally, we have had to wrestle with the issue of curriculum status. This has been a particularly sensitive issue for design & technology in England with a government expert panel suggesting that the subject has 'weak epistemological roots' compared to other subjects more established in the curriculum. We believe that it is imperative to acknowledge the integrity of each subject, respect their ways of knowing and understanding and appreciate their different learning intentions. Vera John Steiner has written about the dignified interdependence that underpins creative collaboration. We fully support her position with regard to the highly creative and collaborative endeavour of teaching science in the light of STEM. Hence, in teaching science in the light of STEM, our advice is not to neglect the importance of regular conversations with colleagues from design & technology and mathematics.

Recommended reading

The following deal with the nature of science

Al-Khalili, J. (2020) *The world according to physics*. Princeton, New Jersey: Princeton University Press.

Gibney, D. (2018a) Towards an ideal chemistry curriculum. *School Science Review*, *100*, 30–35. Herts, Association for Science Education.

Harlen, W. (2010a) *Principles and big ideas of science education*. Hatfield Herts: Association for Science Education.

Layton, D. (1993a) *Technology's challenge to science education*. Buckingham, UK: Open University.

McLeod, L. (2018a) Developing a framework for the biology curriculum. *School Science Review*, *100*, 23–29. Hatfield Herts: Association for Science Education.

Tracey, C. (2018) Guidelines for future physics curricula. *School Science Review*, *100*, 36–43. Hatfield Herts, Association for Science Education.

References

Architecture for Humanity (2006) *Design like you give a damn: Architectural responses to humanitarian crises*. San Francisco: Architecture for Humanity.

Design & Technology Association (2019) *Starting points for designing*. www.data.org.uk/for-education/curriculum/ (accessed 30 April 2020).

Donnelly, J. (2003) A loss of faith in science education? *Education in Science*, *203*, 18–19.

Explain that Stuff (2019) *A Door Bell Explained*. www.explainthatstuff.com/how-electric-door-bells-work.html (accessed 30 April 2020).

Gibney, D. (2018b) Towards an ideal chemistry curriculum. *School Science Review*, *100*, 30–35. Hatfield Herts, Association for Science Education.

Harlen, W. (Ed.) (2010b) *Principles and big ideas of science education*. Hatfield, Herts: Association for Science Education.

Institution of Mechanical Engineers (2017) *'We think it is important but don't quite know what it is': The culture of engineering in schools*. London: Institution of Mechanical Engineers.

Instructions for assembling electric motors from everyday items. www.arvindguptatoys.com/toys/motor.html (accessed 30 April 2020).

John Steiner, V. (2000) *Collaborative creativity*. *England*: Oxford University Press.

Layton, D. (1975) *Science for the people*. London: Allen and Unwin.

Layton, D. (1993b) *Technology's challenge to science education*. Buckingham, UK: Open University.

Lynas, M. (2011) *The God species how the planet can survive the age of humans*. London: Fourth Estate.

McLeod, L. (2018b) Developing a framework for the biology curriculum. *School Science Review*, *100*, 23–29.

Millar, R. (2012) Association for Science Education Presidential Address 2012. Rethinking science education: Meeting the challenge of 'science for all'. *School Science Review*, *93(345)*, 21–30.

National Science Learning Centre (2008) *The STEM framework*. York: National Science Learning Centre.

New Scientist (17 August 2012) *Nutrient-boosted Golden Rice should be embraced*. www.newscientist.com/article/mg21528783.000-nutrientboosted-golden-rice-should-be-embraced.html (accessed 30 April 2020).

Nuffield/York (2011) *Twenty first century science: GCSE biology workbook*. Oxford: Oxford University Press.

Ogborn, J. (2004) Science and technology: What to teach? In M. Michelini (ed.), *Quality development in teacher education and training* (pp. 69–84). Udine: Forum.

Onion, A. (2010) Private communication with the authors.

Osborne, J. (2012) *Sustaining the spirit of Nuffield?* Presentation given at an event to celebrate the 50th anniversary of the Nuffield Foundation's work in curriculum development. www.nuffieldfoundation.org/sites/default/files/files/Jonathan_Osborne_speech_8_May_2012.pdf (accessed 30 April 2020).

Pearce, J. (2012) *Do it yourself and save: Open-source revolution is driving down the cost of doing science.* http://phys.org/news/2012-09-open-source-revolution-science.html (accessed 30 April 2020).

Pilloton, E. & Chochinov, A. (2009) *Design revolution: 100 products that are changing people's lives.* London: Thames and Hudson.

Postman, N. (1993) *Technopoly: The surrender of culture to technology.* New York: Vintage Books.

Pyne, C. (2015) *National innovation and science agenda.* www.minister.industry.gov.au/ministers/pyne/media-releases/agenda-transform-australian-economy (accessed 30 April 2020).

Roberts, G. (2002) *SET for success: The supply of people with science, technology, engineering and mathematics skills.* London: HMSO. https://webarchive.nationalarchives.gov.uk/+/http:/www.hm-treasury.gov.uk/d/robertsreview_introch1.pdf (accessed 30 April 2020).

Royal Society (2007) *S-T-E-M working together for schools and colleges* (unpublished).

Rutland, M. (2009) *An investigation: Is there a need to modernise the secondary school food technology curriculum? Unpublished Report.* Wellesbourne, UK: Design and Technology Association.

Rutland, M. (2011) Food technology in secondary schools in England: Views on its place in a technologically advanced nation. In Kay Stables, Clare Benson, & Marc de Vries (eds.), *Perspectives on learning in design & technology education.* Proceedings of the *PATT 25: CRIPT 8 Conference 2011*, 349–356. London: Goldsmiths University of London.

Science and Learning Expert Group (2010) *Science and mathematics secondary education for the 21st century.* London: Department for Business, Innovation and Skills.

SCORE (2013) *Improving practical work in science.* www.stem.org.uk/resources/collection/3982/score-practical-work-school-science-reports (accessed 30 April 2020).

Smith, C. (2008) *Design for the other 90 per cent.* New York: Smithsonian Cooper-Hewitt National Design Museum.

Snow, C. P. (1959) *The two cultures and the scientific revolution.* Cambridge: Cambridge University Press.

Tracey, C. (2018b) Guidelines for future physics curricula. *School Science Review, 100*, 36–43. Hatfield Herts: Association for Science Education.

Wolpert, L. (1992) *The unnatural nature of science.* London: Faber.

Young, M. (2008) *Bringing knowledge back in: From social constructivism to social realism in the sociology of education.* London: Routledge.

Young, M. & Lambert, D. (2014) *Knowledge and the future school.* London: Bloomsbury.

Teaching design & technology in the light of STEM

The nature and purpose of design & technology

There is little doubt that humanity has behaved technologically since the emergence of the species from Africa. Underpinning this was the development and use of tools. It is a moot point as to whether tools enabled the development of language or vice versa but the powerful combination of tool use and language has defined the development of human civilisations ever since. Jacob Bronowski (1973b: 19) explained this in terms of human's ability to envisage what might be.

> Man is a singular creature. He has a set of gifts, which make him unique among animals; so that, unlike them, he is not a figure in the landscape – he is a shaper of the landscape.

Bronowski (1973b: 116) captured the nature of this accomplishment in three brilliant sentences.

> The hand is the cutting edge of the mind. Civilisation is not a collection of finished artefacts; it is an elaboration of processes. In the end, the march of man is the refinement of the hand in action.

For those devising the National Curriculum for England in 1988, it was important to include a subject that reflected this unique feature of human achievement. David Layton echoed this in an interim report (Department for Education and Science and Welsh Office, 1988: 3) to the government of the time as follows:

> What is it that pupils learn from design & technological activity which can be learned in no other way? In its most general form the answer to this question is in terms of capability to operate effectively and creatively in the made world. The goal is increased 'competences in the indeterminate zones of practice'.

In the early days of the National Curriculum the above statement was puzzling to many teachers, and in Chapter 1 we noted some of the changes that took place in the statutory requirements for the subject as it evolved from its inception to its current form. Put briefly, the subject evolved in a direction that valued procedural competence, which was taught through the activity of designing and making at the expense of defining a clearly articulated body of knowledge to support this activity. This position led to a significant criticism of the subject. During the revision of the National Curriculum in England initiated by the then Minister of Education Michael Gove in 2011, the Expert Panel commissioned to undergo the revision initially decided that design & technology lacked the necessary epistemological roots to merit a statutory programme of study and should be seen as a subject outside the National Curriculum (DfE, 2011). Research by John Williams and John Lockley (2012) explored the views of early career science and technology teachers as to what might be considered 'enduring ideas' within the subjects they taught. Interestingly, this research revealed that while the science teachers had little difficulty in identifying such ideas, this was not the case for the technology teachers. The authors noted that this may be in part due to the extensive place of procedural knowledge in technology but also that technology has no commonly agreed epistemology. After much lobbying, design & technology was accepted as a National Curriculum subject and a statutory programme of study was agreed (DfE, 2013).

We believe that a useful approach to this problem is to adopt that used by Wynn Harlen and her colleagues for science education. They identified important ideas *of* science and important idea *about* science. What would we list as ideas 'of' and 'about' design & technology?

Ideas of design & technology might include:

- knowledge of materials: sources, properties, footprint, longevity;
- knowledge of manufacturing, by: subtraction, addition, forming, construction;
- knowledge of functionality: powering, controlling, structuring;
- knowledge of design, methods for: identifying peoples' needs and wants identifying market opportunities; generating and developing design ideas;
- evaluating design ideas;
- knowledge of critique, for: justice, stewardship.

Ideas about design & technology might include:

- Through design & technology people develop technologies and products to intervene in the natural and made worlds.
- Design & technology uses knowledge, skill and understanding from a wide range of sources especially but not exclusively science and mathematics.
- There are always many possible and valid solutions to technological and product development challenges, some of which will meet these challenges better than others.
- The worth of technologies and products developed by people is a matter of judgement.
- Technologies and products always have unintended consequences beyond intended benefit, which cannot be predicted by those who develop them.

How would these ideas play out in the way the subject will be taught? If we continue to look at the work of our colleagues in science we see that their prevailing pedagogy is based on constructivist thinking encapsulated by Rosalind Driver's seminal work *The Pupil as Scientist* (1983). By analogy we might want a pedagogy based around the 'pupil as technologist', an idea already espoused by Richard Kimbell and David Perry in *Design and Technology in a Knowledge Economy* (2001). Clearly, such pedagogy would include design and making activities but might also include activities in which pupils make without designing, design without making and explore the relationship between technology and society. Barlex and Steeg (2017b) have developed further the idea of *Big Ideas for Design & Technology*, indicating that it is a powerful approach. Their ideas are at a high level of summary and considerable detail will need to be added as teachers devise a curriculum that incorporates these ideas. And it is important that such detail is added in a way that embraces a wide range of approaches to design incorporating, for example, the practices of different cultures in different places and at different times. In this way, an over-emphasis on modern Eurocentric approaches can be avoided and the insights of indigenous peoples can be taken into account.

A corollary to design & technology becoming part of the National Curriculum in England was the development of a single title GCSE to replace the several material area specific GCSE qualifications that existed (DfE, 2015). The following features of this new qualification, taken when young people are aged 16 years old, are particularly relevant:

- The knowledge, understanding and skills that all students must develop are separated into technical principles and designing and making principles.
- Through their work in design & technology students must apply relevant knowledge, skills and understanding from key stage 3 and 4 courses in the sciences and mathematics.
- The assessment requirements involve both a written paper (worth 50 per cent of the marks) and an extended task called a contextual challenge (worth 50 per cent of the marks).
- Food is not a material to be considered or used in design & technology.

The emphasis on technical principles has significant implications for teachers in that all young people studying the subject need to be taught about the Big Ideas identified earlier across a range of materials and technical components. This knowledge and understanding would be assessed in the written paper along with the application of science and mathematics. Also, the contextual challenge was significantly different to the course work assessment of the previous GCSE qualifications. There is an emphasis on user-centred iterative design and the Awarding Organisation (AO) does not set a brief that candidates must tackle as used to happen. Instead, the AO provides a context which candidates must explore and through identifying need and wants in that context develop their own brief. This places much greater decision-making responsibility on the candidate. Barlex (2007) developed an approach to such decision making by means of a 'design decision pentagon' shown in Figure 4.1 involving five key areas of interdependent design decision:

- conceptual (overall purpose of the design, the sort of product that it will be);
- technical (how the design will work);

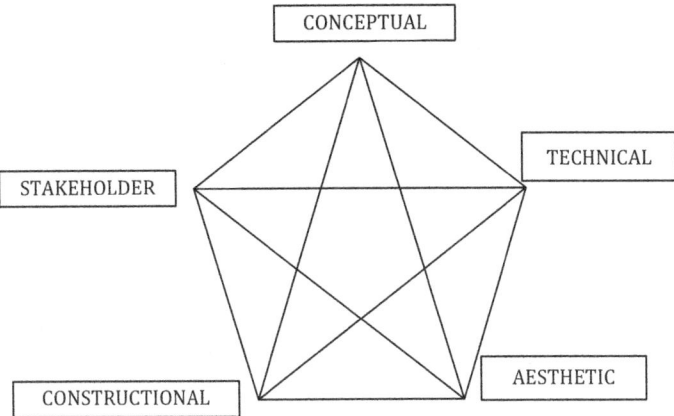

FIGURE 4.1 Design decisions as a pentagon

- aesthetic (what the design will look like);
- constructional (how the design will be put together);
- stakeholder (users, buyers, sellers, manufacturers).

This model of designing as a series of interrelated design decisions at different levels of detail culminating in a design proposal that is realised through making is useful for teaching, learning and assessment in the new single title GCSE. The teacher can use the model to focus a pupil's attention on what it means to design. The model allows the pupil to focus deliberately but not exclusively on particular features of his or her designing. The model allows the assessor to focus on particular features of a pupil's designing without losing the important holistic overview of the design process.

Design & technology and science

In May 2019, David revisited the interview that he had with Torben Steeg in 2011 that discussed the links between design & technology and science. Steeg is a freelance consultant in education and is widely regarded as a national expert in the teaching of electronics, systems and control and modern manufacturing. He also has a strong background in science education having spent the early part of his teaching career as a physics teacher. Initially, Steeg identified the 'usual suspects' of science topics that might be useful in design & technology: electricity, energy, materials, structures, forces, motion, food and nutrition. However, teachers need to be clear that just because these topics have been met in science doesn't mean that students will be able to effortlessly apply them in design & technology. There are two reasons for this; first, as we discussed in Chapter 2, it's well established that students don't easily transfer knowledge from one subject setting to another (Perkins & Salomon, 1992). Teachers need to do some work in making the links between subjects recognisable. Second, even when provided with relevant but 'raw' scientific knowledge, it isn't always clear exactly how it can be used in design & technology (Layton, 1993b). Work needs to be done in turning the scientific knowledge into a form that is useful for informing designing and/or making. Torben used the term 'reconceptualised knowledge' for this. Work needs to be done in turning the scientific knowledge into a form that is useful for informing designing and/or making. He also noted the potential for the use of scientific method or scientific thinking to inform designing and making. That is the ability to approach a question in design & technology with a desire for empirical evidence; the attitude of 'let's find out'. Students need to learn when such an approach is and is not appropriate for design & technology – and that, equally, science teachers might also want them to be able to do this. For example, in deciding technical matters, such as defining the cross-sectional area of a material to give the required strength and stiffness, empirical investigations and the application of material science concepts may be useful and appropriate. It's worth noting that, in reality, a designer is likely to simply look up appropriate cross-sectional values in a table (which is a great example of reconceptualised knowledge). However, there may be pedagogical value in requiring students to undertake investigations of this sort, for example to help make it clear how reconceptualised knowledge, such as the information in a cross-sectional table, is produced.

In contrast, deciding on the overall appearance of a product such that it is 'cool' or has what product designers Dick Powell and Richard Seymour term visceral appeal – 'you want it before you know what it is' – does not rely on a scientific approach.

Torben felt it was valuable to note that science teachers often use examples of applications to illustrate scientific principles. This reveals an interesting difference between the approaches used by science and design & technology teachers. A science teacher might use the example of a drawing pin to illustrate pressure, instructing the pupils to hold the pin between thumb and forefinger and squeeze – but not too tightly – and then ask them to explain what they felt. A design & technology teacher is more likely to ask pupils to take 'user trips' with a variety of drawing pins – different lengths of pin, different surface area of head, different types of head, pushed into different surfaces – and explore how easy it is to use different types of pin and speculate why there are different types. Although the physics of the drawing pin as a pressure multiplier underlies both activities, the science teacher is using the drawing pin to help pupils understand the nature of pressure whereas the design & technology teacher is using the drawing pin to help pupils understand users. In fact, this design & technology activity is likely, inherently, to require no explicit discussion of pressure as force/area, unless there is a cross-curricular imperative to do so. It does seem plausible that having an understanding of drawing pins as pressure multipliers would make a product analysis of pins better informed.

Torben suggested that a good example of a design & technology task that rests on scientific understanding is the use of electric circuits in the design of masks that light up in the dark. There would clearly be a case for considering what pupils might have learned in science about simple circuits. If the light-up elements are in series then the number of elements that can be used is quite small, but if they are arranged in parallel then a greater number can be used. Torben has often used a circuit simulator to allow students to explore different circuit designs while reinforcing the scientific ideas of series and parallel circuits (and ensuring that the circuits adopted do actually work!). If the pupils were required to explain the behaviour of their circuit and its limitations, then science learning would be useful here. If the pupils are required to consider battery life (i.e. work out how long the mask can be used for on a single battery), then the need for technical understanding increases. What kind of scientific understanding is required here is an interesting question. If the explanation of the circuit is couched in terms of current flow then the explanation will revolve around understanding that 'more components' leads to 'higher current flow' and therefore 'the less time the battery will last'. Hence Torben was inclined to use current as the pertinent factor (which can either be measured or looked up from a component's data sheet) and use the fact that most batteries provide an mAh value; this reduces the science understanding demand significantly as there is no need to consider voltage or power. The unit of 'mAh' is another good example of reconceptualised knowledge, allowing different power sources to be easily compared and matched to the current demand of a circuit. This indicates that it might unnecessarily complicate matters to require pupils to understand all the science behind the performance characteristics of the components they are using.

This led Torben to consider the use of chooser charts, developed by the Nuffield Design & Technology Project (2000). An example is shown in Figure 4.2. These are tables that summarise a suite of related reconceptualised knowledge, often drawn from science (e.g. to describe the performance characteristics of components or materials or the usefulness of particular techniques). Their aim is to provide pupils with the information they need to make informed design decisions either unaided or with minimal support from their teacher. An confident pupil can use such charts to make decisions, which he or she can then justify to the teacher. For a less confident pupil, the teacher can ask questions, engaging the pupil with the content of the chart and leading the pupil to make their own decisions.

Electric Components Chooser Chart

What the components might need to do	*Options*	*Symbols*	*Points to check*
To provide a power supply	batteries: • zinc carbon for low current, infrequent use • zinc chloride for medium current, regular use • alkaline for high current, heavy use		Make sure voltage of battery is suitable for components in the circuit.
To make light			
To give a signal	a light-emitting diode 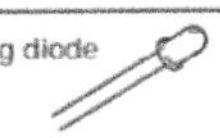		Use protecting resistor. Must be correct way round.
	a flashing light-emitting diode 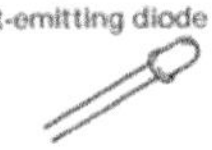		Does not need protecting resistor. Must be correct way round.
To provide illumination	a light bulb 		Must match power source.
To give rotary movement	an electric motor 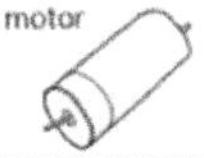		Must match power source. May need 'gearing'.
To make sound			
	a bell		Must match power source.
	a buzzer 		Buzzer must be correct way round.

FIGURE 4.2 Part of a chooser chart from the Nuffield Design & Technology project

Design & technology and mathematics

To gain insight into the links between design & technology and mathematics, David interviewed Celia Hoyles in 2009. Celia has been Professor of Mathematics Education at the Institute of Education, University of London, since 1984 and was the UK government's Chief Adviser for Mathematics between December 2004 and November 2007. She was Director of the National Centre for Excellence in the Teaching of Mathematics from 2007 to 2013. Celia explained that much of the mathematics curriculum for pupils aged 11–14 years is about discerning and expressing structure, pattern and relationships, which include exploring data and appreciating and describing trends. Within this latter activity, there are important concepts that need to be understood if pupils are to be successful: for example, understanding the scale used on axes, the nature of the units used, the gradient of straight-line graphs and how all these might relate to compound units, rates of change and effects over time. Understanding probability and its relationship with assessing risk is also an important area to be explored. Celia identified sustainability as an area of increasing importance across the curriculum. This manifests itself with a concern for dealing with resource depletion (the world is running out of natural resources) and global warming (the impact of increasing carbon dioxide emissions on climate change). Inaction is not an option and despair is counterproductive. So, looking at relevant data using mathematical understanding to gauge the scale and significance of the problem is important. Only then is it really worth giving our pupils the chance to think, just as a professional designer would, about how these problems might be solved. So, a joint venture might be that mathematics teachers and design & technology teachers identify the sorts of data needed for a collaborative venture around sustainability. Such a project could provide a rich context for learning about data, its representation and interpretation in mathematics with the understanding of the data being used to explore designing for a sustainable future in design & technology.

Since measuring is a fundamental part of both mathematics and design & technology, Celia thought it would be an area of exploration likely to be of mutual benefit. The conversation moved quickly on from the hoary bone of contention 'measuring length in millimetres in design & technology versus centimetres in mathematics' to the more positive arena of collaborating over the designing and making of a measuring device of some kind, suitable perhaps for Year 9 pupils. She wondered about pupils designing and making a weighing machine to meet an identified need in school (e.g. a weighing machine that can be used in the school prep room to weigh small animals). Here, the nature of the artefact immediately suggests mathematical thinking: understanding the range of measurement, an appropriate scale, calibrating the device, understanding the need for, and demonstrating reliability as well as other considerations, such as ease of use and comfort/minimal distress for the animal being weighed. Developing such a device might involve calibrating the stretch characteristics of a range of elastic bands such that the device could operate over a wide range of loads. Celia thought this was an example that would be worth mathematics and design & technology teachers discussing. A good starting point for this conversation would be the Nuffield Capability Task Better Weighing (Nuffield Design & Technology Capability Tasks, 2000).

The contextual challenge in the new GCSE design & technology requirements enables pupils to move outside individual materials and tackle mixed media projects. Disaster relief would provide an intriguing context for such a challenge and pupils might respond by designing and making a pack that can be dropped via parachute into an area of natural disaster that survivors could locate easily and then use to provide emergency food, shelter and clothing. Celia thought it would be interesting to speculate on where mathematics might be used to enhance the design decisions made by the pupils, in an authentic way: for example, in maximising the volume and insulation of the pack. However, Celia did raise a word of warning: it would be important not to impose constraints on the design & technology that rendered the task non-authentic. She thought it was very important to be aware that making the mathematics more visible might in some cases be counterproductive for the design as it introduces constraints that are just too artificial.

Finally, Celia made this one point very strongly. An essential requirement is for mathematics teachers and design & technology teachers to work together is time: time for them to initially explore possible mutual benefits that might be achieved through collaborating around a carefully selected design, time to develop the activity and time to actually tackle the 'design and make' assignment for themselves, checking if it works and ultimately experiencing the mutual enhancement of mathematics and design & technology learning. Then they can jointly plan the classroom experience and review it following teaching. If this activity is started towards the end of the academic year it may be possible for mathematics teachers and design & technology teachers to work alongside one another in the classroom. Alternatively, student teachers may be able to work alongside specialist teachers in the complimentary discipline.

The views of Torben and Celia clearly indicate the considerable benefits that are possible if design & technology is taught in the light of knowledge and understanding acquired by pupils in science and mathematics lessons. However, it is important to ensure that the learning in these subjects is not compromised by attempts to form a curriculum relationship between them. This issue is now discussed in terms of maintaining subject integrity.

Maintaining subject integrity

As indicated by both Torben Steeg and Celia Hoyles, in using pupils' learning in mathematics and science to enhance their learning in design & technology, it is essential that the integrity of design & technology be maintained. It is all too easy for the learning intentions to become subverted so that the learning of mathematics or science dominates the proceedings. The simplistic and erroneous definition of technology as 'applied science' can easily lead to situations in which the application of science overrides all other considerations to the detraction of learning in design & technology. Brian Arthur's definition of technology as the 'exploitation of scientific phenomena' is to be preferred because it enables a much wider interpretation as exploitation encompasses far more than application (Arthur, 2009b). This reduces the possibility of important wider influences being ignored. This point is given further weight by David Layton who argued that the knowledge constructed by scientists in their quest for understanding of natural phenomena is not always available in a form that enables it to be used directly and effectively in design & technology tasks. Knowledge that has been conceptualised so that it is useful in providing explanation is not necessarily the knowledge needed to inform the taking of action, although both formulations of knowledge are concerned with the same domain. Indeed, there are examples in the history of science and technology in which the knowledge to take action preceded the knowledge needed for explanation. The classic example is the development of the steam engine by James Watt almost 50 years before the explanation of the underlying thermodynamics by Sadi Carnot in 1824. This is discussed in greater depth in Chapter 3, Teaching science in the light of STEM.

One way to maintain integrity is to plan on the basis of the utility–purpose model proposed by Janet Ainley and colleagues (Ainley et al. 2006). They argue that it is possible to engage the utility of some subjects in pursuing the learning purposes of others. Hence it should be possible to capitalise on the utility of mathematics and science in pursuing the learning purposes of design & technology. If one considers that a fundamental purpose of design & technology is for pupils to learn how to make genuine design decisions then it is not difficult to see how such decisions can, and ought to be informed by learning in mathematics and science. It is important that such decisions are genuine and authentic design decisions and not simply technical decisions contrived to support learning in mathematics and science. Ainley and colleagues also argue that there is mutual benefit in this arrangement. In utilising mathematics and science, pupils will become more adept at these subjects while at the same time enhancing their ability in design & technology.

In this section we have considered the benefits of teaching design & technology in the light of STEM and briefly discussed the importance of maintaining subject integrity. In the following section we will provide examples of design & technology activities that build on the advice from Torben Steeg and Celia Hoyles and exemplify the utility–purpose approach developed by Janet Ainley and her colleagues with particular reference to pupil design decisions.

Examples of teaching design & technology in the light of STEM

The following examples cover the breadth of design & technology and indicate the benefits of teaching in the light of learning in science and mathematics. They take into account both the utility–purpose model and the views of Torben Steeg and Celia Hoyles and consider pupil design decisions.

EXAMPLE 1: TRYING TO EXPLOIT A SCIENTIFIC PHENOMENON IN PRODUCT DESIGN

Consider a unit of work in which pupils aged 14 are required to design and make a device that exploits a scientific phenomenon – echoing Brian Arthur's definition of technology. For example, the phenomenon to be exploited could be the Peltier effect. The Peltier effect is enshrined in a solid-state device that when activated transfers heat from one side of the device to the other side against the temperature gradient. Although this phenomenon is outside the usual science curriculum for 14-year-olds, it is likely that pupils will find the sensation of a device that is 'cold on one side hot on the other side' highly intriguing. And such a new phenomenon provides the opportunity to investigate the extent to which it can be exploited – what do we have to do to get the cold side really cold and the hot side really hot, i.e. how do we maximise the effect? Here is an opportunity for a genuine investigation that will provide information useful to pupils in pursuing a design and make task.

This clearly mirrors the utility-purpose argument proposed by Ainley and colleagues. With appropriate collaboration it would be possible for the investigation to be carried out in pupils' science lessons and the results used in design & technology lessons. This would provide the science teacher with an investigation that is linked to a purpose wider than developing explanation and the opportunity to see how scientific the pupils could be in pursuing such an investigation. If it were not possible to carry out the investigation in a science lesson then pupils could carry out the investigation as part of their design & technology lessons, assuming the design & technology teacher felt confident enough and had the necessary scientific understanding to do this. However, it almost certainly would be a lost opportunity for pupils to see the potential for their learning in science to be related to their learning in design & technology. Torben Steeg has investigated the performance of a Peltier device and found that after a minute the hot side reached 49°C in a room at 20°C. The cold side did not cool as much as the hot side heated, probably due to heat leakage through the device from the hot side; hence the need to remove the heat away from the hot side to get significant cooling. With a heat sink on the hot side plus a fan to draw away the hot air Torben found that after five minutes the temperature of the cool side had dropped 6°C from a room temperature of 20°C. This idea has been developed into a unit of work by Philip Holton, when he was a head of faculty at a school (pupils aged 11–19 years) in South East England. The demands of the task are considerable as demonstrated by the instructions shown in Figure 4.3.

It is noteworthy that understanding the physics of the Peltier device in terms of a scientific explanation is NOT required for this unit of work. Hence although the underlying science will not be taught until pupils are several years older, it is possible for Year 9 pupils to engage with the performance characteristics of the device and that is the knowledge needed to be able to take action. In developing a prototype product, the design decisions made range across those described by the design decision pentagon. The pupils were able to choose the sort of product they would design (conceptual decisions), devise the means to optimise heating and cooling (technical decisions), decide on the appearance of their device (aesthetic decisions), work out how to enclose the Peltier device within their prototype (constructional decisions); all in the light of stakeholder considerations, mainly user requirements. Across this decision making there are opportunities to use knowledge and skills from science and mathematics. In Philip's classroom, pupils have designed and made a variety of cooling devices for different purposes that they consider worthwhile, including a device for cooling drinks and a device for maintaining an organ for transplant at the correct temperature during transportation. Holton's pupils are also introduced to some of the commercial applications for the Peltier effect through considering existing patents.

Peltier Cell project
In this project you are challenged with designing a unique concept product using Peltier Cell technology.
You will need to conduct research into the capabilities of a Peltier Cell; understand current and patented uses of the technology; before under going creative designing of conceptual uses for the cell.
You will need to model your best idea to a level where it can be tested; evaluate your concept; and finish by creating a patent document which describes the unique idea you have developed.

FIGURE 4.3 Instructions for the Peltier Cell project

This is an interesting opportunity for the science teacher to reinforce the idea that science is concerned with explanation involving the use of concepts by questioning the pupils about the working of their finished devices. If a pupil has added metal fins to the hot side of the device, questioning should reveal whether they can explain their function in terms of conduction and convection. There will also be opportunities to probe the extent to which pupils are distinguishing the concepts of heat and temperature. It is extremely worthwhile for the design & technology teacher to sit in on this questioning to give insight into the sorts of conceptual confusion that can arise and the way that the use of language in both design & technology and science lessons for such tricky ideas needs to be consistent. Note that this approach of developing a product that exploits a scientific phenomenon could provide a generic approach to teach design & technology in the light of STEM. Conversations with science colleagues to identify a range of such phenomena would indeed be worthwhile. This would also link strongly to Torben Steeg's point of utilising scientific thinking as an important feature of design & technology.

Moving toys have an intrinsic appeal and even the simplest of such toys have to meet significant technical requirements. Such requirements have to be met in ways that have appeal to potential users. Hence designing and making such toys provides rich learning possibilities. In terms of design decision, the teacher often makes the conceptual decision underlying the activity; the pupils will design and make a moving toy. But that still leaves many other possible design decisions for the pupil. With regard to technical decisions, the teacher might give them three possible choices: a wind-up clockwork motor, a battery-powered electric motor or a ripcord flywheel drive. For aesthetic decisions, the teacher might ask them to opt for a toy that looks like a modern car, or a toy that looks like a vintage car or a toy that looks like an animal. For construction decisions, the teacher might ask them to opt for a frame structure, or a shell structure made from a card net or by vacuum forming (from a range of given formers). For stakeholder decisions, the teacher might ask them to choose between users of three different age groups: 4/5 years, 9/10 years and 13/14 years. If possible, they should find a real person in one of these age groups that they can talk with about what they would like from the toy. This range of design decisions is summarised in Figure 4.4.

The following case study deals with a simplified version of the task suitable for pupils in Year 7.

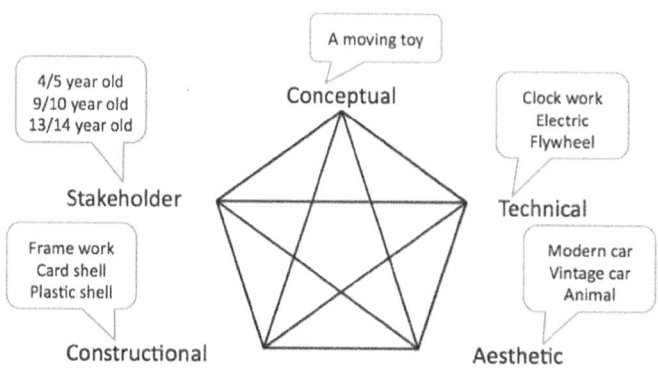

FIGURE 4.4 Design decisions for designing and making a moving toy

EXAMPLE 2: DESIGNING AND MAKING MOVING TOYS

The teacher wants pupils in Year 7 (12-year-olds) to design and make a moving toy but, given their limited experience, she has decided to restrict the means of movement to a single electric motor that produces very high no-load speeds of rotation on the output shaft. Invariably, this high speed of rotation is reduced by a transmission system. The simplest involves elastic band belt drives and pulley wheels. These are inexpensive to resource and forgiving in that they do not require a high level of accuracy to work well. The elastic bands stretch and can easily accommodate an error in locating the drive axle. Conversely, gearing systems have to be located precisely if the teeth in the gears are to mesh in a way that does not bind or slip. Compound gear trains clearly require more accuracy. Often, pupils are provided with a set of wheels and a ready-made transmission system, items that the teacher knows will work, which to some extent guarantees a successful product. However, this approach denies pupils the opportunity to consider how fast they want their toy to move and what factors might affect this speed of travel. How fast the toy moves should to some extent be decided on the needs of the person who will play with the toy. If it will be played with indoors, in a small apartment for example, a slow speed would be preferable but if it will be played with outdoors, in a garden or school playground then a fast speed is required. Pupils can be introduced to the effect of wheel size on speed of movement by providing information as shown in Figure 4.5. Note that the chassis structure is deliberately simple. Pupils may already have carried out such constriction in primary (elementary) school. However, keeping the construction simple enables time to be spent on other features that contribute to the toy's performance.

FIGURE 4.5 Introducing students to thinking about the effect of wheel size on movement

Calculating the speed of each toy is not a trivial task. The speed of travel of the toy depends on both the size of the wheel and its speed of rotation. In the example below, the toy with the slowest speed of rotation (Derek's toy) will travel the fastest because this speed of rotation in combination with the wheel size gives the greatest distance travelled per minute. There are many opportunities for interesting conversations between pupils working in pairs and between the teacher and the pupil about the factors that will affect the speed of travel. And such conversations can reveal any misunderstandings pupils might have about the concept of speed and be used to help pupils address such misunderstanding. Understanding speed is a precursor to understanding acceleration so the science teachers will be interested to know which pupils are having difficulty with the idea of speed. At the same time, the mathematics teacher will be pleased that pupils are gaining practical experience in dealing with compound measures.

It might be necessary to have discs of the different diameters available and so that pupils can experiment with rolling them the requisite number of revolutions to develop an accurate representation of how the different toys will move. From this experience they can decide which toy is the fastest and how to adapt a toy (such as Mary's) so that it becomes the fastest. Other pupils will be able to reach this decision by using annotated sketches. Some pupils might be able to reach the decision intuitively but it will be important to ask them to justify their decision. At the moment this is a theoretical exercise and it is essential to use the understanding achieved in deciding on the motor speed and wheel size for the toys they are designing and making. Usually teachers provide only one sort of motor and most suppliers provide data sheets which will give no load speeds in revolutions per minute. However, once under load, the rate of revolution decreases significantly.

We then need to find out just how fast the motor turns under load and the effect of different sized wheels being turned by the motor on the speed of the toy. Some empirical evidence is required here. Depending on the time and finances available, a teacher could provide a range of motors, pulleys and gears as well as different-sized wheels. Some pupils could pause their investigations and use the results they have gathered so far to predict what might happen with different arrangements of motor, transmission system and wheels. This provides a great introduction to mathematical modelling. If the speed of the motor under load (obtained from the investigation) is known, the ratio of the transmission system (obtained from decisions about pulley or gear size and how they are arranged) and the diameter of the wheel (chosen from the range available) we can work out how fast the toy will go. Pupils can show their understanding of the model by responding to 'what will happen if' questions (e.g. 'You know how fast your toy will travel but what if you make this pulley bigger, this wheel smaller?'). The aim here is to help the pupils discern the patterns of behaviour in the arrangement of components and their understanding of the relationship between their behaviours to make design decisions that lead to a toy that is suitable for a particular user. This approach engages pupils with pattern recognition and the use of relationships, key parts of mathematics, as noted by Celia Hoyles, and vital for carrying out investigations. These are also key parts of science as noted by Torben Steeg. Taken together, the utility of these activities inform the purpose of designing and making a moving toy.

EXAMPLE 3: MODELLING WIND TURBINES

Celia Hoyles indicated that using mathematics could enhance the rigour with which sustainability issues were considered in the curriculum. As many design & technology schemes of work include a consideration of alternative energy sources, there is the opportunity to engage with mathematics. The Science Enhancement Project (SEP) produces a useful wind power booklet to support practical activities using the SEP Wind Turbine and to help explain the science behind wind power. The SEP wind turbine is ready-assembled and can be used to carry out a wide range of experiments on wind power. In some design & technology schemes of work pupils can be required to make simple wind turbines from given components or design and make their own wind turbines. All of these activities enable pupils to develop an understanding of how the energy in the wind can be harnessed to produce either mechanical or electrical energy.

Of course, the unexpected can still occur in terms of the toy's performance. The wheels might fail to grip the surface and slip, thereby reducing the toy's speed. Attempts to make the wheels look attractive (e.g. cutting large holes in them) might reduce their weight so that the motor can turn them faster than predicted increasing the toy's speed. Adding a larger battery to the toy may mean that it can be played with for longer, but it might also result in the toy being too heavy for the motor/transmission system to move. In such cases, the limitations of the model of performance are revealed and this is an important learning point. And, of course, there is a range of other design decisions to be made, including: overall appearance (what sort of vehicle is the toy?); special effects (flashing lights, buzzers) and how these will be controlled; how the motor will be controlled (what sorts of switches and where are they placed); and how will all the different circuits be wired up to fit neatly into the toy?

These activities can be extended to include consideration of wind turbines that might be used to produce significant amounts of energy, reduce carbon footprint and the use of fossil fuels.

Here is an interesting arena for collaboration between design & technology and mathematics. The exploration of relationships involving the use of tables, graphs and background knowledge of direct proportion is extremely useful mathematics learning. Teaching about wind energy in design & technology allows this mathematics learning to be considered in an authentic context supported by a range of practical activities. The mathematics allows the pupils to begin to consider the feasibility of wind energy as a source of power and so enables the engaging practical activities to be extended to include an exploration of technology and society.

For example, it is possible for pupils:

- To consider the relationship between wind speed, wind energy and wind power by plotting appropriate graphs and to use these graphs to estimate the number of different domestic devices that could be powered by different speeds of wind by wind turbines with different areas of sweep.
- To use formulae to calculate the energy and power in winds of varying speeds.

■   To use the Betz limit to calculate the power that can be captures by wind turbines of different diameters.

Without such mathematics, this topic lacks the rigour that will enable pupils to sensibly consider the energy supply issues facing society. The utility of mathematics in making important decisions comes to the fore. Of course, there are issues other than the mathematics of wind and of wind turbines. The situation is complex. Where the turbines are situated and their impact on natural beauty and local wildlife are important factors that need to be considered. The availability and variability of wind on any chosen site need to be taken into account. The cost of setting up and maintaining the wind turbines versus the price for which the electricity generated can be sold has to enable both business and industry to make a profit. The government will be involved in providing incentives to business and industry but unless there is sufficient energy available in the wind that can be extracted by wind turbines then these wider considerations are irrelevant. This approach echoes strongly Celia Hoyle's point that using mathematics to explore data (in this case data about the wind) enables pupils to think in a 'designerly' way about how problems concerning sustainability might be solved. This approach would also benefit strongly from meeting Hoyles' plea that mathematics and design & technology teachers spend time together exploring the activities so that they are comfortable with each other's learning requirements and can see how to support them in their own lessons. The resulting plan of action might require the mathematics of wind energy to be considered in mathematics lessons either before or at the same time as the introductory practical activities in design & technology so that the wider discussions about using wind energy in particular situations can be informed by this mathematical understanding. If this were not possible, and it was decided that the design & technology teacher would teach the mathematics of wind energy within the design & technology lessons, then it would be very important for her to liaise strongly with the mathematics department on how best to approach this.

There are, of course, many links with science possible in teaching about wind energy in design & technology and investigations of the performance of small wind turbines pupils have constructed from given parts or designed and made provide many opportunities. For example, measuring the mechanical power of the output shaft by timing how long it takes to lift a mass through a metre is a very direct way. Alternatively, the output shaft can drive a small electric motor that acts as a generator to light up LEDs. The greater the number of LEDs lit, the better the performance of the turbine. Within these activities there are opportunities for pupils to use the concepts of energy, force, work and power and to talk about their meaning. As with the conversations about heat and temperature in the Peltier effect project, discussion with pupils will reveal both understanding and misconception. Hence, if possible, it would be useful for the science teacher to discuss their investigations with pupils in ways that require pupils to use relevant concepts correctly. Here, we have an inverse use of the usual suspects identified by Torben Steeg – their use in design & technology being scrutinised by the science teacher to reveal possible misconceptions.

Clearly, there are strong links here with the 'usual suspects' identified by Torben Steeg. The physics of the semiconductor materials that produce light in LEDs is

EXAMPLE 4: LIGHTING DESIGN

In many design & technology courses pupils aged 14–16 years are required to design and make simple lighting devices. Such tasks provide interesting opportunities to explore the way the technologies we use in daily life change over time and may change in the future. A comparison between filament lamps and light emitting diodes (LEDs) gives an interesting starting point for the way the provision of lighting is undergoing change. The way in which filament lamps work is relatively easy for pupils to understand and it is not difficult for them to appreciate how inefficient such lamps are in that only a fraction of the energy consumed is used in providing light. Most of the energy is used in bringing the filament up to the temperature at which the filament begins to glow. And, of course, pupils can feel filament lamps becoming hot. LEDs, on the other hand, do not rely on a heating effect to produce light.

probably not taught in science courses to pupils under the age of 16 years but it is relatively straightforward for pupils to carry out an investigation comparing the energy consumption of small filament lamps and LEDs. At the time of writing, filament lamps for domestic lighting have almost been being phased out in the UK and being replaced by lower energy consumption fluorescent bulbs which, whilst saving energy, do present environmental problems with regard to disposal because of their mercury content. Research and development activity into the design of LEDs that are suitable for domestic lighting has come to fruition. Hence although the major part of a unit of work on lighting would be the designing and making of an LED-based light for, say, task or mood lighting, it is possible to support this with science-based investigations into filament lamps and LEDs and an exploration of the research that led to the development of LED lighting for domestic use.

An alternative to using electricity to generate light is to consider how living organisms generate light; so-called bioluminescence. Such developments are an example of the emerging field of biomimetics – adopting and adapting biological systems for use in technologies. The company Glowee (Glowee, 2019) is using a combination of biomimicry and synthetic biology to explore ways of developing bioluminescence as an alternative to traditional lighting. They genetically engineer bioluminescent micro-organisms to make them more efficient in terms of light production (intensity, stability, capacity). Table 4.1 shows the advantages they assign to this form of lighting. In planning such a topic to exploit the links with science as indicated above, it will be important to have conversations with the relevant science teachers. Of course, the physics teacher is likely to be interested in the filament lamp LED comparison and may be inclined to carry out the investigation as part of the pupils' physics course. If this were the case, then it would be useful for the design & technology teacher to observe some of the lessons. She might also suggest that the pupils carry out additional investigations into the illumination provided by their light design proposals. Her guidance on this would be useful.

The biology teacher is likely to be interested in the bioluminescence part of the unit and may well be able to suggest practical activities in which pupils grow cultures of such bacteria. However, this is very specialist territory with particular health and

TABLE 4.1 Advantages of bioluminescence

1. Reduce the Environmental Footprint of Lighting	2. Improve Comfort and Well-Being	3. Offer New Design Possibilities for Lighting
A raw material that can grow indefinitely	A cool and natural light to reduce light pollution	A raw material that can take different shapes and states
A light powered from waste rather than electricity	A soft light to reduce visual pollution	Infinite possibilities to adapt this to different needs and uses of lighting
100% organic by-products that can be neutralised or revalorised	An hypnotic light with relaxing and soothing properties	

safety issues, so it would be wise for such activities to be taught by the biology teacher preferably in a science-teaching laboratory. The utility of scientific knowledge and understanding is clearly important in developing the investigations and this should be apparent to pupils as they pursue their lighting design and make tasks and explore the way the technologies we use for lighting have changed and may change in the future. This approach supports Brian Arthur's view that technology may be interpreted as the exploitation of scientific phenomena.

Of course, in designing and making a light there are a wide range of design decisions other than those concerned with the production of the light. The light may need to be directed by a reflector to provide task lighting or diffused to avoid dazzle, the appearance of the light will need to appeal to the user and be appropriate for the setting, the various parts will need to be assembled to provide visual appeal and ease of use.

Whilst it is important for pupils to design and make items of worth, there is a case to be made for designing a product and associated services without making the product. Although there is no physical artefact 'to take home' research (Murphy, 2003), Murphy has shown that this does not deter pupils and that they respond enthusiastically to such tasks especially if they are required to work collaboratively and make short presentations justifying their design proposals. In many cases, 'not requiring to make' was welcomed by the pupils, as it released them from the constraints of the materials and equipment available in their school workshops. It is important to scaffold such designing and the Young Foresight project (Barlex, 2012) identified four factors that teachers should encourage their pupils to take into account.

1 The technology that is available for use. This should be a new and/or emerging technology and be concerned primarily with how the new product or service will work. Pupils should not concern themselves with manufacture.

2 The society in which the technology will be used. This will be concerned with the prevailing values of the society, and what is thought to be important and worthwhile. This will govern whether a particular application of technology will be welcomed and supported.

3 The needs and wants of the people who might use the product or service. If the product does not meet the needs and wants of a sufficiently large number of people, then it will not be successful.

4 The market that might exist or could be created for the products or services. Ideally, the market should be one with the potential to grow, one that will last, and one that adapts to engage with developments in technology and changes in society.

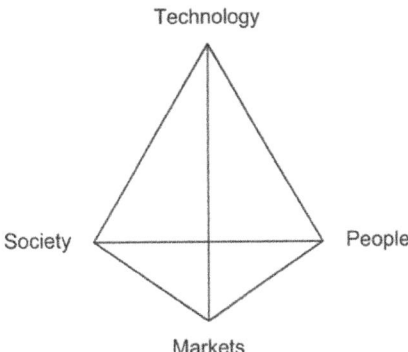

FIGURE 4.6 The Young Foresight 'Tetrahedron' describing four factors to take into account when designing products and services for the future

Clearly, these factors interact with one another and influence the sorts of products and services that can be developed and will be successful. Using this way of thinking, unencumbered by the necessity of making the proposed designs, enables pupils to be creative and develop highly original, conceptual design proposals. This framework for designing was represented diagrammatically as a tetrahedron as shown in Figure 4.6.

In developing and justifying the design proposal, pupils have to visit each vertex of the tetrahedron. However, there is no prescribed starting point.

An important decision for teachers is the order in which they ask pupils to tackle a task. One way is to start with a particular new technology and ask a sequence of questions like this:

- What sorts of things can we use this technology for?
- For each of these, what needs will they meet?
- Will meeting these needs be seen as important and worthwhile in society in the future?
- Will people want products or services to meet these needs?
- What sort of market is there likely to be for these products and services?

If teachers adopt this approach, it will be important to be wide-ranging in answering the first question.

Another way to start is by asking pupils to construct a scenario of what a future society will be like, and what life will be like for particular groups of people in that society. Pupils can then explore a sequence of questions like this:

- What needs exist in that society?
- What products and services will people want to meet these needs?
- What sort of market is there likely to be for these products and services?
- What technology do I need to make the product or service work?

This is a much more demanding approach, but it offers more scope for considering the nature of a future society and the impact of technology on that society. It is an approach that is more likely to stall, as the starting point is much less concrete than a particular technology. However, it does have the potential for developing some really big ideas.

The following case study considers using QTC (quantum tunnelling composite) as the new and emerging technology as the starting point.

EXAMPLE 5: DESIGNING WITH A NEW AND EMERGING TECHNOLOGY IN MIND

QTC is a stress sensitive conductor. In a relaxed state it is good insulator but the more it is stressed by compressing, stretching or twisting, the more it conducts. Pupils aged 14 were challenged to use this property of QTC as the basis for their designing. Some of the products they devised included the following:

1. Clothing that changes colour as you dance
2. Car tyres that sense their internal pressure
3. An epileptic fit detector
4. A self-weighing suitcase
5. An arthritis treatment device
6. Keep fit apparatus
7. Depth sensitive submersible
8. Internal heart beat monitor

In all cases the pupils had to justify the feasibility of their proposals and these required that they understand the nature of force and how force acting over an area gives rise to pressure, which would affect the conductivity of QTC. They had to explain this in their presentations and indicate how the change in current flow and/ or potential difference enabled their proposed devices to work. The explanations provided by the pupils would make interesting listening for their science teachers as they would reveal the extent of their understanding of current and potential difference given that they needed to reconstruct this in devising their proposals.

EXAMPLE 6: RADIO DESIGN

Designing a radio receiver circuit is beyond most school pupils but the experience of making a radio receiver that has been designed by someone else is a very worthwhile activity. Any teacher who has taught this will remember the expressions of surprise and delight on pupils' faces when they hear a local radio station on their own radio. There are several radio kits available from educational suppliers consisting of a printed circuit board, components and assembly plans to support this activity. There are different approaches to organising the assembly. Some teachers prefer to structure the activity on a step-by-step basis giving the class precise instructions for the identification and placement of each component. These teachers argue that this approach guarantees each pupil a working radio. Other teachers prefer to organise pupils in pairs, provide illustrated step-by-step assembly instruction and instruct each pair to produce two working radio circuits with each pupil in a pair being responsible for checking the other pupil's work. These teachers argue that this approach encourages the pupils to take more responsibility for their learning and enhances their collaboration and communication skills. Their position is that the few mistakes that cause circuits to malfunction can easily be identified and rectified, and the increased learning more than justifies this approach. In terms of design & technology learning, this making activity will enable pupils to learn how to identify a range of components, orientate components according to a layout diagram and soldering skills.

This learning can be extended to include a consideration of how the circuit actually works. This provides a useful opportunity to use a systems approach to describing circuits and to overlay the various components in the circuit onto the system blocks. In terms of links with science, this also provides the opportunity to consider the electromagnetic spectrum. This is an important and demanding idea that many pupils find difficult. Hence it will be important to liaise carefully with science teacher colleagues to ensure that the discussions in design & technology lessons about how the circuit works do not lead to conceptual confusion. Although it is possible for quite young pupils to assemble a working radio from given components, here it would probably be inappropriate to consider the electromagnetic spectrum. However, it can be used as a motivating starter activity to a design & technology electronics course for pupils aged 14–16 years and at this age it is likely that, in their science courses, they will be learning about the electromagnetic spectrum (either in physics programmes or applied science programmes dealing with communication). Ideas about frequency and wavelength will almost certainly be considered. So, it is possible that the science teacher could use the radios made by pupils in their design & technology lessons as a starting point for considering the electromagnetic spectrum.

It is likely that as part of their design & technology courses pupils will be required to produce a housing or enclosure for the radio circuit they have made. There are a variety of design decisions to be made in this activity and some of them can involve mathematics. For example, pupils will study nets in their mathematics lessons. Nets are two-dimensional shapes that can be folded to three-dimensional forms. These are sometimes studied in design & technology where they are called 'surface developments'. In mathematics, pupils may investigate the relationship between the surface area and enclosed volume, and they may also link their study of nets to geometry, relating a variety of three-dimensional forms to the variety of nets from which they might be constructed. So, it is possible that pupils will have at their disposal knowledge of a wide range of possible forms and associated nets to use for the radio enclosure. The net has to accommodate a variety of features and be large enough to accommodate the circuit and battery. These features include an on/off switch, a tuning dial and a volume dial and, if the radio is sufficiently complex, an AM/FM switch. All these features need to be arranged on the net to give user convenience and the overall appearance, which may include graphics, should have visual appeal. And, of course, the net should enable access for repair (e.g. wires coming loose) and maintenance (e.g. changing batteries). Hence designing a successful enclosure is not a trivial task. An example of a radio enclosure made from a net is shown in Figure 4.7.

If the pupils can use CAD software to draw the required net with places to insert the various features, then they can use their CAD files to drive a laser cutter to produce the required net from thin sheet material such as card or polypropylene, complete with creases to enable folding up around the circuit and battery to form the enclosure. The range and variety of enclosures formed will, to some extent, depend on pupils' initial knowledge of nets and it is here that conversations with their mathematics teachers can pay dividends. If the radio task can be timed to take place just after the pupils have studied nets, then the mathematics teacher can contextualise the nets topic by using the radio enclosure design in the mathematics lessons. There is the possibility here of using Celia Hoyles' example of considering the general case by which the volume for a particular enclosure can be maximised. This will enable the design & technology teacher to capitalise on the utility of the taught mathematics as

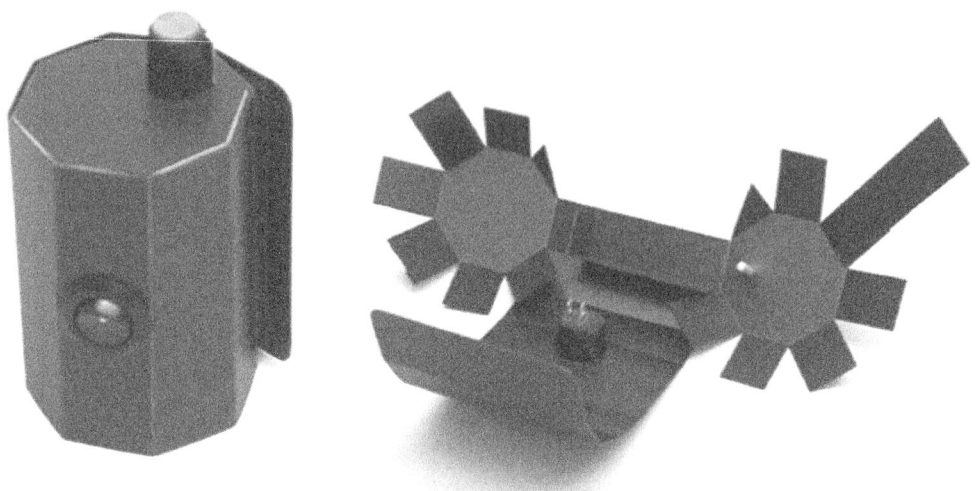

FIGURE 4.7 A radio enclosure constructed from a net

pupils produce the enclosures. Even if such juxtaposition and contextualisation are not possible, the design & technology teacher can still support pupils' designing tasks by helping them to remember what they have previously learned in mathematics and hence illustrate the usefulness of mathematics for design purposes.

EXAMPLE 7: PROTECTIVE TEXTILES

In many design & technology courses concerning textiles, pupils are required to design and make items concerned with protection. This provides the opportunity for pupils to consider the many different situations in which there is the need for protection that can be provided by the use of textiles. The following examples indicate the wide range of situations pupils might consider.

1. Trawler fishermen keep warm and dry by wearing clothing made from waterproof fabric with welded seams and flaps over fastenings.
2. Soldiers avoid being seen by the enemy by wearing clothing that is randomly coloured, causing the figure to merge in the background.
3. Mechanics keep clean by wearing overalls made from densely woven fabric that does not allow grease or dirt to penetrate.
4. American football players avoid being hurt by wearing padding that protects by absorbing impact.
5. People keep warm in the snow by wearing coats made from thick fabric that traps air between clothes and coat to provide insulation.
6. People out in the sun keep cool by wearing clothes made from thin fabric that allows perspiration to be absorbed and evaporate keeping the wearer cool.
7. Cyclists maintain visibility by wearing brightly coloured, light-reflective fabric that enable them to be seen.

In each of these examples there is the potential for pupils to make a wide range of design decisions that will interact with one another as the eventual design unfolds. All these situations provide the potential for investigations into fabric performance and underlying these investigations is the important idea of 'properties'. It would capitalise on both of Torben's categories of usefulness – the 'usual suspects' – properties of materials and scientific thinking. Within the arena of textiles this is complex because the properties of fabrics depend on both the nature of the fibre and the structure of the fabric. To give the investigations purpose, the pupils will need to consider the following or similar questions:

- What properties are important to achieve the protection required?
- Which fabrics or fabric combinations have these properties?
- What investigations can I carry out that will help decide which fabrics might be suitable?
- What other factors should I consider (e.g. cost, availability or appearance)?

Pupils will already have been introduced to the idea of materials having properties and how these are established by investigation, giving rise to tables of data describing such properties. It is possible for pupils to identify the fabrics that might be useful by using such tables. However, this requires a sophisticated understanding of the properties under consideration and initially it will almost certainly be necessary for pupils to devise and carry out simple investigations for themselves to understand the nature of relevant properties. Such investigations can be designed to give a rank order of materials with regard to a particular property, e.g., increasing ability to resist wear, or to give values of properties in particular units, such as the tensile strength of a fabric in kg/cm. Clearly, conversations with science teachers will be valuable here not least to ensure that the ideas concerning fair testing and measurement of properties that are taught in science lessons are utilised and built upon in the design & technology lessons. Science teachers might also use the investigations as part of their science teaching as in the Peltier device investigation described earlier in this chapter.

As the pupils become familiar with a wide range of fabric properties through investigations, their ability to use information in tables of properties will increase and they will be able to justify their choice of fabric without necessarily carrying out investigations. In those cases where pupils will design and make a textile item, it is likely that the choice of fabric will be limited by cost. However, in some cases it would be appropriate for pupils to develop their ideas to a detailed design proposal only and stop short of actually making a finished article. In such cases, a mock-up in an inappropriate fabric, supported by details of the actual fabric to be used in the final article, would suffice. This would allow pupils to consider the use of very modern textiles unavailable to schools – Kevlar is an obvious example. Comparison of the properties of Kevlar compared with other textiles that might be used for protection purposes soon indicates how unusual and useful it is. However, without the preliminary understanding of the properties of materials, learning about Kevlar and considering possible uses will lack a 'wow' factor. It is not difficult to extend this approach to include pupils speculating about the uses of cutting-edge materials being developed by science-based research and development. Spider silk is such a material: five times stronger than steel, tougher than Kevlar and highly elastic, it is potentially extremely useful – if only it could be manufactured. It has proved impossible to farm spiders in the same way as silk worms, so scientists have been trying to get the best of both worlds – super-strong silk in industrial quantities – by transplanting genes from spiders into worms. Recent successful research represents a step towards the commercial production of a combination of silk and spider silk spun by silkworms. Currently, it is thought that the main applications for spider silk will be in the medical sector creating stronger sutures, implants and ligaments. But the GM spider silk could also be used as a 'greener' substitute for toughened plastics, which require a lot of energy to produce.

Encouraging pupils to speculate about possible uses of genetically modified materials provides an interesting way of raising pupils' awareness of the way biomimicry and biological manufacturing are likely to become important in the future. Conversations with science teachers about how to relate this teaching in design & technology to the teaching of genetics in biology classes are an important part of ensuring that the utility-purpose argument developed by Ainley and colleagues is on a sound footing and can be extended to activities involving exploring technology and society.

After considering a wide range of examples of teaching design & technology in the light of STEM, the following section revisits the issue of 'maintaining subject integrity' with a short discussion on the importance of ensuring that the learning in design & technology, mathematics and science isn't in anyway undermined by 'teaching in the light of STEM'.

Ensuring continuity of learning across the subjects

Having considered a wide range of examples in which design & technology can be taught with regard to links to pupils' learning in science and mathematics, it is important to ask to what extent might this approach compromise the learning in the interacting subjects? It is vital that there is sufficient mutual benefit to the subjects involved to ensure that the not inconsiderable effort required for the interaction is worthwhile. Celia Hoyle's warning that the linking process should not impose constraints on design & technology that render the tasks non-authentic is key. And it is also important to ensure that the process of interacting does not confuse pupils by giving mixed messages about learning in the interacting subjects. In developing approaches to teaching design & technology with regard to pupils' learning in science and mathematics, one must start with one or two activities and then build in evaluations to give some sense of the costs and benefits of the exercise. Ideally, the interaction between the subjects should enhance the learning across the interacting subjects. Hence the science and mathematics teachers should be able to see improvement in their pupils' learning as a result of the interaction. Similarly, the design & technology teacher should be able to see improved learning in design & technology through pupils' use of science and mathematics. Once a few successful 'teaching in the light of STEM' activities have taken place, it will be easier to develop further effective examples.

Conclusion

So, what are we to make of this chapter? If we define a knowledge base for design & technology, will it become more acceptable as a subject suitable for a place in the national curriculum? Will such a definition cause the subject to become less concerned with procedural competence to such an extent that it loses the essence of its initial rationale, 'enabling competence in the indeterminate zone of practice'? Adopting pedagogy around designing and making would certainly make this less likely. Where does this leave teaching design & technology in the light of STEM? Torben Steeg and Celia Hoyles were in no doubt as to the advantages of pupils of being encouraged if not actually required to use their science and mathematics learning to enhance their learning in design & technology. And what of the danger of teaching design & technology in the light of STEM resulting in the legitimate learning requirements of design & technology becoming submerged and merely subservient to meeting the learning requirements of science and mathematics? The utility-purpose model proposed by Janet Ainley and colleagues goes some way to mitigating against this difficulty in that there is benefit to all the collaborating subjects only if design & technology can pursue its 'designerly purpose' which lies at the core of its learning requirements.

The examples of teaching design & technology in the light of STEM used in this chapter were developed in part to show that in taking such an approach the integrity of design & technology would not be compromised. We provided examples to illustrate that teaching design & technology in the light of STEM is not only possible but really worth exploring. If you can use our examples that is all to the good, but if you find them inappropriate for your situation it is our view that this should not be a barrier to developing your own examples that are appropriate. Indeed, we would urge you to develop your own ways to teach design & technology in the light of STEM. Of course, in tackling this task it will be important to avoid giving pupils mixed messages that will confuse rather than enhance their understanding. Developing a coherent appreciation of important ideas across the STEM subjects will only be achieved through on-going conversations between all those involved in the teaching. This will require time – something that the evaluation of the STEM Pathfinder Programme (2012) indicated was seen by teachers as the scarcest and most valuable resource needed for teaching collaboratively across the STEM subjects. So, in teaching design & technology in the light of STEM, our advice is don't neglect the importance of regular conversations with colleagues from science and mathematics.

Recommended reading

The following deal with the nature of design & technology:

Arthur, W. Brian. (2009a) *The nature of technology*. London, England: Allen Lane.

Barlex, D. & Steeg, T. (2017a) *Big ideas for design & technology*. https://dandtfordandt.wordpress.com/working-papers/big-ideas-for-dt/ (accessed 12 June 2020).

Bronowski, J. (1973a) *The ascent of man*. London: British Broadcasting Corporation.

Dakers, J., Hallstrom, J., & de Vries, M. (eds) (2019) *Reflections on technology for educational practitioners*. Lieden: Brill Sense.

Layton, D. (1993a) *Technology's challenge to science education*. Buckinghamshire, UK: Open University.

References

Ainley, J., Pratt, D., & Hansen, A. (2006) Connecting engagement and focus in pedagogic task design. *British Educational Research Journal*, 32(1), 23–38.

Arthur, W. Brian. (2009b) *The nature of technology*. London, England: Allen Lane.

Barlex, D. (2007) Assessing capability in design & technology: The case for a minimally invasive approach. *Design and Technology Education: An International Journal*, 12(2), 9–56.

Barlex, D. (2012) The Young Foresight Project: A UK initiative in design creativity involving mentors from industry. In B. France & V. Compton (ed.), *Bringing communities together: Connecting learners with scientists or technologists*. Rotterdam: Sense.

Barlex, D. & Steeg, T. (2017b) *Big ideas for design & technology*. https://dandtfordandt.wordpress.com/working-papers/big-ideas-for-dt/ (accessed 12 June 2020).

Bronowski, J. (1973b) *The ascent of man*. London: British Broadcasting Corporation.

Department for Education (2011) *The framework for the national curriculum: A report by the Expert Panel for the National Curriculum review*. London: Department for Education.

Department for Education (2013) *The national curriculum in England framework document*. London: Crown Copyright.

Department for Education (2015) *Design and technology GCSE subject content*. London: Crown Copyright.

Department for Education and Science and Welsh Office (1988) *National Curriculum Design and Technology Working Group interim report*. London: HMSO.

Driver, R. (1983) *The pupil as scientist*. Milton Keynes: Open University Press.

Glowee (2019) Bioluminescence. www.glowee.eu/uses (accessed June 12 2020).

Kimbell, R. & Perry, D. (2001) *Design and technology in a knowledge economy*. London: Engineering Council.

Layton, D. (1993b) *Technology's challenge to science education*. Buckinghamshire, UK: Open University.

Murphy, P. (2003) The place of pedagogy. In D. Barlex (ed.), *Creativity in crisis? Design & technology at KS3 and KS4*. Wellesbourne, UK: Design and Technology Association.

Nuffield Design & Technology Capability Tasks (2000) Capability tasks. https://dandtfordandt.files.wordpress.com/2013/07/ks3-ct-better-weighing.pdf (accessed June 12 2020).

Nuffield Design & Technology Chooser Charts (2000) https://dandtfordandt.wordpress.com/resources/nuffield-ks3-dt-resources/chooser-charts/ (accessed June 12 2020).

Perkins, D. N. & and Salomon, G. (1992) *Transfer of learning*. International Encyclopedia of Education, Second Edition Oxford, England: Pergamon Press.

Science Enhancement Project Wind Power booklet. Wind power booklet. www.stem.org.uk/resources/collection/3379/wind-turbines (accessed 12 June 2020).

STEM Pathfinder Programme(2012) The Pathfinder Programme. www.stem.org.uk/elibrary/resource/26103 (accessed June 12 2020).

Williams, P. John & Lockley, J. (2012) An Analysis of PCK to elaborate the difference between scientific and technological knowledge. In Thomas Ginner, Jonas Helstrom, & Magnus Hulten (eds), *Technology education in the 21st century: Proceedings of the PATT 26 Conference 2012* (pp. 468–477). Stockholm: Linkoping University.

5

Teaching mathematics in the light of STEM

A subject some people just can't do?

In the UK there is a cultural view that some people 'just can't do maths' and this is a great impediment to enabling many young people to be successful at mathematics. Charlie Stripp, National Director at the National Centre for Excellence in teaching Mathematics (NCETM) suggested to the authors that the problem might lay with the way mathematics has often been taught in the UK in that it can seem to students like an ever-growing collection of rules to be remembered, rather than a logical, connected discipline that can be understood and used as a powerful, flexible problem solving tool and that it will be important to encourage teaching for connected understanding, rather than remembering lists of seemingly unconnected rules. The popular image of the mathematician is not dissimilar to the popular image of the scientist mentioned in Chapter 1: male, elderly, unfashionable, untidy, withdrawn into world that only he, and I stress he, is interested in or understands and which he can't explain to others in everyday language. Yet, mathematics is the product of the human mind. Unfortunately, the mathematics we learn at school tells us little if anything of the mathematicians who produced it. Vera John Steiner and Reuben Hersh write compellingly about the 'life mathematical' enjoyed by those who commit to mathematics. They acknowledge that it is certainly not an easy life. It is full of intellectual struggle accompanied by a roller coaster of emotional highs and lows as ideas which seem promising turn out to be false and must be discarded in an ever more ruthless pursuit of truth. Among mathematicians there is fierce rivalry as well as intense friendship and loyalty, played out within a domain that few others can appreciate. However, Steiner and Hersh do acknowledge that for many the 'life mathematical in school' is a very different affair. They write, 'People aren't born disliking math. They learn to dislike it at school!' This is perhaps an oversimplification in that a cultural view endorsing not being able to do mathematics will influence learners studying mathematics in schools. This is not to decry the efforts of teachers but to acknowledge that the content of current mathematics courses in conjunction with their significance in high-stakes testing and examination success needed to gain access to college or university courses puts a very heavy burden on students who find the subject bemusing. This is not a particularly new insight. In 1907, Betrand Russell wrote:

In the beginning of algebra, even the most intelligent child finds, as a rule, very great difficulty. The use of letters is a mystery, which seems to have no purpose except mystification. It's almost impossible, at first, not to think that every letter stands for some particular number, if only the teacher would reveal what number it stands for. The fact is, that in algebra the mind is first taught to consider general truths, truths which are not asserted to hold only this or that particular thing but of any one of a whole group of things. [...] Usually the method that has been adopted in arithmetic is continued: rules are set forth, with no adequate explanation of their grounds; the pupil learns to use the rules blindly, and presently, when he is able to obtain the answer that the teacher desires, he feels that he has mastered the difficulties of the subject. But of inner comprehension of the processes employed he has probably acquired almost nothing.

(Russell, 1907: 60)

In fairness, we should acknowledge that Bertrand Russell's attempts at school teaching were not successful and the school he set up with Ludwig Wittgenstein was a complete failure. But we cannot deny that there is considerable concern over many young peoples' dislike of school mathematics and the resultant poor levels of attainment. Hence in the rest of this chapter we will consider

■ causes for concern;

■ responding to the concern;

■ examples of teaching mathematics in the light of STEM to address the concern;

and

■ revisit the life mathematical in school.

Causes for concern

Writing in the *Washington Post*, Joe Hiem (2016) reports that in the latest Program for International Student Assessment (PISA) measuring math literacy in 2015, US students ranked 40th in the world. The US average math score of 470 represents the second decline in the past two assessments – down from 482 in 2012 and 488 in 2009. The US score in 2015 was 23 points lower than the average of all of the nations taking part in the survey. Although 6 per cent of US students who took the test had scores in the highest proficiency range, 29 per cent of US students did not meet the test's baseline proficiency for math. In response to these findings the then US Education Secretary John B. King Jr (2016) said on the morning the results were announced,

We're losing ground – a troubling prospect when, in today's knowledge-based economy, the best jobs can go anywhere in the world. Students in Massachusetts, Maryland, and Minnesota aren't just vying for great jobs along with their neighbors or across state lines, they must be competitive with peers in Finland, Germany, and Japan.

Some experts dismiss the value of the PISA results and say they ignore integral aspects of education. Yong Zhao, a professor in the School of Education at the University of Kansas is reported as saying

> The results basically tell us how well these students took the test, that's all. Whether that performance has anything to do with real life or the quality of education, we don't know. There's no other evidence. We don't have to really jump on this, let alone try to borrow policies or ideas from other places. I disregard all these tests because no test actually measures exceptionality. But an economy, especially today, is driven by individual exceptionality. Entrepreneurship, entertainment, inventiveness, creativity – no tests can measure that.
>
> (Ibid.)

However, Marc Tucker (2016), the then president of the National Center on Education and the Economy described the results as a 'Sputnik moment' for the US leaders and educators saying:

> We're living in a world that is highly integrated. And the United States cannot long operate a world-class economy if our workers are, as the OECD statistics show, among the worst-educated in the world.

He described how the Chinese had achieved their success as follows:

> They have redesigned their schools to take advantage of very highly educated and trained teachers. They have organized their schools so that teachers work together in teams in a very disciplined way to get better and better at teaching and to constantly improve the performance of their students.
>
> (Ibid.)

It is worth noting that the Trends in International Mathematics and Science Study (TIMSS) collects data concerned with the performance of students in primary (elementary) and secondary (junior high) schools. These tests are more closely aligned to the school curriculum and recently schools in England and America have shown improved performance. However, a comparison of the score trends for secondary schools affords little comfort to politicians. Although there has been improvement over the past 20 years, when this is compared with the trends from other higher performing countries this improvement is at best lacklustre.

The current administration in America has voiced similar concerns in the speeches of US Secretary of Education Betsy DeVos. In addressing the American Enterprise institute in January 2018, she commented on the PISA results as follows:

> And, you know this too: it's not for a lack of funding. The fact is the United States spends more per pupil than most other developed countries, many of which perform better than us in the same surveys.
>
> (DeVos, 2018a)

In addressing State Chiefs in March 2018, DeVos endorsed the previous administration's attempt to move on from No Child Left Behind and its introduction of the bipartisan Every Student Succeeds Act (ESSA). She saw this act as recognising that federal overreach had failed, and that ESSA was enacted to give State Chiefs the flexibility and opportunity to address their state's unique challenges. But there was a sting in the tail. She said, 'The trouble is … I don't see much evidence that you've yet seized it.'

Put bluntly she is saying emphatically, 'The ball is in your court'. And it appears that she is open to highly individual, context dependant innovative approaches as she continues,

> Because the imperative to do something shouldn't have to come from Washington. It shouldn't have to come from your state capital. The imperative to do better comes from students. We are accountable to them! That's why your ESSA plans aren't a ceiling. There is no ceiling. There is no ceiling on what students can achieve. That's why we must pursue a paradigm shift … a fundamental reorientation … a rethink. I've called for this nation to rethink school, and I want to make sure you're clear what I mean. 'Rethink' means we question everything to ensure nothing limits a student from pursuing his or her passion, and achieving his or her potential. So, each student is prepared at every turn for what comes next. Question everything. At every school. In support of every student's success. So, I ask each of you: What are you going to do to rethink education in your state?

In England, the most recent Ofsted Report ('Mathematics Made to Measure', 2012) into the teaching of mathematics offers little comfort. As with all such reports the sample is limited. Inspectors visited 160 primary and 160 secondary schools, and observed more than 470 primary and 1200 secondary mathematics lessons, but there is little reason to suspect that the findings are not typical of the wider picture. The report highlights the failure of much teaching to develop pupils' conceptual understanding alongside their fluent recall of knowledge, and a lack of confidence in problem solving indicating that too much teaching concentrated on the acquisition of disparate skills that might have enabled pupils to pass tests and examinations but did not equip them adequately for the next stage of education, work and life. The report emphasised the problems of a poor start which doubtless reinforces the cultural view that some people 'just can't do mathematics':

> The 10% who do not reach the expected standard at age 7 doubles to 20% by age 11, and nearly doubles again by 16.
>
> (Ofsted, 2012: 4)

The report noted that this is compounded by the lack of curricular guidance and professional development in enhancing subject knowledge and effective pedagogy. In his introduction to the report, the then Chief Inspector of Schools Sir Michael Wilshaw returns to the instrumental theme:

> Our failure to stretch some of our most able pupils threatens the future supply of well-qualified mathematicians, scientists and engineers.
>
> (Ofsted, 2012: 4)

Four years on since the data was collected for Made to Measure, the keynote presentation made by Jane Jones, HMI, the National Lead for mathematics at the Better Mathematics Conference in 2015 indicated that in the light of recent data collected by Ofsted little had actually changed (Jones, 2015). While emphasising that the best teaching develops conceptual understanding alongside pupils' fluent recall of knowledge and confidence in problem solving, she noted that conceptual understanding and problem solving are under emphasized. Teaching tended to focus on exams, relying on pupils' short-term memory, rather than on progression and development of understanding and that teachers were not clear enough about progression, so teaching was fragmented and did not link concepts.

Responding to the concern

Mathematical habits of mind

Given the concern expressed about developing conceptual understanding and the lack of problem solving, the approach developed by Cuoco et al. (1996) in articulating mathematical habits of mind might bear fruit. He and his co-workers identified mathematical habits of mind as shown in Table 5.1.

Cuoco suggests that it would be more useful if the curriculum was built around the habits of mind used by mathematicians when they think about problems and how they set about solving them. Cuoco (1996: 401) concludes that is possible to design courses that:

> meet the needs of students who will pursue advanced mathematical study, at the same time as serving those who will not go on to advanced mathematical study but who will nevertheless use these ways of thinking in other researchlike domains such as investigative journalism, diagnosis of the ills of a car or a person, and so on.

TABLE 5.1 Mathematical habits of mind

Students who think like mathematicians should be:

Pattern sniffers – always on the lookout for patterns and the delight to be derived from finding hidden patterns and then using shortcuts arising from them in their daily lives.

Experimenters – performing experiments, playing with problems, performing thought experiments allied to a healthy scepticism for experimental results.

Describers – able to play the maths language game, for example, giving precise descriptions of the steps in a process, inventing notation, convincing others and writing out proofs, questions, opinions and more polished presentations.

Tinkerers – taking ideas apart and putting them back together again.

Inventors – always inventing things – rules for a game, algorithms for doing things, explanations of how things work, or axioms for a mathematical structure.

Visualisers – being able to visualize things that are inherently visual such as working out how many windows there on the front of a house by imagining them, or using visualization to solve more theoretical tasks.

Conjecturers – making plausible conjectures, initially using data and increasingly using more experimental evidence.

Guessers – using guessing as a research strategy; starting with a possible solution to a problem and working backward to achieve the answer.

Other researchers have extended the idea of mathematical habits of mind by focusing on particular aspects of mathematics. Ersen and colleagues (Ersen et al. 2018) have explored geometric habits of mind with 10th grade students. Their study, which took place over 15 weeks, considered reasoning with relationships, generalising geometric ideas, investigating invariants and balancing exploration and reflection. The teaching environment consisted of the students being given unfamiliar geometric problems to solve using the software GeoGebra with the teacher providing limited guidance and encouragement to use geometric habits of mind. They concluded that the teaching environment developed improved the students' geometric habits of mind in a lasting way. Eroglu and Tanish (2017) explored algebraic habits of mind. Their study was with a class of Grade 7 students and was much shorter, lasting only three lessons. The students were set the sum of consecutive numbers task shown in Panel 5.1. The researchers concluded that providing rich classroom environments (that includes teacher–student and student–student interaction, using multiple representations, questioning, functional thinking, thinking independently) could develop students' algebraic habits.

Interestingly, both these more focused studies on mathematical habits of mind use problem solving as a dominant pedagogy that resonates with Ofsted findings (2012 and 2015) that in England problem solving is underemphasised indicating that perhaps the use of mathematical habits of mind as a tool for both planning and implementing the curriculum might pay dividends. Problem solving will be considered in more detail in Chapter 6.

Teaching for mastery

Recently, the idea of teaching for mastery has been gaining traction as a means of overcoming the difficulties experienced by many learners with regard to mathematics. The National Centre for Excellence in Teaching Mathematics (NCETM, 2016b) explains as follows:

- Mastering maths means pupils acquiring a deep, long-term, secure and adaptable understanding of the subject.
- The phrase 'teaching for mastery' describes the elements of classroom practice and school organisation that combine to give pupils the best chances of mastering maths.

7=3+4
9=2+3+4
22=4+5+6+7
Above, there are some numbers written as the sum of consecutive numbers. Seven is written as the sum of two consecutive numbers; nine as the sum of three consecutive numbers; and 22 as the sum of four consecutive numbers. Accordingly:
a) Find all the ways that can be used to write each number between 1 and 35, as the sum of two or more consecutive numbers.
b) Find a rule that can be used to write the numbers as the sum of two or more consecutive numbers.

Panel 5.1 The sum of consecutive numbers task

■ Achieving mastery means acquiring a solid enough understanding of the maths that's been taught to enable pupils to move on to more advanced material.

Teaching for mastery can be considered as the interaction of five big ideas drawn from research evidence, underpinning teaching for mastery. Figure 5.1 shows how these ideas bind together.

A brief description of these ideas is as follows:

Coherence
Lessons are broken down into small connected steps that gradually unfold the concept, providing access for all children and leading to a generalisation of the concept and the ability to apply the concept to a range of contexts.

Representation and structure
Representations used in lessons expose the mathematical structure being taught, the aim being that students can do the maths without recourse to the representation.

Mathematical thinking
If taught ideas are to be understood deeply, they must not merely be passively received but must be worked on by the student: thought about, reasoned with and discussed with others.

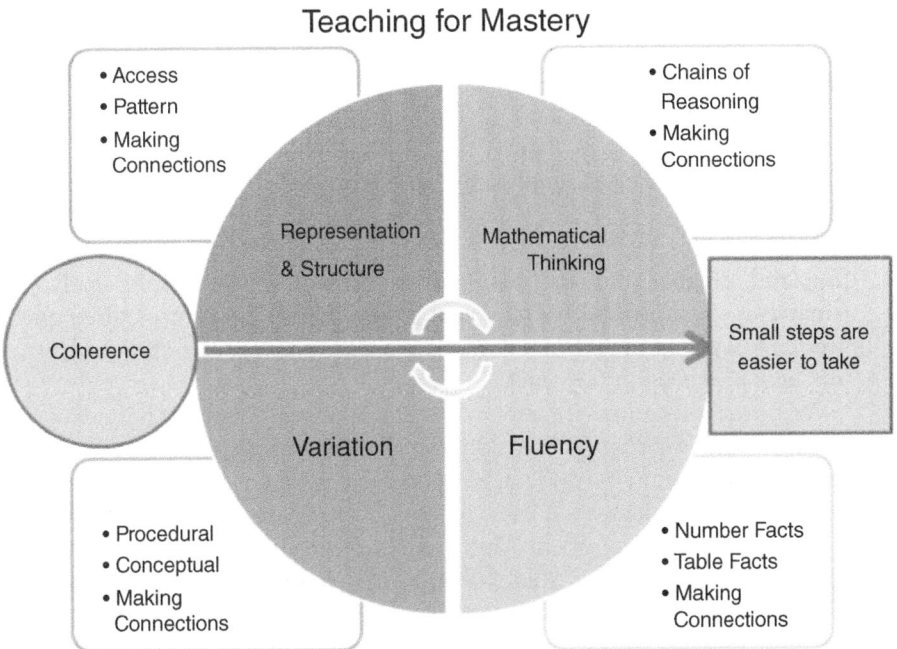

Figure 5.1 The five big ideas informing teaching for mastery
Source: NCETM website

Fluency

Accurate and efficient recall of facts and procedures and the flexibility to move between different contexts and representations of mathematics.

Variation

Variation is twofold. It is first about how the teacher represents the concept being taught, often in more than one way, to draw attention to critical aspects, and to develop deep and holistic understanding. It is also about the sequencing of the episodes, activities and exercises used within a lesson and follow-up practice, paying attention to what is kept the same and what changes, to connect the mathematics and draw attention to mathematical relationships and structure.

The guidance on teaching for mastery produced by the NCETM (Maths Hub, 2017b) makes three important points:

- It is recognised that practice is a vital part of learning, but the practice is intelligent practice that aims to develop students' conceptual understanding and encourage reasoning and mathematical thinking, as well as reinforcing their procedural fluency.
- Significant time is spent developing a deep understanding of the key ideas and concepts that are needed to underpin future learning.
- The structures and connections within the mathematics are emphasised, which helps to ensure that students' learning is sustainable over time.

If mathematics is to be used as part of the learning in other STEM subjects, it will be important that the way students have been taught for mastery is not compromised.

Mathematics post-16

The review of post-16 mathematics (Smith, 2017: 6, 7) describes the state of mathematics in England as significantly different from that in most other advanced nations.

> England remains unusual among advanced countries in that the study of mathematics is not universal for all students beyond age 16. Almost three quarters of students with an A★–C in GCSE mathematics at age 16 choose not to study mathematics beyond this level.
>
> With the exception of mathematics degrees, more than 40 per cent of English 19 year olds studying STEM subjects in UK universities do not have a mathematics qualification beyond GCSE. This increases to over 80 per cent for students on non-STEM degree courses, many of which have a significant quantitative element. A lack of confidence and anxiety about mathematics/statistics are problems for many university students; and many have done little or no mathematics pre-university for at least two years.

The report made a strong economic case for an increased uptake of mathematics:

> There is strong demand for mathematical and quantitative skills in the labour market at all levels and consistent under-supply, reflecting the low take-up of and achievement in 16–18 mathematics in England relative to other developed countries. The quantitative demands of university courses in both STEM (science, technology, engineering and mathematics) and non-STEM subjects are increasing and are set to increase further.
>
> Higher levels of achievement in mathematics are associated with higher earnings for individuals and higher productivity. Increased productivity is a key determinant of economic growth.
>
> The increasing sophistication of technology is driving change to the economy and the nature of work. This is increasing the demand for mathematical and quantitative skills.
>
> (Smith, 2017: 6)

The report made a wide range of recommendations the most significant with regard to teaching mathematics in secondary schools and colleges were the recommendation for a core mathematics qualification to be made available to post-16 students as shown in Panel 5.2.

The government welcomed the report with Nick Gibb (2017), the Minister for Schools, writing to Adrian Smith as follows:

> I agree with your conclusion that government, employers, schools and colleges must take greater action to encourage and support more young people to choose mathematics post-16, particularly in areas where take-up is low.

In the wake of the Smith Review support for a post-16 Core Maths programme for those not studying conventional A level mathematics courses has grown. Considering the importance of Core Maths, Charlie Stripp the Director of NCETM wrote in 2019:

Recommendation 1: The Department for Education should seek to ensure that schools and colleges are able to offer all students on academic routes and potentially students on other level3 programmes access to a core maths qualification.
Recommendation 2: The Department for Education and Ofqual should consider how the core maths brand could be strengthened with the aim of improving awareness and take-up of the qualification.

Panel 5.2 Recommendations for core post-16 mathematics

Maths education is so important that 16-year-olds shouldn't be pondering 'Should I take maths post-16?' Instead they should be asking 'Which maths should I take?'

Historically, if AS/A level maths was not the right choice for students who had 'passed' GCSE maths, no suitable alternative post-16 maths option existed. Now Core Maths, which is specifically designed to fill this qualification gap, is establishing itself as an excellent option for these students.

I believe we should aspire to an annual cohort of over 100,000 students choosing to take Core Maths within the next 10 years.

The Nuffield Foundation (2010) had highlighted some of the issues the Smith Review sought to address with the publication of *Is the UK an Outlier? An International Comparison of Upper Secondary Mathematics Education*, which reported that in England, Wales and Northern Ireland fewer than one in five students study any mathematics after the age of 16 (Scotland does slightly better). In 18 of the 24 countries, more than half of students in the age group study mathematics; in 14 of these, the participation rate is over 80 per cent; and in eight of these countries *every* student over 16 studies mathematics. The authors of the report concluded that when it comes to the mathematics education of its upper secondary students, the UK is out on a limb and recommended that there should *be a review of post-16 mathematics policy in the UK and alternative models for post-16 mathematics* should be developed. The commissioning of the Smith Review, its recommendations and results can be seen to stem from the 'outlier' publication.

In 2017 the Nuffield Foundation has commissioned research into the early take up of core mathematics (Homer et al. 2018). This project aims to assess the early success (or otherwise) of this new qualification. It also aims to make suggestions for how the government and other agencies can best act to ensure its long-term success. The researchers will quantify the uptake of Core Maths in its first three years of existence, and gather views on how successful Core Maths has been in widening post-16 mathematics participation in England. The project ran from 1 March 2017 to 29 February 2020 and reported some early findings (Homer et al. 2018) generally indicating support for the wider policy imperative of ensuring more students study mathematics post-16 but identifying two main challenges for schools and colleges: the logistics of positioning Core Maths within the curriculum framework and funding conditions now characterising the post-16 sector, and whether to target certain students or allow students to opt in.

Whether schools and colleges are able to overcome these challenges remains to be seen, although it is worth noting that the government in England is committed to meeting an aspiration voiced in the recent review of post-16 mathematics (Smith, 2017) that in ten years' time all students will be studying some mathematics post-16 (Gibb, 2017) . It is also worth noting that a student's experience at both Key Stage 3 (ages 11–14 years) and Key Stage 4 (ages 14–16 years) might not necessarily predispose them to take mathematics post-16 even if these experiences had led to public examination success. While the answer to the question, 'Should all students study some mathematics post 16?' may appear to be a resounding 'Yes' from the perspective of policy makers with an eye to productivity and economic success, overcoming the influence of the unhelpful cultural view may not be quite so simple. The response of the government to the Smith Review acknowledges this when the minister notes, 'The need to address negative cultural perceptions of mathematics' (Gibb, 2017). In considering the response of young people to the possibility of studying mathematics post-16, it is worth asking how their learning in science and design & technology might be used to enhance both their learning *and* enjoyment of mathematics before they make a decision as to whether or not to continue studying mathematics post-16? It is this we will consider in the following section. It is also worth noting that science and design & technology are not the only subjects that benefit from competence in mathematics, the social science, humanities and business studies are all enhanced by mathematical understanding, particularly that regarding data analysis and statistical inference.

How might teaching mathematics in the light of STEM help?

Our approach will be to identify topics in science and design & technology in which understanding and application of mathematics are required and then scrutinise this using mathematical habits of mind as a lens and taking into account how teaching for mastery might also be involved.

Deriving chemical formulae

One of the great advances in chemistry was the idea that when a material burns it combines with oxygen and the resultant material has a greater mass than the starting material. This is counterintuitive as many everyday materials 'go up in smoke' when they burn, giving the impression that the results of combustion will be materials of less mass than the starting material. Early on in chemistry courses learners will have the opportunity to burn various materials and identify the increase in mass.

BURNING MAGNESIUM IN AIR

A common approach is for learners to be given instructions to burn a weighed amount of magnesium ribbon in a lidded crucible and then weigh the amount of magnesium oxide formed. This is a tricky experiment to perform well. First, the crucible and lid should be weighed, then the magnesium ribbon added to the crucible in the form of an open coil so that it can burn and the whole assembly reweighed. Then, the crucible must be heated to cause the magnesium to ignite and the lid must occasionally be raised ever so slightly to allow more air into the crucible to ensure complete combustion. But it is important not to allow any of the oxide being formed to escape. Once the reaction appears to be complete then the whole assembly must be left to cool before it is reweighed. The weighing data can be used to show that the oxide is indeed heavier than the starting magnesium and to find the mass of oxygen that has combined with the starting mass of magnesium. Ultimately, the challenge is to turn this data about combining masses into a formula for magnesium oxide.

How might this experiment be reworked in ways that support mathematical habits of mind?

First is to start with burning some ribbon in the open air and ask the learners to be *guessers*. Is the white stuff likely to be heavier or lighter than the starting magnesium? Second is to move on to ask the learners to be *conjecturers*. Is it likely to be heavier or lighter? How might we find out? Third is to move on to ask the learners to be *experimenters*. They are tasked with devising a good way to find out. With some prompting, class discussion and peer-to-peer discussion, an experimental procedure similar if not identical to that described above will be arrived at and the learners can work in pairs to carry out the experiment and record their results. Once the results are in and there is general agreement that the material produced when magnesium is burned weighs more than the starting magnesium, the challenge for the class is to think about how the data they have collected might be used to calculate the formula for magnesium oxide. Here they are moved on to be *visualisers* in imagining the various possibilities. If one atom of magnesium combined with one atom of oxygen, the formula would be MgO. If one atom of magnesium combined with two atoms of oxygen, the formula would be MgO_2. If two atoms of magnesium combined with one atom of oxygen, the formula would be Mg_2O. And perhaps the next step is for them to become *pattern sniffers* in presenting the data in a way that enable it to be used to distinguish between the different possibilities. Any calculation has to take into account the differing atomic masses of magnesium and oxygen. Magnesium has an atomic mass of 24 whereas oxygen has an atomic mass of 16. Hence for a formula of MgO, the ratio of combination would be 24:16, equivalent to 3:2, whereas a formula of MgO_2 would require the ratio of combination to be 24:32, equivalent to 3:4. Given that pupils will be handing masses of magnesium between 0.5 and 1.5 g, the arithmetic can become complicated. So perhaps a way forward is to try and present the data graphically, involving the learners in being *conjecturors* in deciding what the graphic presentation of the data means.

A way forward here is to plot the data obtained on a chart that shows line graphs for different possible formulae of magnesium oxide. Such a graph is shown on Figure 5.2. Pupils in the class can share their results and then plot them on the graph. The result will be a scatter of points and the position of this scatter should enable the class to decide which of the three possible formulae is correct according to their experiments. The scatter of the points should align more closely with MgO line leading the learners to *conjecture* that this is the formula. Of course, the chemistry teacher will want her pupils to know and remember the correct formula of magnesium oxide (MgO) but the point of this experiment is to reinforce the idea that formulae are derived from experimental data. Later in the chemistry course the pupils will learn about atomic structure and valency and be able to use these ideas to explain why the formula of magnesium oxide is MgO and not Mg_2O or MgO_2.

We can see the data manipulation and presentation as a series of mathematical operations. Each operation is not complicated but the overall sequence of operations is demanding. This requires a clear understanding of the purpose of the endeavour, experimental skill in obtaining the required data, competence in presenting the data graphically, an understanding of ratios and the ability to interpret the data once presented graphically. It is a worthwhile opportunity for pupils to use their mathematics in a chemistry lesson. Using the approach outlined above enabling learners to use various mathematical habits of mind will almost certainly take longer than a simple

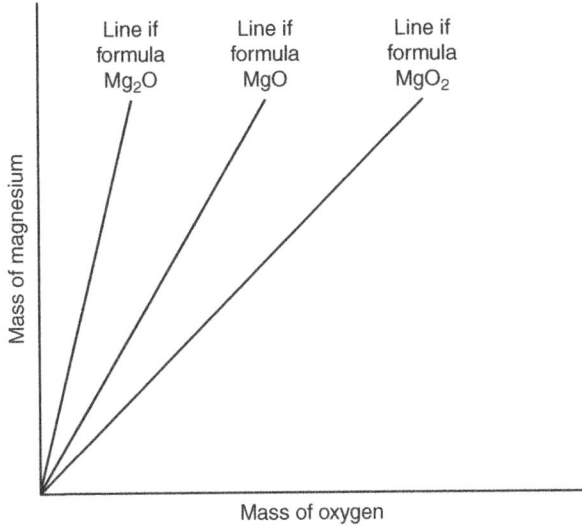

Figure 5.2 Graphical representation of possible formulae for magnesium oxide

'follow the given instructions' approach and its merits will needed to be gauged against the extent to which it is worthwhile in the long game of developing such habits of mind and enabling learners to use mathematics learning to support in science learning. The approach described meets some of the key principles for teaching mastery (Maths Hub, 2017b). There is a high level of teacher–learner and learner–learner interaction where all learners are in the class are thinking about, working on and discussing the same mathematical content and the learners are revisiting important ideas purposefully, which can be seen as intelligent practice with the time spent on this developing a deeper understanding of these ideas. Conversations between the chemistry teacher and the mathematics teacher are necessary to ensure that learners are encouraged to use mathematical thinking and justify their procedures as described above as opposed to simply following instructions. It might even be possible for the mathematics teacher to include the 'work up' of the results in a mathematics lessons. This would require a deliberate intervention to teach mathematics in the light of STEM. If the mathematics teacher felt that the learners had sufficient mathematical knowledge and skill before such an intervention then the 'work up' lesson can be seen as an opportunity for revision and assessment of previous learning. If the learners are going to be using unfamiliar mathematics then the work up session provides a novel way to introduce such topics.

Calculations from equations

CONSIDERING NEUTRALISATION OF ACIDS

In attempts to show the application of chemistry in everyday life, teachers often consider simple medicines such as indigestion remedies. The basis of such remedies is sometimes the neutralisation of excess stomach acid. Such products are often called 'antacid tablets' and the key ingredient will be a substance that reacts with the acid in the stomach. In some cases, this is a carbonate. The reaction of the carbonate with the acid causes the production of carbon dioxide so that the reduction of acid content in the stomach is accompanied by the formation of a gas. In such cases, those taking the tablets often 'burp'. In other cases, the key ingredient is a hydroxide. Here the neutralisation of the acid is not accompanied by the production of carbon dioxide.

For those producing indigestion remedies, it is important to know how much of the antacid ingredient to put in each tablet. Too little and the remedy will fail; too much and the reduction of acid in the stomach will be so great that food cannot be properly digested. The acid in the stomach is hydrochloric acid and the equation for the reaction of magnesium hydroxide with hydrochloric acid is as follows:

$$Mg(OH)_2(s) + 2HCl(aq) \rightarrow MgCl_2(aq) + 2H_2O(l)$$

The challenge is to use this equation to calculate how much magnesium hydroxide is needed to neutralise some of the acid in the stomach. Typically, an adult's stomach contains about 200 ml of gastric juice with a concentration of 0.1 mol/l. To ensure that not too much of the acid is neutralised, we can assume that only half the acid in the stomach should be neutralised. What are the steps in the calculation?

1   Decide on the amount of acid to be neutralised.

2   Use the equation to decide on the number of moles of magnesium hydroxide and hydrochloric acid that are reacting together.

3   Use the answer from (1) to decide on the number of moles of hydrochloric acid that need to be neutralized.

4   Use the answers from (2) and (3) to decide on the number of moles of magnesium hydroxide needed to neutralise the hydrochloric acid.

5   Convert the answer to (4) into grams of magnesium hydroxide needed for a single antacid tablet.

Underpinning this lengthy calculation are the important ideas of ratio and proportion. For the chemistry teacher, it is important that the learners become fluent in such calculations not losing sight of the chemistry because the mathematics appears over-complicated. For the mathematics teacher, the chemistry lesson might provide opportunities for learners to further develop mathematical habits of mind. It is clear that a conversation is necessary to explore these requirements and develop an approach in which the learning in both chemistry and mathematics is enhanced.

Let us now move on to physics.

Understanding waves

Waves are a fundamental concept in physics and the behaviour of waves is used to explain a wide variety of phenomena. These include earthquakes, (seismic waves), sound waves, the electromagnetic spectrum and the properties of light (propagation, refraction and diffraction). Fundamental to understanding the behaviour of waves is the wave equation

Wave speed   = Frequency × Wavelength
(metres per second, m/s)   (hertz, Hz)   (metres, m)

The unit hertz refers to cycles per second, a unit named in honour of the physicist Heinrich Hertz who discovered radio waves. This is sometimes abbreviated to

$$v = f \times \lambda$$

where

$$v = \text{wave speed,}$$
$$f = \text{frequency,}$$
$$\lambda = \text{wavelength}$$

This can be shown diagrammatically as a sine wave (See Figure 5.3). The amplitude of the wave can change without any change to the wavelength or frequency.

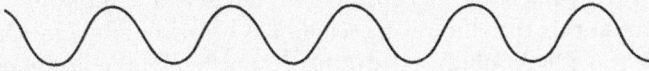

FIGURE 5.3 Sine wave
Source: Adapted from 21st Century Science Physics

It is possible to use this wave equation to calculate any one of the features in the equation if values are known for the other two features. The equation can be used to explain the colour of visible light. The speed of light can be treated as a constant, hence the frequency and wavelength are also constant. But if the frequency is decreased then the wavelength must increase. Similarly, if the frequency is increased then the wavelength must decrease. In the visible spectrum, blue light has a higher frequency and lower wavelength than red light. The numbers get scary because of the units. Blue light has a frequency of 606–668 THz and a wavelength of 450–495 nm. Red light has a frequency of 400–484 THz and a wavelength of 620–750 nm. T stands for tera which means x million million – 1 000 000 000 000 or 10^{12}, so the frequency of blue light is in the region of 460 million million cycles per second (or Hz); n stands for nano, which means one thousand millionth – 0.000000001 or 10^{-9}, so the wavelength of red light is in the region of 700 thousand millionths of a metre.

The potential for learner confusion is high so bringing mathematic habits of mind to bear on understanding such large and small numbers might well be beneficial. How might a physics teacher and a mathematics teacher approach this problem? One way might be to play a card game.

- Start with a small set of cards in which on each card is the name and definition of one of the mathematical habits of mind, say 32 cards so there are four cards for each habit of mind.
- Shuffle the cards and take the one from the top. Whichever one it is the teachers ask themselves the question, 'What part of a lesson on waves might encourage this habit of mind?'
- Then return the card to the pack, shuffle the cards and again take the one from the top.
- Whatever the habit of mind, ask the same question again.

If there are too many repeats simply move on to another shuffle but it's probably worth allowing for a habit of mind to come up two or three times. Over several shuffles a set of lesson parts will emerge that can be assembled into a lesson that teaches about waves in a way that utilises mathematical habits of mind. The random nature of the emerging habits of mind should encourage creative thinking. It will be necessary to keep the idea of teaching for mastery in mind and on this occasion it is perhaps fluency in dealing with indices that is an important goal.

Estimating the size of a molecule

Practical Physics, developed by the Institute of Physics and the Nuffield Foundation, has a useful experiment to estimate the size of a molecule using an oil film (Nuffield Foundation, 2007). The diameter of a tiny droplet of olive oil is measured and the droplet placed on a water surface covered with lycopodium powder. The drop spreads out to form a circle and it is assumed the circle is one molecule thick. The basis for this assumption is that the oil molecule has a hydrophilic end that is attracted into the water and a hydrophobic chain that is repelled by the water and stands up out of the water (see Figure 5.4a). The diameter of the circle is measured (see Figure 5.4b).

Mathematical thinking is required to turn the measurements into an estimate of molecular size.

- The volume of the oil drop is proportional to the diameter cubed.
- The oil drop spreads out to form a cylinder one molecule thick.
- The volume of the cylinder is given by area of the circle times its depth.
- The area of the circle is proportional to the diameter squared.
- If the diameter of the circle is D, and the diameter of the oil drop of d, and the length of the molecule is l, then $d^3 = D^2 \times l$.
- A typical diameter of the initial oil drop would be 0.5 mm.
- A typical diameter of the film would be 250 mm.

Hydrophilic end which is attracted by water

Hydrophobic end which is repelled by water

The result is that the molecules stand on end side by side with the hydrophilic end in the water.

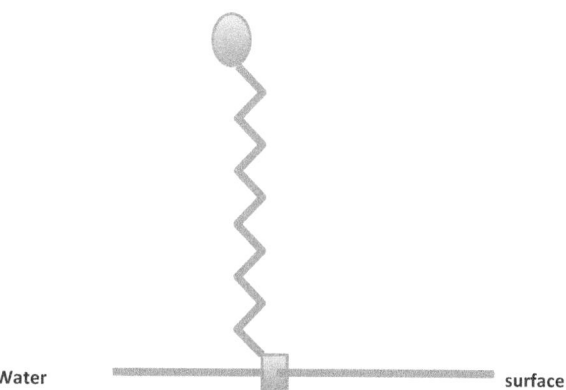

Water   **surface**

FIGURE 5.4A The oil molecule and its behaviour in water

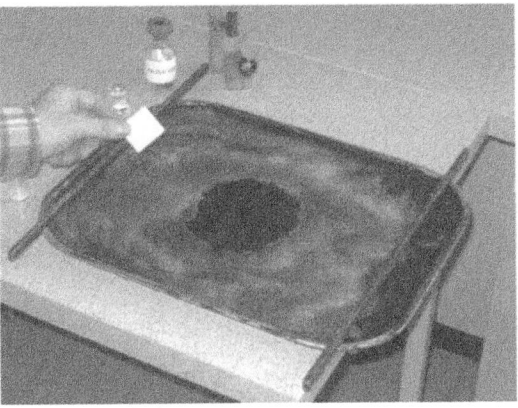

FIGURE 5.4B The Practical Physics oil drop experiment

Hence

- The volume of the drop is given by $0.5 \times 0.5 \times 0.5$ mm^3.
- The volume of the film is given by $250 \times 250 \times l$ mm^3 where l is the length of the molecule.

The volume of the drop is the same as the volume of the film, hence

$$0.5 \times 0.5 \times 0.5 = 250 \times 250 \times l$$

So, we can find l by rearranging as follows

$$\frac{0.5 \times 0.5 \times 0.5}{250 \times 250} = 1 \text{ (the length of the molecule in mm)}$$

Using a calculator:

The length of the molecule is 0.000002 mm or 0.000000002 m.

Using standard notation this becomes: 2 and 10^{-9} m.

There are approximately 12 atoms in the olive oil chain the size of an atom is given by 2 and 10^{-9} ÷ 12.

Hence the approximate size of an atom in the molecule is 1.7 and 10^{-10} m.

This can be written as 0.000000000017 m, 0.00000000002 m or 0.2 nanometres.

The above is, of course, an approximation, as the precise formulae for the volume of the initial drop and the cylinder have not been used in order to keep the calculation relatively simple. The result is, however, of the correct order of magnitude, the size of a carbon atom being 0.7 and 10^{-10} m.

So, what sort of conversation might the mathematics teacher have with the physics teacher about this experiment? A discussion on accuracy of measurement and the impact of inaccuracies on the estimate of molecule size is a possibility. The formulae for the volumes of spheres and cylinders might feature along with the possible effect of the approximations in the calculations shown above. The arithmetic involved in the calculations, the use of power of ten notation, and the prefix nano meaning one billionth might all be supported in mathematics lessons. However, with so much varied mathematics embedded in the activity, it is unlikely that the mathematics teacher can use the experiment to introduce these topics in the mathematics curriculum. But if the mathematics teacher *could* find time to discuss the experiment with learners, she would gain an insight into to the extent that they had mastery of the underlying concepts. Discussion of the results of the experiment and how to derive an estimate for molecular size would reveal where pupils were comfortably confident and where they were experiencing difficulties. It might provide the opportunity for further practice. The conversation might be extended to consider the mathematical habits of mind that the learners were using in the physics lesson and whether any changes of approach to this might be beneficial. Let us now move on to biology.

Estimating population size

Deciding on the number of particular animals or plants in a particular location is a challenge for professional biologists. In most situations it is impossible to count the actual number of flora or fauna present so experimental procedures for estimating the population from a small sample have been devised. Pupils are introduced to these procedures in most biology courses and these estimation procedures are underpinned by mathematical understanding.

Using quadrats

This procedure is relatively simple and involves using a metal or wooden frame called a quadrat. It is usually used to count plants but can be used for slow moving insects. The most basic approach is as follows:

a   Placing the quadrat on the habitat under investigation.

b   Counting the number of a particular plant present inside the quadrat.

c   Estimating how many quadrats are needed to cover the habitat.

d   Multiplying the answer to (c) by the answer to (b) to estimate the number of the particular species in the habitat.

It is possible that the number of plants within the quadrat will vary from place to place in the habitat. A habitat could be variable in terms of abiotic factors (e.g. light and shade, exposure to wind, availability of water and minerals) and biotic factors involving competition from other organisms. Hence the procedure often involves placing several quadrats at different positions in the habitat in order to improve the estimate and avoid bias (see Figure 5.5).

   Part of a typical examination question might include the results of such an experiment as follows:

Quadrat	Number of dandelions
1st	5
2nd	1
3rd	0
4th	2

■   Each quadrat has an area of 0.25 m^2.

■   The total area of the habitat, a playing field, is 20,000 m^2.

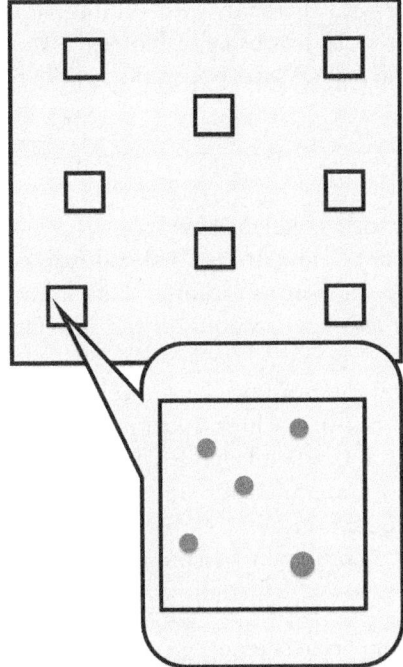

FIGURE 5.5 Diagram showing the use of quadrats shown as small squares to estimate the number of particular plants shown as dark green circles in a field

Question: Estimate the total number of dandelion plants in the playing field.

If the candidate realises that the sum of the area of the four quadrats is 1 m², the task becomes simple. Simply add the number of dandelions in each quadrat to find the number of dandelions in 1 m² and then multiply the result by 20,000. Surprisingly, teachers report that many pupils find this sort of question confusing and use inappropriate arithmetical techniques – multiplying the areas and dandelion numbers instead of adding them giving ten dandelions in .0039 m²; averaging the number of dandelion in a quadrat and then miscalculating the number of 0.25 m² in the field. It is as if the requirement to 'be mathematical' presses a panic button.

A conversation between the biology teacher and the mathematics teacher to explore why pupils tried such apparently illogical approaches would be a start to overcoming pupil's poor responses. A further step would be for a quadrat exercise to be undertaken as part of a mathematics lesson and the results then considered in a biology lesson. If this took place the mathematics teacher might deliberately orchestrate the lesson to reinforce mathematical habits of mind and require intelligent practice leading to fluency.

Using mark, release, recapture

Mark, release, recapture is a common approach to estimating the size of an animal population in a habitat. A portion of the population is captured, marked, and released. Later, another portion is captured and the number of marked individuals within the sample is counted. This method assumes that the study population is 'closed'. In other words, the two visits to the study area are close enough in time so that no individuals die, are born, move into the study area (immigrate) or move out of the study area (emigrate) between visits. It is usual in biology texts book to simply provide the following formula, which allows the population of the particular animal to be estimated.

$$N = \frac{MC}{R}$$

where

N = Estimate of total population size

M = Total number of animals captured and marked on the first visit

C = Total number of animals captured on the second visit

R = Number of animals captured on the first visit that were then recaptured on the second visit

I must admit that I did not find the formula easy to understand intuitively. I wasn't sure why it worked. Some teachers try to provide insight by talking pupils through a situation in which ten labelled ladybirds are released in a greenhouse and on recapture of ten ladybirds sometime later, only one is marked indicating that the ladybird population in the greenhouse is 100. However, I do wonder whether this insight is sufficient to provide genuine understanding.

Then I found out how the formula was derived.

The method assumes that in the second sample, the proportion of marked individuals that are caught (R ÷ M) should equal the proportion of the total population that is caught (C/N).

This can be written

$$\frac{R}{M} = \frac{C}{N}$$

This can be rearranged to give

$$N = \frac{MC}{R}$$

This is the formula often given to the pupils without explanation. I wonder whether it is worth pupils understanding where the formula comes from (i.e. how it is derived) to better understand the logic behind the technique. Before attempting to do this with a class it would be wise for the biology teacher to have a conversation with the mathematics teacher. The result of this conversation might well be that the mathematics teacher would be happy to use the derivation of the formula as a means of teaching or reinforcing algebraic manipulation. This would fit in with using mathematical habits of mind and teaching for mastery. Once the formula had been derived in mathematics, pupils could ideally use it in actual investigations of populations of small creatures in a local environment (e.g. wood lice or ground beetles). If this were not possible then second-hand data from some original investigations could be used. Either way, the pupils would be in a stronger position to understand the mathematical basis of the experimental procedure. Before moving on, let us look at another example in biology.

Probability and genetics

Some diseases are genetic. They are passed from the parents to their children. For example, the causes and probability of a child having the genetic disorder cystic fibrosis is taught in most biology courses. The explanation requires pupils to understand

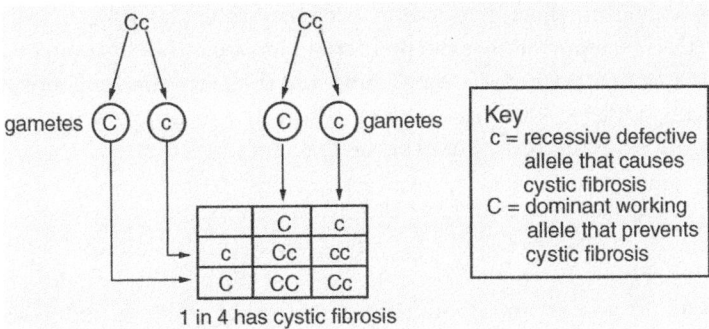

FIGURE 5.6 An explanatory diagram showing the probability of inheriting cystic fibrosis
Source: Adapted from Willams, Biology for you

several concepts: dominant and recessive alleles, faulty alleles, the production of two sets of alleles by each parent, and the combination of alleles during sexual reproduction, some of which give rise to the disease, some of which don't. These concepts are embedded in explanatory diagrams like the one shown in Figure 5.6.

So, if both parents carry the recessive defective allele (c) and the dominant working allele (C), there is a one in four chance that their child will inherit the disease because only one in four of the possible combinations gives two recessive defective alleles.

Parents whose genetic make-up is Cc are known as carriers as they can pass on the defective allele but do not themselves have the condition. It is possible to test an embryo in the womb to discover if it has two defective alleles, in which case the child would suffer from cystic fibrosis. It is now possible for parents to have their DNA tested to discover if they are carriers. But even if they find they *are* carriers, the chance that their child will be born with cystic fibrosis is only one in four. And probability has no memory. Hence if couples who are carriers have three children, all of whom are healthy, the chance of their next child having cystic fibrosis would still be one in four. And a couple who were carriers might have just one child and that child could have cystic fibrosis even though the chance of that child having the disease is one in four. Without some understanding of probability, it seems likely that pupils could become confused with regard to the factors effecting parent's decision making. Hence a conversation with the mathematics teacher might not go amiss. Indeed, the mathematics teacher might consider using recessive/dominant allele combination as a way of teaching probability. A difficulty with this as an introductory approach might be that the science terminology gets in the way of understanding the probability. If this occurs, then it might be preferable for the mathematics teacher to use this as a revision exercise in probability once pupils are familiar with the science. As with the previous example, this would fit in with using mathematical habits of mind and teaching for mastery. Now let us move on to design & technology beginning with food technology.

Considering the energy content of food

The government in England is interested in the teaching of food in schools as part of its concern with regard to the nation's health and the impact of poor dietary choices on the cost of the National Health Service (NHS). As early as 2007, Foresight, a division within the UK government Department of Innovation, Universities and Skills (DIUS), reported that the predicted increase in obesity was a ticking time bomb as far as health service costs were concerned (Department for Innovation, Universities and Skills (DIUS), 2007a). These extracts from the summary of key messages (DIUS, 2007b) indicate the seriousness of the situation:

■ In recent years, Britain has become a nation where overweight is the norm. The rate of increase in overweight and obesity, in children and adults, is striking. By 2050, Foresight modelling indicates that 60 pe cent of adult men, 50 per cent of adult women and about 25 pe cent of all children under 16 could be obese. Obesity increases the risk of a range of chronic diseases; particularly type 2 diabetes, stroke and coronary heart disease and also cancer and arthritis. The NHS costs attributable to overweight and obesity are projected to double to £10 billion per year by 2050. The wider costs to society and business are estimated to reach £49.9 billion per year (at today's prices).

■ The obesity epidemic cannot be prevented by individual action alone and demands a societal approach.

■ Tackling obesity requires far greater change than anything tried so far, and at multiple levels: personal, family, community and national.

■ Preventing obesity is a societal challenge, similar to climate change. It requires partnership between government, science, business and civil society.

This provides a stark warning and now prescriptions for type 2 diabetes caused to a large extent by lifestyle choices leading to being overweight and obese are costing the NHS in England more than £1 billion a year (Ives, 2018).

Many school food technology programmes deliberately educate pupils about the nature of consumer products developed by the food industry with a view to informing individual's food choices. Although this deals with only a minor contribution to the overall obesogenic environment, it is significant in that it empowers pupils and their families to make decisions about their personal eating. Understanding the extent to which the ingredients in a product might contribute to an obesogenic environment relies on significant scientific and mathematical understanding. From the science perspective there is the nature of the ingredients and their ability to act as energy dense food. From the mathematics perspective there is the quantification of the scientific perspective. Overall, the total number of calories in a portion will be important and products that are high in sugar and fat will be energy dense and contribute large numbers of calories. Fats in the diet, depending on their nature, may also lead to arteriosclerosis. As an example, let us consider the energy intake from breakfast cereal. Calculating the calorie intake from a week's consumption of breakfast cereals is not a trivial task. It requires the pupils to take information from the packaging and consider it in the light of a typical portion size and the number of days per week it is eaten. It would be instructive to compare different cereals. This would be an intricate arithmetic exercise requiring the ability to perform a sequence of calculations and comment on the significance of the results. Enabling pupils to use arithmetic fluently is a requirement of many school mathematics courses and this is often seen as important for success in other subjects. In England, for example the National Curriculum for Mathematics states 'Teachers should use every relevant subject to develop pupils' mathematical fluency. Confidence in numeracy and other mathematical skills is a precondition of success across the national curriculum' (Department for Education 2014). Using arithmetic to compare food energy content in the context of the looming obesity crisis provides the basis for collaboration between the mathematics teacher and the food technology teacher. This would also enable consideration of the questions 'What if I ate this instead of that?' Mathematical habits of mind would surely be engaged in this exercise and there are opportunities to develop mastery. Note that this task is made less demanding in terms of arithmetical manipulation if the pupils use a spreadsheet.

Ending hunger

In considering hunger in the world, it is important that learners appreciate how the poorest in the world acquire their food and how this might be changed. The most basic form of food production is subsistence farming by which a family or community grow just enough food for them to be able to eat with little if any surplus. Any disruption of this endeavour quickly leads to hunger and starvation. In 2015, about 2 billion people (slightly more than 25 per cent of the world's population) in 500 million households living in rural areas of developing nations survived as 'smallholder' farmers, working less than 2 hectares (5 acres) of land (Rapsomanikis, 2015). A step up from subsistence farming is the production of sufficient food for those producing the food to be able to have enough for their own needs and a surplus, which they can sell to others through local markets. A further increase in the scale of food production involves farmers producing food for sale only and using their earnings to buy food from other food producers or from shops and business that sell food. This sort of farming can feed into regional, national and global markets. The Food and Agriculture Organisation of the United Nations (2009) warns that the world population will have reached over 9 billion by 2050, from its current population of some 7.4 billion. This will place a significant burden on food production. For example, the report warns that it is estimated that by 2050 developing countries' net imports of cereals will more than double from 135 million metric tonnes in 2008/09 to 300 million in 2050.

Foresight within the Department for Business, Innovation & Skills produced a report *The Future of Food and Farming: Challenges and Choices* (DBIS, 2011) for global sustainability. Chapter 6 (2011: 116) of the report deals with the problem of hunger, which is significant. The report notes:

> Today, there are an estimated 925 million people hungry, and perhaps an additional one billion who are not hungry in the usual sense but suffer from the 'hidden hunger' of not having enough vitamins and minerals. Hunger is the antithesis of human development. It is important for policy makers to take a broad view of the nature and causes of hunger and its many impacts, including the severe and long-lasting nature of the effects that hunger and under-nutrition can cause, particularly in children.

Throughout this chapter there is a wide range of graphical information depicting various features relevant to world hunger, some of which are shown in Figures 5.7a, 5.7b and 5.7c. All school mathematics courses teach the accurate interpretation of graphs. For many pupils this can become an abstract exercise of little interest unless the information encapsulated by the graphs is of some interest to them. Few pupils will be unmoved or disinterested in the plight of hungry people, and the opportunity

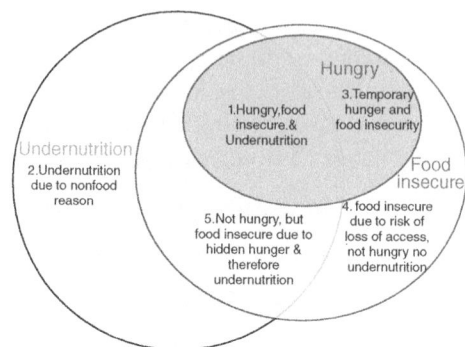

FIGURE 5.7A Unpacking the concept of hunger through a Venn diagram
Source: Haddad (2010)

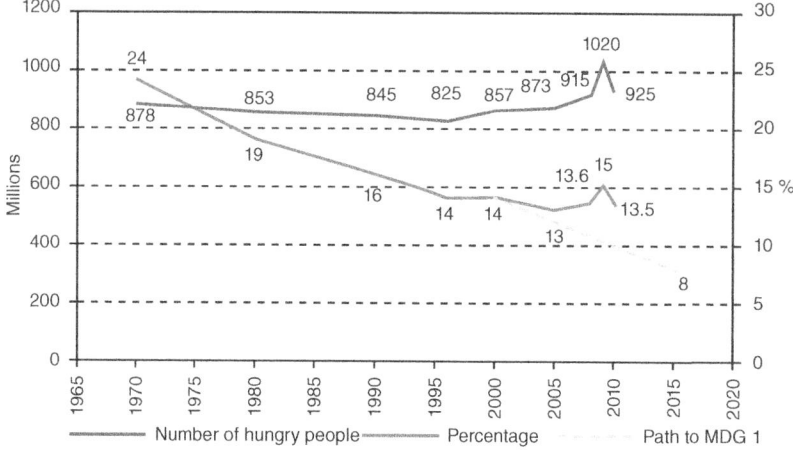

FIGURE 5.7B Showing progress towards reducing hunger through line graphs
Source: Oxfam (2010), data cited from FAO Hunger Statistics (from 1969 to 2006); UN (2009)

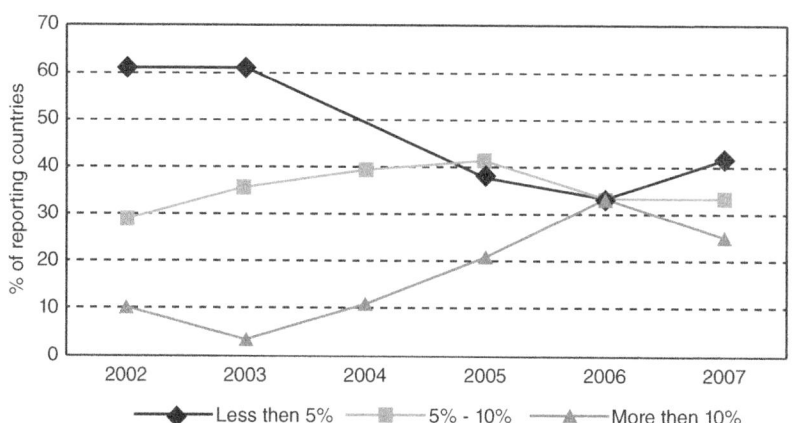

FIGURE 5.7C Showing progress in investing in agriculture
Source: Fan et al., ReSAKSS (2009)

to use graphs dealing with this important issue provides the basis for collaboration between the mathematics teacher and the food technology teacher. It would require some effort on the part of both teachers to analyse the contents of Chapter 6 of the report and extract a range of graphical material that told a coherent story about world hunger, which pupils would only be able to understand by interpreting the graphs. However, this would provide an interesting opportunity to show how this aspect of mathematics is a useful communication tool and to provide a contrast to the situation in the UK and USA where most people have more than enough to eat.

MDG refers to the Millennium Development Goal to halve the proportion of people who are 'undernourished' from 16 per cent in 1990 to 8 per cent in 2015.

For the mathematics teacher there is the opportunity to engage mathematical habits of mind in exploring the graphs particularly tinkering and conjecturing plus being able to use the exploration to gauge the extent to which learners had mastery with regard to graph interpretation. And by including this in food technology it will be possible to begin to engage pupils with the complex global system of food production/consumption and help them consider the moral dimension of a world where there is much hunger and the role that food technologies might play in tackling this problem. This is discussed further in Chapter 12.

Choosing materials

Choosing which material to use for the components of a design is always a challenge. In most of the products designed and made by pupils in schools the choice is inevitably limited. Often, the choices made are based on a combination of precedent – what others have used when they have designed similar products – and availability – what the school has in stock or can afford to purchase. This experience, while defendable on grounds of practicality, does not engage pupils with serious thinking about material choice with regard to matching the required physical characteristics with those of available materials.

Questions concerning both strength (will the part break?) and stiffness (how much will the part deform?) are important, as poor choice of material will lead to poor product performance. Investigations into the properties of materials can give pupils insight into the behaviour of materials. The results of such investigations can be presented graphically, and the interpretation of such graphs requires mathematical thinking. The simplified stress versus strain graph for a metal under tension shown in Figure 5.8 provides a good example. The behaviour of the metal in the linear part of the graph shows elastic behaviour. In this part of the curve the metal will stretch under load and when the load is removed return to its original size. In the non-linear part of the curve the metal deforms but when the load is removed the metal stays permanently deformed. The metal becomes thinner in the final downward part of the curve (known as 'necking') and eventually breaks when the loading exceeds the strength of the metal.

Conversations between the design & technology teacher and the mathematics teacher are essential here if pupil interpretation of such graphs is to be sound. Indeed, it might be possible to go further than a conversation. The mathematics teacher could use stress–strain graphs for different materials as a means of teaching 'interpretation of graphs' and the understanding of compound measures – stress has units of force/area, strain has no units as it is a ratio of extension: original length. The stress/strain ratio

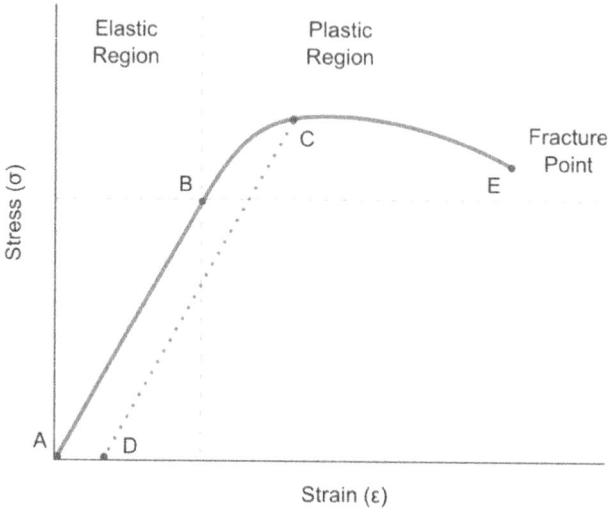

FIGURE 5.8 Simple stress strain graph for ductile metal

(slope of the line in the linear region) gives Young's modulus for the material and is a measure of the materials elasticity. Because strain has no units, the units for Young's modulus are the same as the units for stress. Hence the units for elasticity have the same units as strength – a cause of confusion for many pupils. Of course, the above consideration is an oversimplification of the thinking required to decide on which material to use for a component but it does open the way for the design & technology teacher to ask questions, such as how strong does it need to be? How stiff does it need to be? How will your design decisions about form and material ensure that the component is strong enough and stiff enough? When pupils become intrigued by such questions and how they may be answered they are beginning to appreciate STEM as an holistic approach to designing. As with the interpretation of graphs concerned with hunger, there are opportunities to engage mathematical habits of mind and teaching for mastery.

Designing mechanisms

There is no shortage of mathematics embedded in the design of mechanical systems, but some design & technology teachers have questioned the contexts into which such designing is embedded. Their position is summarised by the question 'Just how many pupils in the twenty-first century really want to make a mechanical toy or point of sale device?' They also argue that, technically, the results are generally unsophisticated and use technology from the nineteenth if not eighteenth centuries – which should not be the hallmark of modern technological learning. While I have some sympathy with this argument, I am reminded of a conversation I had with a friend and colleague who trained as a mechanical engineer. 'You know David,' he said, 'four-bar linkages are bloody amazing!'

    This comment made me wonder about the intrinsic interest there might be in some mechanisms and that a more purist approach might pay dividends. What if one were to consider a mechanism as just an item of intrigue and did not worry too much about a context for use? I realise that this flies in the face of the conventional wisdom that the context for designing is of paramount importance and provides both authenticity and significant motivation for the learner, but when I read a little more about four-bar linkages

I became convinced that this almost reactionary idea might have some worth. I found out about Grashof's rule. Franz Grashof was a distinguished nineteenth-century German engineer who, in 1883, came to the conclusion that with regard to four-bar linkages:

> If the total length of the shortest and longest bars is equal to or shorter than the lengths of the remaining two bars, then the shortest link can make complete revolutions.
>
> (Hartenberg & Denavit, 1964: 77)

This rule struck me as a having great mathematical potential and the possibility of simple practical work involving card strips and split pin paper fasteners. I can envisage mathematics lessons in which the teacher takes four bar linkages as a topic for practical and theoretical investigation with a view to pupils formulating Grashof's rule for themselves. If this investigation took place at a time when pupils were being asked to learn about mechanisms in design & technology and develop products that used mechanical systems, this would provide them with a new mechanism to consider and a mathematical way of considering its design.

Is such formulation feasible in secondary school mathematics lessons? It would certainly require mathematical habits of mind. Just imagine learners discussing what the sentences mean and trying to write mathematical expressions to describe them. It might depart from teaching for mastery in a narrow didactic sense but it would certainly offer the opportunity for students who grasp ideas quickly to engage in deeper analysis and apply what they were learning to different four-bar linkage arrangements. An exploration of Grashof's rule is shown in Panel 5.3.

If the total length of the shortest and longest bars is equal to or shorter than the lengths of the remaining two bars, then the shortest link can make complete revolutions.

In this card model

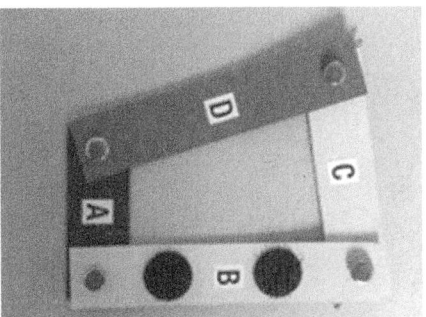

The longest bar is B
The shortest bar is A
The total length must be A + B
The lengths of the remaining bars must be C + D

If A+B = C+D then A can make a complete revolution
If A+B < C+D the A can make a complete revolution

Panel 5.3 Grashof's rule exploration

In a delightful book, *Mathematics Meets Technology* (1991) the author Brian Bolt describes four-bar linkage that obeys Grashof's rule: the crank and rocker, which is one of those shown in Figure 5.9. In musing on mechanisms, I recalled strandbeests, amazing creatures designed and constructed by Theo Jansen, which wander Dutch beaches powered only by the wind and using a combination of rotary motion and linked levers. To achieve this, he wrote a special computer program with a genetic algorithm on an Atari computer for this in 1990 (see Figure 5.10). I find it difficult to image that a mathematics teacher and a technology teacher couldn't make lessons leading to understanding such mechanical wonders irresistible.

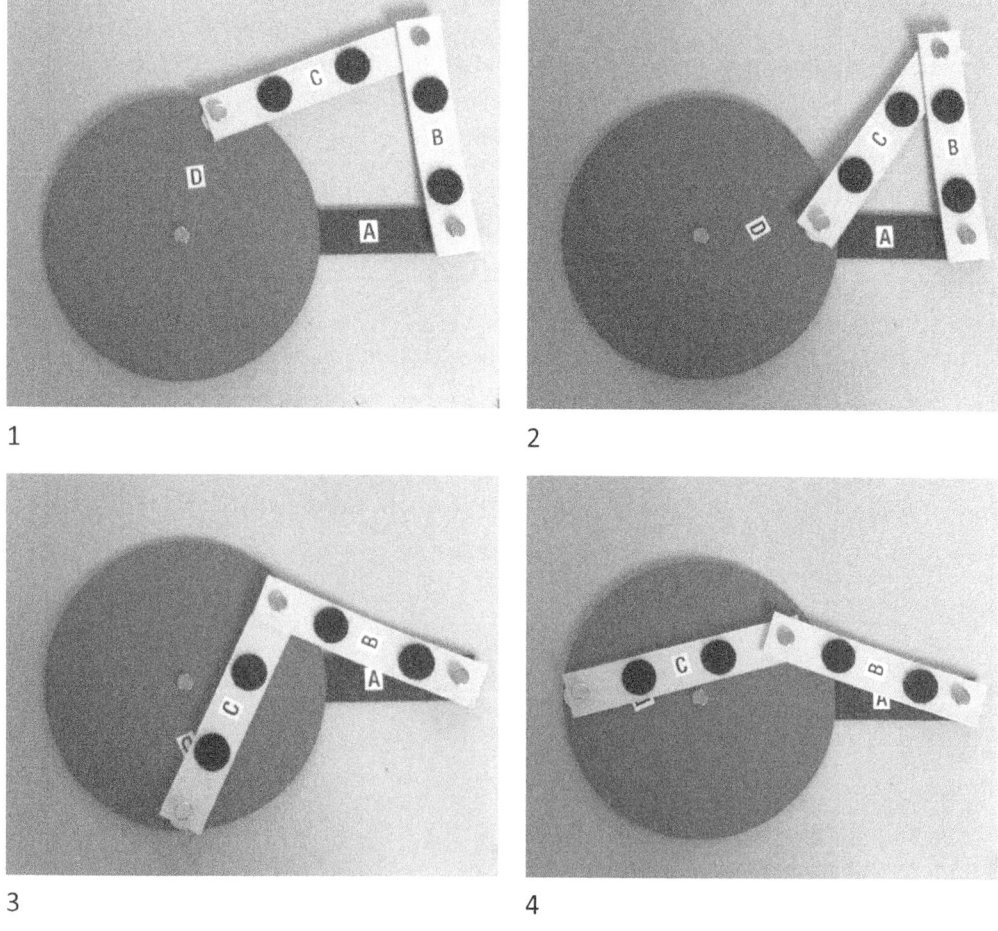

FIGURE 5.9 A sequence showing a four-bar linkage acting as a crank and rocker

FIGURE 5.10 Theo Jansen with the strandbeest Plaudens Vela, which means flapping sails. The animal had a lot of sails so that it could move forward even in gentle breezes.
Source: Animaris Plaudens Vela (2013); image: Marco Zwinkels, © Theo Jansen

Considering robots

Robots are seldom out of the news as concern over the impact of automation on employment grows. Learners perceptions of robots are often limited to 'little metal men' and underpinned by iconic and disturbing figures from the popular culture such as the Terminator. Politicians are using the employment issue as platforms for election. For example, Andrew Yang made automation the key theme of his bid for the White House in 2020 and is reported as saying

> All you need is self-driving cars to destabilize society. … [W]e're going to have a million truck drivers out of work who are 94 percent male, with an average level of education of high school or one year of college. That one innovation will be enough to create riots in the street. And we're about to do the same thing to retail workers, call center workers, fast-food workers, insurance companies, accounting firms.
>
> (Quoted in Roose, 2018)

So, there is an important opportunity in technology education to consider the reality of robots and their likely impact on society. Much of the information about impact on employment is presented in graphical form and requires mathematical understanding for its interpretation. Some are relatively straightforward and easy to understand while others require significant mathematical knowledge and skill to understand. Good examples of this variation can be found in the work of Michael Osborne and Carl Frey (undated) The use of such graphs in design & technology lessons in the exploration of the consequences of technology is fraught with difficulty if the learners are mathematically inept with regard to their interpretation. Hence a conversation between mathematics and design & technology teachers about the interpretation of graphs would be worthwhile.

Such conversations would enable the design & technology teacher to use the correct language when both providing explanations and asking questions and contribute to developing mastery and reinforcing mathematical habits of mind.

Simple robots feature in many design & technology curricula and an important aspect of working with robots is exploring the different ways they can be programmed to perform their various functions. This will involve both mathematical and computational thinking and will be considered in Chapter 9 Computing, digital competence, computer science, TEL and STEM. The physical design of robots, as opposed to their programming, can involve significant mathematics. Consider three basic types of robots: Cartesian robots, cylindrical robots and spherical robots (shown in Figure 5.11). The working envelopes of the robots are an important design feature. This describes the space in which the end effector of the robot can operate. In the case of the Cartesian robot, the position of the end effector at any one point will be described by three numbers – the x coordinate, the y coordinate and the z coordinate. The design of the robot governs the possible sizes of these coordinates and hence the size and shape of the working envelope, which is a cuboid. In the case of the cylindrical robot the radial arm can extend and retract horizontally, rotate about a vertical axis and the entire arm can be raised and lowered. The position of the end effector at any one

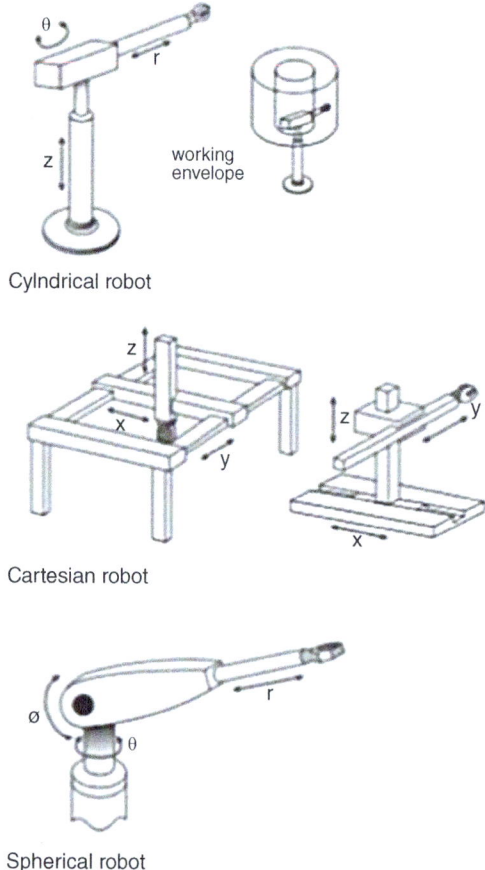

Cylndrical robot

Cartesian robot

Spherical robot

FIGURE 5.11 Three types of robot arms: Cartesian, cylindrical and spherical
Source: Adapted from Bolt and Brian, Mathematics meets technology

point is described by its distance along the horizontal axis (r), the angle it has rotated about the vertical axis (Θ) and the distance it has moved along the vertical axis (z). The design of the robot governs the possible sizes of these cylindrical coordinates and hence the size and shape of the working envelope, which is cylindrical. In the case of the spherical robot the arm can extend along its length (r), rotate about a vertical axis (Θ) and can be rotated about a horizontal axis (Φ) to elevate it above or below the horizontal. The design of the robot governs the possible sizes of these spherical coordinates and hence the size and shape of the working envelope, which is a spherical. The design & technology teacher can engage pupils with the design of these different types of robot quite easily with the use of construction kits such as Lego or Fisher Technic and the mathematics teacher can give such work a significant mathematical dimension by introducing pupils to three different ways of defining points in three-dimensional space. There are many places in which robots are being used in society. Robots are finding their way into homes as domestic cleaning machines, into hospitals to perform surgery, in care homes for the elderly, in military operations such as bomb disposal, in search and rescue, in autonomous transport, in teaching, in manufacturing and in data collection. There will almost certainly be mathematical dimensions to their design in these different situations.

Surface decoration

Surface decoration plays a large part in many textile courses. Repeat patterns of various sorts are one of the main ways of achieving surface decoration and, of course, the mathematics of symmetry underpins such pattern generation.

Starting with a simple geometric shape, a triangle or half circle perhaps it is a relatively simple exercise to use basic transformation operations such as translation, reflection and rotation to produce a variety of different patterns. More complex transformations such as glide translation and helical translation can be added to the mix of operations. Assigning colours as the result of particular sequence of transformations can add even more visual interest, e.g. every time there a shape is reflected there is a colour change from red to blue or blue to red, but when a shape is rotated to give the next shape there is no colour change (see Figure 5.12).

So, it is possible for pupils to write algorithms of transformations to generate patterns. Traditionally, such algorithms can be applied to fabric by using block printing techniques with the block undergoing a particular set of transformations between the making of each print on the fabric. And it is possible to carry out this activity pattern-generating activity on screen and then use sublimation printing to produce the patterned fabric. The basic unit of the pattern need not be confined to a simple geometric shape. Suitable shapes can be derived from natural forms via observational drawing and simplification, abstract forms by assembling a variety of curved and straight lines into an enclosed shape. However, the shape is derived, the way it can be used to produce a repeat pattern can be developed using transformations. School mathematics courses often include an introduction to symmetry and transformation geometry. The generation of patterns for use as surface decoration in a textiles component of a design & technology programme provides an interesting context for the application this mathematics. Hence a conversation between the mathematics teacher and textiles teacher would seem to be in order. The motivation for learning the somewhat abstract ideas of symmetry transformations can be enhanced if the teaching by the mathematics teacher acknowledges explicitly with the class that they will be able

FIGURE 5.12 Pattern generated through reflection, rotation and colour change

to use the learning in their textiles lessons. Indeed, the development of the algorithms to produce patterns in the mathematics lesson can be seen as an essential first step in the overall textiles task of designing and making a patterned fabric. It is, of course, important that the ultimate use for the patterned fabric is considered so that the pattern is appropriate for the garment or furnishing that is being designed. An interesting possibility is that the mathematics teacher, having introduced the pupils to pattern design using transformations, consults with the class as they develop patterned textile products. In this way the teacher can use the students' efforts in design & technology to assess their understanding of transformations and where appropriate intervene to help pupils overcome misunderstandings. As with all the examples of collaboration between mathematics teachers and design & technology teachers, this will contribute to developing mastery and reinforcing mathematical habits of mind.

Having presented a range of examples concerning the teaching of mathematics in the light of STEM we will revisit the 'life mathematical' in school.

Revisiting the 'life mathematical' in school

Enabling success in international tests as a key requirement of the life mathematical in school has been questioned. We saw earlier that Yong Zhao was doubtful and in an article in the *New Scientist* (2013) MacGregor Campbell challenges the conventional wisdom that PISA and TIMMS test scores are important in gauging how well pupils will be in working effectively in a knowledge-based global economy. He argues that several researchers have shown that there is little if any correlation between test scores and measures of economic success such as per capita GDP, Growth Competitiveness Index and entrepreneurship. As a specific example, he notes that Japanese students have always been near the top of the TIMMS and as such you might expect such students to go on to drive a high-flying economy. Yet the Japanese economy stagnated throughout the 1990s and 2000s. He concludes that fixating on international tests as a way to promote the importance of mathematics (and science) is likely to prove counterproductive and that more emphasis should be placed on developing creativity and initiative. Being able to use mathematics fluently in subjects other than mathematics, as proposed and advocated in the above examples, will support the creative use of mathematics and help pupils show initiative in bringing mathematics to bear in learning both science and design & technology. It has also been strongly suggested that this can be aligned with developing mathematical habits of mind and achieving mastery. It is, of course, unlikely that government ministers will ignore PISA and TIMMS rankings, but engaging pupils with the utility of mathematics as indicated is likely to develop a more positive attitude overcoming the 'I'm one of those who just can't do it' disposition and leading to better overall mathematical confidence and attainment.

Clearly, the authors of this book value highly the links between mathematics and science and design & technology and perhaps it would not be going too far to argue that if a school deliberately forges such links then the mathematics teaching overall is likely to become more successful. Underpinning many of the activities that embody such links is the idea of the conversational classroom in which pupils actively discuss their approaches to using mathematics, the difficulties they are experiencing and how they overcome them. Such conversations are regarded as essential for developing mathematical habits of mind and achieving mastery. One can see the failure to develop conceptual knowledge and the ability to solve problems, noted by the most recent Ofsted report into mathematics teaching, as being compounded by the classroom of certainty in which tentative attempts to understand and use difficult ideas through discussion find no place.

The development of Core Math for all students post-16 while at an early stage is, of course, to be welcomed but it is unlikely to be embraced with enthusiasm by students who, although successful in pre-16 mathematics studies have not found their learning enjoyable and are looking forward to dropping mathematics post-16.

David Willets (2019: 119) has an interesting response to this problem. He acknowledges that the pre-16 curriculum often leaves young people not only disenchanted with mathematics but lacking in key skills so that …

> a student arrives at university to study politics or history and find themselves reading research using qualitative techniques that are beyond them. That is the point at which they can be reintroduced to Maths and given extra training. And they may be more likely to do it, if it enables them to understand analysis of voting behavior or whatever else they are focused on rather than being told it is a generic skill they are obliged to learn in the abstract.

He also notes …

> For this to work universities need to do their bit by carefully designing courses and adding training in these generic skills as the means to study a subject effectively.
> (Ibid.)

Eva Jakobla and Karen Skilling (2018) make some telling observations in discussing the nature of numeracy and mathematical literacy. They asked beginning teachers to consider the contexts in which mathematical skills might be deployed and the sensitivity needed in using particular contexts given that 'people's private, professional, social, occupational, political and economic lives not only represent a multitude of different requirements but also different social conditions and identities'. They also note that when numeracy or mathematical literacy is developed in cross curricular projects 'there is a better chance of including and developing both mathematical and other relevant skills and knowledge, and coordinating a range of perspectives'. Teaching mathematics in the light of STEM as we have suggested would seem to provide non-contentious contexts and embrace, to some extent, cross curricular projects.

So perhaps if schools can teach mathematics in the light of STEM pre-16, predisposing more young people to study core mathematics post-16 and the universities do 'keep up the good work' as suggested by David Willets then cultural perception of mathematics that signals to students that they are one of those who, for whatever reason, 'just can't do mathematics' will finally fade away.

Conclusion

So, what are we to make of this chapter? There is no doubt that in both the USA and the UK there are serious concerns about the response of many young people to the teaching and learning of mathematics. Perversely, it might seem, those schools that acknowledge the conceptual struggle involved in learning mathematics and enable pupils to articulate this struggle have more success than schools that, with the best intentions, over-simplify and fragment the learning, denying pupils the opportunity to construct their own personal robust understanding and in the process gain significant mathematical skills. We advocate an approach that supports the importance of conceptual struggle, and suggest that one way to achieve this is to take the mathematics necessary for learning science and design & technology into the mathematics classroom. We see that this will support mathematics habits of mind and develop mastery.

We have developed the examples of teaching parts of the mathematics curriculum in this way to illustrate that it is both possible and worthwhile. If you are able to use these examples we will, of course, be delighted, but if you find them inappropriate then we believe that this should not deter you from developing ideas that will work for you in your situation. Our conversations with science and design & technology teachers have led us to the view that there is no shortage of possible examples that require the use of mathematics in those subjects. Our foray into presenting such examples is, of necessity, limited but we are convinced that conversations between mathematics teachers and those teaching science and design & technology will be able to identify many more examples, and, importantly, examples that will be successful in the individual circumstances of their particular schools. We have noted the importance of conversations in this endeavour: the initial and probably short conversations between teachers exploring possibilities; the subsequent more detailed and time-consuming conversations required to outline what might be done to respond to the possibilities and a consideration of the conversations to take place in the classroom between pupils as they learn mathematics through using it for other STEM subjects. It is this final set of conversations that will both enable and reveal the effectiveness of our suggested approach. Hence we suggest that you discuss with colleagues what sort of conversations you want to happen, how you might support such conversations and how you might monitor the conversations that do take place with the view to improving their effectiveness. In this way, we believe that you will be able to make a significant contribution to the life mathematical in your school, one which pupils will value and enjoy. Hence in teaching mathematics in the light of STEM, our advice is don't neglect the importance of regular conversations with colleagues from science and design & technology.

Recommended reading

These provide more information about mastery and connections between mathematics and technology.
Bolt, B. (1991a) *Mathematics meets technology*. Cambridge: Cambridge University Press.
Maths Hub (2017a) *Secondary mathematics teaching for mastery: Some themes and key principles*. www.ncetm.org.uk/files/69314786/Secondary+Teaching+for+Mastery+December+2017.pdf (accessed 5 May 2020).
NCETM (2016a) *Mastery explained*. www.ncetm.org.uk/resources/49450 (accessed 5 May 2020).

References

Bolt, B. (1991b) *Mathematics meets technology*. Cambridge: Cambridge University Press.
Campbell, Macgregor (2013) *West vs Asia education rankings are misleading*. www.newscientist.com/article/mg21728985-800-west-vs-asia-education-rankings-are-misleading/ (accessed 5 May 2020).
Cuoco, A., Goldenberg, E. P., & Mark, J. (1996) Habits of mind: An organizing principle for mathematics curricula. *Journal of Mathematical Behaviour*, 15, 375–402.
Department for Business, Innovation & Skills (2011) *Foresight –the future of food and farming*. www.bis.gov.uk/foresight/our-work/projects/published-projects/global-food-and-farming-futures/reports-and-publications (accessed 5 May 2020).
Department for Education (2014) *The national curriculum in England: Key Stages 3 and 4 framework document*. www.gov.uk/government/publications/national-curriculum-in-england-framework-for-key-stages-1-to-4 (accessed 5 May 2020).
Department for Innovation, Universities and Skills (2007a) *Tackling obesities: Future choices – project report*. https://assets.publishing.service.gov.uk/government/uploads/system/uploads/attachment_data/file/287937/07-1184x-tackling-obesities-future-choices-report.pdf (accessed 5 May 2020)
Department for Innovation, Universities and Skills (2007b) *Tackling obesities: Future choices – summary of key messages*. https://assets.publishing.service.gov.uk/government/uploads/system/uploads/attachment_data/file/287943/07-1469x-tackling-obesities-future-choices-summary.pdf (accessed 5 May 2020).
DeVos, B. (2018a) *Speech to American Enterprise Institute*. www.ed.gov/news/speeches/prepared-remarks-us-education-secretary-betsy-devos-american-enterprise-institute (accessed 5 May 2020).
DeVos, B. (2018b) Speech to State Chiefs www.ed.gov/news/speeches/devos-state-chiefs-we-can-we-must-do-better-students (accessed 5 May 2020).
Eroglu, D. & Tanish, D. (2017) Integration of algebraic habits of mind into the classroom practice. *Elementary Education Online*, *16(2)*, 566–583, https://pdfs.semanticscholar.org/821c/cca5313a8fb6f4a59b400e74f6b829c75d8c.pdf (accessed 5 May 2020).
Erşen, Z. B., Ezentaş, R., & Altun, M. (2018) Evaluation of the teaching environment for improve[ing] the geometric habits of mind in tenth grade students. *European Journal of Education Studies*, 4(6), 47–64.
Food and Agriculture Organisation of the United Nations (2009) *How to feed the world in 2050*. Rome, Italy: FAO. www.fao.org/fileadmin/templates/wsfs/docs/expert_paper/How_to_Feed_the_World_in_2050.pdf (accessed 5 May 2020).
Gibb, N. (2017) *Letter responding to the Smith Review*. https://assets.publishing.service.gov.uk/government/uploads/system/uploads/attachment_data/file/631009/Letter_responding_to_Smith_review.pdf (accessed 5 May 2020).

Hartenberg, R. & Denavit, S. (1964) *Kinematic synthesis of linkages*. McGraw Hill: New York.

Hiem, J. (2016) *On the world stage U.S. students fall behind*. www.washingtonpost.com/local/education/on-the-world-stage-us-students-fall-behind/2016/12/05/610e1e10-b740-11e6-a677-b608fbb3aaf6_story.html (accessed 5 May 2020).

Homer, M., Mathieson, R. Tasara, & Banner, I. (2018) Increasing post-16 mathematics participation in England: The early implementation and impact of Core Maths. In J. Golding, N. Bretscher, C. Crisan, E. Geraniou, J. Hodgen, & C. Morgan (eds.), *Research proceedings of the 9th British Congress on Mathematics Education* (3–6 April 2018, University of Warwick, UK). www.bsrlm.org.uk/bcme-9/ (accessed 5 May 2020).

Ives, L. (2018) *Diabetes prescriptions now cost NHS £1bn*. www.bbc.co.uk/news/health-46139595 (accessed 5 May 2020).

Jacobla, E. & Skilling, K. (2018) Numeracy, mathematical literacy and mathematics. In Maguire, M., Gibbons, S., Glackinn, M., Pepper, D. and Skilling, K. (eds.), *Becoming a teacher*.

Jones, J. (2015) *Understanding ways forward*. www.slideshare.net/Ofstednews/better-mathematics-keynote-spring-2015 (accessed 5 May 2020).

King, J. B. (2 December 2016) *US highschoolers fall behind world peers in math* in Reuters report. www.businessinsider.com/r-us-high-schoolers-fall-behind-world-peers-in-math-study-says-2016-12?r=US&IR=T (accessed 29 September 2020).

Maths Hub (2017b) *Secondary mathematics teaching for mastery: Some themes and key principles*. www.ncetm.org.uk/files/69314786/Secondary+Teaching+for+Mastery+December+2017.pdf (accessed 5 May 2020).

NCETM (2016b) *Mastery explained*. www.ncetm.org.uk/resources/49450 (accessed 5 May 2020).

Nuffield Foundation (2007) *Oil drop experiment*. https://spark.iop.org/estimating-size-molecule-using-oil-film#gref (accessed 5 May 2020).

Nuffield Foundation (2010) Is the UK an outlier? An international comparison of upper secondary mathematics education. Nuffield Foundation Report. London: Nuffield Foundation. www.nuffieldfoundation.org/project/is-the-uk-an-outlier-in-upper-secondary-maths-education (accessed 5 May 2020).

Ofsted (2012) Mathematics: made to measure. Ofsted report ref 110159. www.ofsted.gov.uk/resources/mathematics-made-measure (accessed 5 May 2020).

Osbourne, Michael A. & Frey, Karl Benedict. (undated) *The future of employment*. https://futureoflife.org/data/PDF/michael_osborne.pdf

Rapsomanikis, G. (2015) *The economic lives of smallholder farmers: An analysis based on household data from nine countries*. Rome: Food and Agriculture Organisation of the United Nations. www.fao.org/3/a-i5251e.pdf (accessed 5 May 2020).

Roose, K. (2018) His campaign message: The robots are coming. *New York Times*. www.nytimes.com/2018/02/10/technology/his-2020-campaign-message-the-robots-are-coming.html (accessed 5 May 2020).

Russell, B. (1907) *The Study of Mathematics, reproduced in Russell, B. (1957) The study of mathematics in mysticism and logic*. New York: Doubleday.

Smith, A. (2017) *Report of Professor Sir Adrian Smith's review of post-16 mathematic*. https://assets.publishing.service.gov.uk/government/uploads/system/uploads/attachment_data/file/630488/AS_review_report.pdf (accessed 5 May 2020).

Stripp, C. (2019) *The importance of Core Maths and the support available*. www.ncetm.org.uk/resources/42295 (accessed 5 May 2020).

Tucker, M. (6 December 2016) On the world stage, US students fall behind. Report in *The Washington Post*. www.washingtonpost.com/local/education/on-the-world-stage-us-students-fall-behind/2016/12/05/610e1e10-b740-11e6-a677-b608fbb3aaf6_story.html (accessed 29 September 2020).

Willets, D. (2019) Technology at work v4.0: Navigating the future of work. Citi GPS Global Perspectives and Solutions. https://ir.citi.com/%2Bsi3%2BYKA2e3WrSalzmOchzHQqP-UAersOy9%2BRj9AQRfQk%2Bhsikx7zf5aSLAsAXNWO26TTlD49IYM%3D (accessed 5 May 2020).

6

Project-based learning and STEM

Introduction

In Chapter 1 we saw how STEM was considered important by politicians as it had implications for their country's economy. We also looked at the processes involved across the STEM subjects and 'problem solving' was picked out as a key feature that the subjects have in common. In other chapters when 'looking sideways', we have tended to focus on the teaching of the *content* of a subject 'in the light of STEM' – and how knowing what is taught and when it is taught can be exploited by other STEM colleagues. This mutual support in teaching a common topic can lead to better learning for pupils. We will see in what follows that for employers the *processes* are just as important – and due to the unpredictable nature of future work opportunities – possibly more important than any specific content:

> We are currently preparing students for jobs that don't yet exist, using technologies that haven't been invented, in order to solve problems we don't even know are problems yet.
>
> (Gunderson et al., 2004)

> We are in a uniquely exciting time. We understand how to engage kids. We need to give them real-world challenges, have them work with other kids, and provide them with the right kind of adult support. Project-based learning is how people work in the real world. We need to let our kids create portfolios of joy.
>
> (Doug Lyons in Dintersmith, 2018: 18)

> In a survey by the National Association of Colleges and Employers (2017), more than two-thirds of employers reported that they look for employees who demonstrate strong creative problem-solving, teamwork, and communication skills. For the United States to remain competitive in the 21st century, our citizens must be equipped with creative problem-solving skills.
>
> (Duyar et al., 2019: 2)

This chapter takes this further and looks in detail at how the contextual teaching of STEM, the contrasting teaching styles or different pedagogy of colleagues when doing project work, could also be shared, and in doing so pupil learning may be enhanced.

Interest in the importance of teaching problem solving and using practical work to enhance learning is not new. Early in the twentieth century (and before), practical work was seen as a route to learning and separate from vocational preparation. For example, John Dewey (1916: 70), working in Chicago, set up an innovative school in 1896 where cookery and carpentry were seen as important in providing insights into natural materials and processes. He said:

> Give the pupils something to do, not something to learn; and the doing is of such a nature as to demand thinking; learning naturally results.

'Hands-on as well as Minds-on' has a long history, but teaching using problem solving has become mainstream in most countries more recently through what have been termed 'twenty-first-century skills'. The argument is, as stated above, that the future will be very different from the present and that for all citizens mental agility and adaptability will be key. Wagner (2008) in *The Global Achievement Gap*, for example, advocated seven 'survival skills' for the twenty-first century:

- *critical thinking* and problem solving;
- *collaboration* across networks and learning by influence;
- agility and adaptability;
- initiative and entrepreneurialism;
- effective oral and written *communication*;
- accessing and analysing information;
- *curiosity and imagination*.

But in the third decade of the twenty-first century are these still appropriate?

> Although some educators have grown weary of the term '21st-century learning,' the drive to transform education matters more today—a lot more—than when we started the conversation.
>
> (Kay, 2020)

During the 'advocacy phase' in promoting the teaching of twenty-first-century learning, the competencies were condensed into a more memorable set – the '4Cs':

- critical thinking;
- collaboration;
- communication;
- creativity.

You notice that 'critical thinking and problem solving' lead the list, which also includes 'creativity' both of which are characteristics of project-based work. An additional consideration is the 'authenticity' that can be the context of project work – where the project is based in a real-world problem. Engaging in authentic practice involves situations that are real to the pupil, to their lives, and to situations they may encounter in their community.

But how do we move from a set of catchy slogans to actual classroom practice?

To help sort this out, and to investigate how project work can be done successfully, this chapter will consider the following questions:

- What is project work (project-based learning) and why is it important?
- How are successful projects and related tasks organised?
- What is the relationship between project work and assessment?

What is project-based learning – and why is it important?

Features of project-based learning

If teachers were asked to describe the characteristics of project work, their ideas would probably include the following:

- the choice of a project is based upon a need which the pupil can see and identify with, and is based on an authentic 'real-life' situation – a project may well be chosen by the pupil;
- the pupil takes responsibility for the conduct of the project as much as possible – pupil directed and teacher facilitated;
- the exact outcome is open-ended and unpredictable;
- a range of skills, knowledge and concepts are required to complete a successfully, as technological problems in particular do not respect subject boundaries project;
- it is a time limited.

But there is an assumption here that all STEM teachers see that is self-evident that project-based learning is worthwhile. In a number of countries teachers are held to account by parents and governments in ways that do not necessarily match preparing pupils for life after school. Rather, the 'school game' is to perform well compared to other schools by scoring high in tests and examinations. In the USA, for example, Dintersmith says, 'Our national K–12 education mantra is "College and Career Ready", but the phrase is misleading. In reality, schools prepare students for their college application, not college. Career is strictly an afterthought' (2018: 57). Dintersmaith argues that testing in America is centred on retaining knowledge and it is extensive – those graduating from US high schools may take a Preliminary SAT (PSAT)/National Merit Scholarship Qualifying Test, SAT (standardised tests) and AP (Advanced Placement tests). AP is a program that offers college-level curricula and examinations to high school students and American colleges and universities may grant placement and course credit to students who obtain high scores on the examinations. In England, the English Baccalaureate (EBacc) is a set of subjects examined at the age of 16 and requires the passing the following subjects:

- English language and literature;
- Maths;
- the sciences;
- geography or history;
- a language.

As in the USA, schools in England consider that pupils gaining success in examinations enhance their prestige in relation to other schools, and the EBacc subjects, like PSAT and AP in the US, are prioritised by schools. There are consequences. In the US, often school technical facilities (shop) have closed down. In England, the EBacc itself excludes 'creative' subjects like design & technology, art and music. However, although these creative subjects are not included in the EBacc, student performance in these subjects can be taken into account when reporting on the school accountability measures though so called 'Progress 8 and Attainment 8' measures of school effectiveness.

However, there are important exceptions. In the USA, despite initiatives like 'No Child Left Behind' and 'Race to the Top (RTTT)', which have funding attached, some schools in each state have rejected the associated focus on testing and have prioritised entrepreneurship and creative thinking for a career – 'Career Ready'. In England, T Levels introduced in phases from September 2020 to September 2022 for high school graduates are equivalent to three academic 'A levels'. These two-year courses have been developed in collaboration with employers and businesses so that they meet the needs of industry and prepare students for work. T Levels offer students a mixture of classroom learning and 'on-the-job' experience during an industry placement of at least 315 hours (approximately 45 days). They are designed to provide the knowledge and experience needed to open the door into skilled employment, further study or a higher apprenticeship. (See T Levels, 2020.)

Project-based learning in science

Project-based learning, therefore, emphasises learning activities that are student-centred and bring together a 'real' worthwhile problem with relevant techniques to solve it. Teachers are, of course, a resource to the students but act more as a facilitator and students are required to organise their own work and manage their own time, and will need to communicate and collaborate with one another.

EXAMPLE 1: PROJECT WORK IN SCIENCE

Fresh milk is heat treated to 132oC for one minute before it is sold to customers. This is known as ultra-heat treatment.

A milk manufacturer is trying to find ways to cut their production costs and has suggested that temperatures lower than 132oC might achieve the same results.

You are a microbiologist investigating the effect of temperature on the levels of bacteria in milk samples. You will report your findings to the milk manufacturer.

Or

Research has shown that the foods and drinks that are given to young children can affect their development and how well they progress at school later on in childhood and adolescence.

Fruit juices can seem like a healthy option for young children but there are concerns by the National Health Service because different juices contain differing amounts of vitamin C, sugar, acidity and fibre.

You are a food analyst working for the Food Standards Agency (FSA).

You have been asked to investigate five different fruit juices.

Your investigation could include:

- pH tests;
- food tests;
- vitamin C content;
- acid concentration;
- mass of suspended matter.

You should write a report on your findings, which could be used by the NHS to help parents, nurseries and child minders choose the best fruit juice for toddlers and young children.

Suggested examples for the Assessment and Qualifications Alliance (AQA Applied Science GCSE in the UK.

These examples match some of the general characteristics of project-based learning as described above and some concepts borrowed across STEM subjects would have to be appropriately applied to produce the outcome.

Project-based learning in technology and engineering

An example of project-based learning in the area of technology and engineering is offered by the UK Design and Technology Association. The significant driving question posed here is: What can nature offer Architectural Design?

In this case, the issue of significance is sustainability and the project-based learning unit of work is focused on education for sustainable development. A project-based learning approach will allow pupils to turn their critical faculties on the way buildings do or do not achieve sustainability, and if they do not what might be done about it. They will not be able to do this without the requisite knowledge, understanding and skill and it is important that the pedagogy used in this unit of work provides this. There are some suggested 'big tasks' or projects and some contributory 'small tasks' from across the STEM subjects that might be useful in undertaking the big task. The following is taken from the teachers' guide.

EXAMPLE 2: PROJECT WORK IN DESIGN AND TECHNOLOGY OR ENGINEERING

The students can formulate their project through a big task, for example:

- Re-designing an existing building in order to make it more sustainable.
- Designing a new building on a given plot of land to ensure that it is sustainable.
- Taking a department from within the school and redesigning it to make it more sustainable.

■ Identifying an existing building with sustainable credibility. Developing a presentation to explain how it achieves this sustainability.

The projects at the centre of this unit of work pose considerable challenges for students. To support the students in meeting these challenges there are a range of small tasks that provide structured opportunities for learning that will be useful in tackling the big tasks. There are three sets of tasks, one set for each of the contributing STEM subjects

For design & technology the small tasks are concerned with:

■ architecture informed by nature;
■ sustainable architecture;
■ product life cycle analysis.

For science, the small tasks are concerned with:

■ reducing heat loss;
■ wind power;
■ materials;
■ forces and structures.

For Mathematics, the small tasks are concerned with:

■ exploring the Fibonacci Series and the Golden Ratio;
■ scale drawings, plans and elevations;
■ power from the wind and calculating wind power.

(DATA, 2013)

Here are two more examples in engineering. They were developed jointly between a UK examination board, the Assessment and Qualifications Alliance (AQA) and the charity Practical Action.

EXAMPLE 3: PROJECT WORK USING APPROPRIATE TECHNOLOGY

You have been asked to design a treadle pump for use in remote areas. You will need to carry out concept development to produce visual design solutions. You must use CAD software to create a 3D model of the treadle pump that can be shared with potential sponsors who will fund the manufacture and the engineers who will work locally to produce it.

Or

In this task you will use modelling to carry out the optimisation of a design for a bicycle trailer, then consider the technologies that could be used to make it and plan its manufacture. You must also consider its environmental impact, both during manufacture and across the product life.

(AQA/Practical Action Level 3 Technical Level Design Engineering Specimen examples, 2019)

Project-based learning in mathematics

Although practical work is often a group activity in science as equipment and apparatus usually have to be shared, project-based learning in the teaching of science and in technology/engineering tends to be a solitary endeavour as it follows an individual pupil's ideas. Mathematics, traditionally, has been particularly resistant to group work and so projects that can be done in a group and that are constructed around real-life problems, ill-structured, open-ended and ambiguous deserve careful consideration.

An approach to mathematics common in schools is Model-Eliciting Activities (MEAs), which are designed to help pupils apply the mathematical procedures they have learned to create mathematical models. To initiate MEAs, typically, the teacher sets up a context for the pupils and this is often a simulated newspaper article about the real-life topic to be considered and pupils respond to some questions based on the article. The problem is then posed to the pupils who work in groups to model possible solutions. Often, such problems have multiple solutions or the students are working towards a 'best fit'. This example is given by Chan, Chun Ming (2008).

EXAMPLE 4: PROJECT WORK IN MATHEMATICS

The hiring problem
MISSION
   Your group is in charge of hiring some workers to help clean, paint, and move furniture in the school. These workers must complete the job within four days.
CONDITIONS

1.   You can hire only from one company once, and you have to accept the number of workers for that company.
2.   You have a worksite supervisor who can only supervise at most 12 workers per day, so you try to hire as many as 12 per day. Assume that each worker to be hired works the same amount of time, and produces the same amount of work per hour.
3.   You need at least: 14 workers for moving furniture, 14 workers for painting, and 14 workers for cleaning within the 4 days.

PRESENTATION
   You have to present your case to your class. Show in full detail (with different solution options) how you arrive at hiring the workers. Show your productivity index and use it to make your decision.

THE HIRING PROBLEM DATA

Cleaning services:

Company	A	B	C	D	E
No. of workers	4	2	6	3	5
Cost $	160	76	270	120	175

Painting services:

Company	F	G	H	R	J
No. of workers	3	6	7	4	5
Cost $	114	240	315	160	210

Moving services:

Company	K	L	Q	N	T
No. of workers	7	4	3	6	4
Cost $	245	160	135	225	140

Productivity index is calculated as follows:
(Total no. of workers / Total Cost) × 100 { Give index to 3 decimal places }
Are you getting value for money? The larger the index the better.

From a teacher's point of view, some of the managerial problems such work presents may be evident too: the teacher has suggested the task – not the pupil, who may or may not see it as relevant to them; and if the pupil lacks appropriate knowledge, that knowledge must be gained somehow. These issues need to be addressed, but will be considered later.

Tamara Moore and Gillian Roehrig (2019) at the University of Minnesota believe that MEAs allow for a more *thoughtful and inclusive approach* to gauging student understanding of STEM subjects, which is impossible with a textbook-based approach, and set out five characteristics of MEAs:

- **Model-eliciting** – meaning that students are required to develop a model to not only solve the problem at hand, but also others like it. This usually looks like a step-by-step method for *how* to solve the problem, rather than just an answer to one question. This is important because it helps students understand the mathematical structure of the problem.
- **Self-assessable** – meaning the individual or student team can critique their own work for accuracy and effectiveness.
- **Open-ended** – to allow for creative and thoughtful interpretation of the lesson. Rarely in the real world is there one way to solve a complex problem, and you can't find the answer in the back of a textbook! MEAs let students develop their own ways of thinking about the problem, in that they design the model for the problem based on their own prior knowledge and experiences, thus improving their problem-solving capabilities.
- **Realistic** – to connect students with familiar topics, like solar energy or paper airplanes. MEAs illustrate how STEM subjects can help solve the problems, big and little, of the world.
- **Generalisable** – in that MEAs are useful tools for *all* STEM disciplines: science, technology, engineering and math.

Why is this type of work important?

Project-based learning is particularly valuable in that it enables pupils to:

■ integrate skills (in applying knowledge; speculative thinking; communication skills; ability to manipulate ideas and materials; etc.) and knowledge from a variety of sources in the process of developing useful outcomes;

■ become more autonomous through taking increasing responsibility for the direction of their own work.

The aim of encouraging pupils to become autonomous – able to plan, investigate and research aspects of their own learning – has long been part of the rationale for many of the STEM subjects.

A balanced, practical-based curriculum will include many activities such as teacher demonstration, discussion work and also 'focused activities' to teach specific skills or aspects of knowledge or wider 'resource tasks', which are also specific inputs that are pertinent to the work in hand and matched to the programme of study. In one sense, 'projects' could be considered to be just one teaching technique among many, but the qualities pupils require to solve problems and engage in successful project work cannot be inculcated by teacher-directed activities alone. It has been argued that project work is able to encourage people to 'create and do' rather than just 'know and understand'.

Such capability is important in many aspects of life and particularly, it is argued, in industry and commerce. Barlex (1987: 7) reports that one school so highly valued the attributes promoted by project work that it included the following on its references for pupils:

Employers please note:

The qualities engendered in students who successfully complete a project are of value to Britain's industrial needs and your firm:

a. the capacity to acquire new skills when they are needed;
b. industriousness over a long period of time;
c. perseverance in the face of disappointment and problems;
d. research skills necessary to become familiar with established ideas in the fields related to the project, be they technical, scientific or aesthetic;
e. the ability to use such ideas in the new and unique context of the project;
f. the ability to communicate clearly and effectively the development and final outcome of the project in both written and graphic form.

It is clear that project-based learning enables pupils to express their capability in a way that written papers alone cannot do. Teachers' enthusiasm for project work has influenced its inclusion in schools across the US, UK, Europe and beyond but the characteristics of open-ended project work – so laudable for small groups of older pupils – must be examined more critically in the light of managing the capabilities and larger numbers of pupils in the lower secondary school.

How is successful project-based learning and related tasks organised?

The ability of a teacher (or a team of teachers) to ensure that the learning of pupils engaged in project work is progressing while using *open-ended* projects is severely stretched by the characteristics of an unpredictable outcome with unpredictable knowledge needs, and facilitating the concept and skill requirements of a class of 20 pupils on a 'when needed' basis.

Technology, engineering and design-and-make assignments

A 'design-and-make assignment' is a type of project work in engineering or design & technology that conforms to the following general characteristics:

- the *exact* outcome is unpredictable (although the framing of the task reduces the possible number of outcomes);
- the pupil takes responsibility for the conduct of the project as much as possible;
- it is based upon a need which the pupil can see and identify with, and is a 'real-life' situation.

Consider the following example. This is an identified task chosen by the teacher and presented to the class, but the outcome is only loosely specified and further work is needed to identify the likely learning outcomes.

EXAMPLE 5: PROJECT WORK IN DESIGN & TECHNOLOGY IN THE LOWER SECONDARY SCHOOL (11–14 YEARS)

Context: Safety in the home.
Task: To design and make a device that will give warning of potential hazards in the home.
Outcomes: Warning devices for: intruders, overflowing vessels, high temperature, needless energy loss.
Materials and components: Wood, metals, composites, range of switches and/or sensors, resistors, LEDs, ICs, batteries, connectors

This task could be presented to the pupils in different ways to recognise the differ-ences of individuals and also the depth of knowledge and skills pupils possess at age 11 compared with age 14. A range of project briefs might result:

- Task 1: Design and make an intruder alarm that turns on when someone treads on a mat.
- Task 2: Design and make an intruder alarm that is tripped on (and stays on) when someone enters a room.
- Task 3: Design and make a device that will give warning of an intruder in the home.

Depending on the age and experience of the pupil, these tasks for lower high school students could be very different to the science investigations set out above, in particular in the ways the teacher structures the activities to help and direct the work, basing the structuring on the previous experiences of the pupils. The organisation of activities will be considered next.

The starting point in organising successful project work depends, to some extent, on the *level* of that organisation. Before considering the organisation at a teacher-in-classroom level, the higher level of planning a 'scheme of work' (the collection of projects and associated activities) for a whole stage of learning (say 11–14-year-olds) should be thought through by all the teachers involved. If learning is to be meaningful, the work done must be:

- **Differentiated** – able to be tackled at a number of levels so that individual pupils understand what is expected of them and the work makes appropriate demands.
- **Build progressively on previous activities** – a new project must offer new challenges that, at least at a general level, are supported by previous tasks; activities must not become a treadmill where pupils 'go through the motions' but learn few new skills or ideas.
- **Relevant to pupils** – pupils must see the point of the project, particularly if it is more open-ended and steered by the enthusiasm of the individual.

Considerable planning is required to ensure that this happens in practice.

When planning project work it is important that you are clear on the following:

1 The capabilities, resources and awareness that pupils are likely to bring with them to the project.
2 The resources the pupils will reinforce and develop by means of the project.
3 The capabilities the pupils will be required to demonstrate by means of the project.
4 The awareness that will be highlighted by the project.

Here the term 'resources' includes knowledge and skills, and 'awareness' means the way the project affects and impinges on our everyday lives, with the consequent consideration of values.

The rest of this section looks at the important issues to consider when planning projects:

- teaching techniques to 'open up' the real-life problem;
- teaching knowledge when needed, or as structured development;
- problem ownership and motivation;
- the relative importance of skills.

Teaching techniques to open up the real-life problem

This is an extract from *Teaching Problem Solving in the Digital Era* by Moshe Barak (Barak, 2020b) where he introduces the technique of systematic inventive thinking:

SYSTEMATIC INVENTIVE THINKING (SIT)

Systematic Inventive Thinking (SIT) is a method of finding solutions to problems by making systematic alterations or manipulations with a system's components and attributes, rather than searching randomly for ideas using methods such as brainstorming.

Among the principles or tools learned in the SIT course are:

- *Unification:* solving a problem by assigning a new use or role to an existing object;
- *Multiplication:* solving a problem by introducing a slightly modified copy of an existing object into the current system;
- *Division:* solving a problem by dividing or cutting an object or subsystem and reorganizing its parts;
- *Change relationships between variables (attribute dependency):* solving a problem by adding, removing or altering relationships between variables;
- *Removal:* solving a problem by removing an object (with its main function) from the system; and
- *Inversion:* solving a problem by inverting the structure or functions of components in a system.

Following is an example of use of the SIT method by students in a final exam during an inventive problem-solving course held at Ben-Gurion University of the Negev (Barak & Albert, 2017). One of the exam questions was 'Suggest a method of how to encourage youngsters to use special night buses for going out on weekends instead of driving a car (especially for those who drink ...)'.

Examples of conventional solutions to this question are providing free buses or having more police on the roads. To find an inventive solution using the SIT method, we first make a list of the components in the world of the problem: **youngsters, cars, police, buses, pubs, alcohol, music, etc.**

We then try to find a solution by carrying out systematic manipulations in the system's components according to SIT principles. One of the SIT principles or 'tools' is **unification**: assigning a new use or role to an existing object in the system. A solution some students suggested was **serving drinks and playing music on a bus.** In terms of the SIT method, this solution assigns the role of a pub to a bus.

(Barak, 2020b)

Teaching knowledge when needed, or as structured development

Teachers can manage project-based learning with small tasks that provide learning necessary FOR tackling an open-ended problem, and in using this specific small-task learning, the students will achieve learning THROUGH the problem-solving activity giving a cognitive shift with regard to their understanding achieved in the learning FOR tackling the problem. In other words, a strategy is to reduce the unpredictability, as is shown above, by designating specific problems and assignments for pupils to tackle.

Pupils may know what they want to do but not be able to realise their solution because they do not have the required knowledge or skills. For instance, the example illustrated above indicates a need for knowledge of switches and transducers before the tasks can be successfully accomplished. More critically, when planning their work pupils may not consider certain approaches to a problem because they are ignorant of the existence of equipment or a technique that might help them. For these pupils, 'problem solving' is doing little more than applying their common sense.

So, what is the best approach? Should pupils learn skills in isolation, which might prove useful later but for which they perceive little immediate value? Should pupils learn skills 'as needed' within projects when they appreciate the usefulness of what they are learning but without a coherent structure and without realising that there *was* something new that they should know, to transfer to future work? The best approach is probably to steer a middle line as is illustrated by the 'big tasks' and 'small tasks' in the DATA example above. A carefully planned selection of shorter projects or small tasks emphasises particular aspects of the programme of study, skills and techniques, together with the longer, more open task (big task), which allows pupils to develop their capability by drawing on their accumulated experiences. In these longer big tasks, new skills and knowledge will have to be covered, just as the shorter small tasks will need to be meaningful and situated in an appropriate context to make sense.

Problem ownership and motivation

If pupils choose a project themselves, they may be more motivated to work independently and with interest but they may have insufficient knowledge and skill to complete it successfully. A teacher-decided project may be better suited to build progressively on the pupil's previous work, be more controlled in the materials and equipment needed to resource it, and easier to manage as part of a whole class's work; however, students may not be so interested in what they have been asked to do. This issue assumes a great importance as the pupils progress through the school and are engage in more open project work, but the issue is still relevant in earlier stages. The careful introduction of the project is vital and ways in which the pupils can themselves identify a need to investigate and work on is important.

Organising project work in the classroom

The word 'classroom' here is used generically to denote any space where STEM education takes place. It has already been suggested that much of the strategic planning of project work should be done at a department team level to satisfy the statutory requirements of each stage and the examination boards, tempered with important education issues concerning the individual pupil.

Revisiting the 'process' in projects

We have already considered the importance of the *process* in project work and what many call the 4Cs of twenty-first-century skills. There are as many different interpretations and critics of process as there are different subjects in STEM! In design & technology and engineering, the criticisms centre on the simplistic use of process as a linear movement from 'identification of need' to 'ideas' to 'specification' to 'product' to 'evaluation of product'. People do not actually follow a neat step-by-step process to design like that – the process is much more varied. Similar criticisms can be made about techniques for solving problems in mathematics. The thinking process is not linear but a complex activity where new possible solutions and evaluations of current ideas continually circle back and permeate every part of the activity at every stage. The over-emphasis on particular aspects of the process, perhaps because of a need to award marks for identified stages, can be unhelpful and leads to such distortions as pupils inventing 'initial ideas' after their design is finished!

While accepting the shortcomings of the process descriptors, many technology/engineering projects will contain the following activities:

- **Researching** – finding out information from books, magazines etc.
- **Investigating** – experimenting with equipment, materials, processes etc.
- **Specifying** – stating clearly the criteria that the chosen solution has to meet.
- **Developing ideas** that might make a contribution to the chosen solution.
- **Optimizing ideas** to formulate the details of a chosen solution.
- **Planning the making** or organisation of the chosen solution.
- **Making** – manufacturing the artefact.
- **Evaluating** – assessing the output against the specification.

The skill of the teacher is to integrate these activities within the constraints of the resources, materials and equipment available and the timetable restrictions. It is also necessary to create a learning environment where pupils can work independently and gain support from each other (see Chapter 10). Niall Seery (Seery, 2020) has written about the importance of the social interaction in project-based learning. Niall's contention is that, to the uninitiated, a technology workshop in which young people are tackling a range of designing and making projects appears chaotic. The suggestion that 'doing' is taking precedence over learning will come as no surprise to those of us who have taught in such situations. However, we know from our experience that this is not the case, and that underlying this apparent chaos is a network of social interactions that 'grows' the knowledge available to the students enabling them to make difficult decisions about the details of their emerging but as yet unresolved design proposals. This knowledge is not evenly distributed among the students; serendipity plays its part in who knows what, but the skilful teacher orchestrates the social interaction to ensure that the workshop is a place in which communication between students is the norm, invariably on task and beneficial. This requires the teacher to develop trust between herself and the students, and between the students themselves. Establishing the technology workshop as such a collaborative creative community of practice takes time and the teacher will need to nurture this over a significant period.

Within this there is a place for specific teaching of matters relevant to making sound design decisions and Niall would argue that this is essential if the students are to be able to find out more on an 'as needed' basis. Such teaching forms a springboard for independent activity, but if ignored puts the students in a position where they have to learn everything relevant to a designing and making task from scratch which is difficult and inefficient. As students become adept at finding out more for themselves, such specific teaching becomes less imperative. However, it is probably worth revisiting important ideas at regular intervals in the light of the learning that has taken place across the class in their designing and making. Imagine asking students, 'OK, now what more do we know and understand about X or Y given our experience and conversations in the designing and making project we've just tackled?' This is not an easy ask but it would make explicit the learning that is taking place within the class and provide the opportunity to make this more widely available. It could also be used to convince those who have a limited view of the learning that takes place in technology education lessons, often seeing it as limited to just 'making' with little, if any, cognitive gain.

The best lessons feature some or all of the following:

- Pupils are taught safely by specialist teachers who were confident and familiar with the media, tools and equipment being used, and who knew the standards they should expect.
- The work is well planned, with systematic teaching of skills, knowledge and techniques.
- Teachers provided a good range of resources and materials, and encouraged pupils to use these to investigate, design, make, test and evaluate their work.
- Specialist teachers use a variety of teaching techniques, and provided pupils with a good balance of activities both as part of an individual lesson and as part of their long-term planning.
- Teachers know the pupils for whom they are responsible, know the most appropriate moment to intervene, and are able to respond flexibly to the requirements of individual pupils.
- Teachers set a brisk pace and provide work which is realistic and interesting to pupils.

However, a well-planned scheme of work, a lively introduction, carefully prepared resources for skill enhancement and teacher inputs, and 'a good balance of activities' will still produce disappointing results if there is insufficient attention given to the allocation of short-term targets within the long task. There should be a clear purpose to each lesson.

By helping pupils to know what they need to have accomplished by strategic points throughout the project, they can be guided to a successful outcome. This does not mean that all pupils should do exactly the same thing in a rigid undifferentiated way, but we should be aware of the way pupils can get side-tracked by a particular facet of the work and lose sight of the whole task or problem presented.

What is the relationship between project-based learning and assessment?

In his comments about social interactions in lessons, Niall Seery notes that, in England, it is ironic that the current assessment arrangements set out to deny this powerful learning through social interaction. The contextual challenge set by Awarding Organisations requires students to explore a context, identify a need or want that needs addressing and then design, make and evaluate a prototype solution. Indeed, it has been observed for some time that the influence of assessment can alter the nature of STEM project work (see Banks, 2009). The assessment criteria of examinations are important, but sometimes teachers can insensitively force pupils through a research or evaluation process for the sake of gaining marks on a particular scheme, rather than to help them develop their own ideas in a natural way. In contrast, and when assessing design & technology project-based learning, Barlex (2007: 53) advocates a 'minimally invasion' approach:

> What is required is the means to allow pupils to reflect on and reveal their progress in making design decisions as the task progresses. Essentially the assessment exercise has to probe and record chronologically the pupil's thinking. Such probing must take place as a pupil moves through the design task. I suggest that probes are required at three junctures in any design and make activity.
>
> The first probe will be used when a pupil has developed his or her first ideas for a product. A pupil will be asked to consider whether his or her proposals meet the requirements of the brief and to clarify and justify the design decisions made so far. The pupil will also be required to review these decisions and consider whether what he or she is proposing is likely to be achievable in relation to resources of time, materials, equipment and personal skills.
>
> The second probe will be used when most of a pupil's design decisions have been made through sketching, 3D modelling, and experimenting. This will be at the point where making is imminent or has just started. Again, the pupil will be asked to clarify and justify the design decisions made so far. Again, the pupil will also be required to review these decisions and consider whether his or her design fully meets the requirements of the brief and whether his or her plans for making are achievable.
>
> The third probe will be used when the product is complete and will include an evaluation against the brief and the specification.
>
> These probes will be used by pupils working in pairs or small groups under structured guidance with their work on the design task available for reference. The probes will provide a script through which pupils can reflect on and justify their design decisions.

Assessment needs to be integrated naturally into lessons.

- Assessments need to be made during a project as well as at the end. The burden of assessment needs to be spread out; indeed, some important attributes can only be assessed as the work is being done. Such assessment opportunities need planning.
- Pupils can help in recording assessments by noting points in their books, project folder or design portfolio; for example, the outcome of a discussion that led to a decision, a new idea or modifications to a design.
- Internal moderation between teachers is necessary to come to a shared meaning of what is required for specific levels. A collection of evidence in the school will help to establish common agreement.
- Several projects are needed to build up a view of the capability of a pupil as different projects bring different types and levels of responses from pupils. This is a case where 'looking sideways' not only helps with the construction of the curriculum but also an understanding of the capability of the pupils.
- The aims and purposes of project work should not be unduly affected by the assessment process. As described above, certain procedures will be strongly

suggested by teachers because 'it earns marks in the exam', but the relationship between the pupil and the teacher and the desire of the pupil to take ownership of the task should not be compromised.

A complementary and in some ways contradictory approach, however, has been advocated by Richard Kimbell. His long-term look at assessment rejects criteria-based assessment as a fallacy – the criteria gets more and more detailed teachers still don't agree that the end 'result' is valid, so fiddle the contributory criterion marks to make it right. For example, his 'e-scape project' worked up the concept of a six-hour structured activity (two consecutive three-hour sessions) in which pupils take a design task from its starting point up to the point of a working prototype, but instead of constructing a separate portfolio of assessment 'evidence' the pupils use a hand-held device – such as a smartphone or iPad – to create an e-portfolio in real time. Kimbell (2007: 68) says:

> The clever bit of this project (at the classroom end) lies in the fact that the e-portfolio is unlike anything that currently exists by that name. Typically such things are second hand re-constructions of real designing – in PowerPoint (PP) or some other sequential software. The construction of the e-portfolio is typically a different task to the designing that it seeks to illustrate. First do your designing – then tell the story in your PP e-portfolio. By contrast the e-scape system uses hand-held digital tools directly in the nitty-gritty of the designing activity in workshops and studios. As learners do their thing, the hand-held digital tools up-link the work dynamically into a secure web-space, where their e-portfolios emerge before their eyes as they work through the activity. These are real-time design e-portfolios.

Another smartphone app that takes a similar approach to 'one-the-hoof' portfolio construction is 'kapture8', which has the capability to upload work in a variety of media, including video, audio and CAD files (see kapture 8, 2020).

Conclusion

Many of the issues to consider when supervising project-based learning are context- dependent. The school environment, the subject traditions of the teachers in the STEM subjects, the whole school timetable, and the financial resources delegated to the different departments – all are highly influential on successful project work. Consumables on projects, for example, may have to be paid for by pupils or their families, but sponsorship from industry or assistance from a parent support group may help. The type of projects tackled may reflect the expertise of the staff and be hampered by the class size which is determined by the school management team.

The most important factor, however, is general to all teachers in every school. Project and investigation work will be most successful when pupils are matched to a task which they find challenging but manageable, and which is relevant to a need which they can perceive. This means that teachers need to know the pupils' background – both personal and in their subject. Some of this will be on record cards in the different departments, but another way to find out is to hold STEM staff meetings regularly. The crucial information to help a teacher in project work may well be in a colleague's head.

In more ways than one, the key to successful projects is team work. As we have said before, it is important to engage in regular conversations with colleagues.

Recommended reading

What is project work (project-based learning) and why is it important?

Scott, L. (2020) STEM Projects Toolkit, British Association, www.stem.org.uk/resources/collection/3926/stem-projects-toolkit (accessed 1 May 2020).

How are successful projects and related tasks organised?

Barak, M. (2020a) Teaching problem solving in the digital era. In P. John Williams & David Barlex (eds), *Pedagogy for Technology Education in Secondary Schools*. Switzerland: Springer.
Brahier, D. J. (2020) *Teaching secondary and middle school mathematics*. Abingdon: Routledge.
Fahrman, B., Norström, P., & Gumaelius, L. (2020) Experienced technology teachers' teaching practices. *Int J Technol Des Educ*, 30, 163–186.
Vogelzang, J., Wilfried, F., Admiraal, W. F., & Van Driel, J. H. (2020) A teacher perspective on Scrum methodology in secondary chemistry education. *Chem. Educ. Res. Pract.*, 21, 237–249.

What is the relationship between project work and assessment?

Griffin, P., McGaw, B., & Care, E. (eds.) *Assessment and teaching of 21st century skills*. Heidelberg: Springer.

References

AQA (2019) *Tech-level engineering*. www.aqa.org.uk/subjects/engineering/tech-level/design-engineering-2016 (accessed October 2019).
Banks, F. (2009) Research on teaching and learning in technology education. In A. Jones & M. Vries (eds), *International handbook of research and development in technology education*. Rotterdam, Netherlands: Sense Publishers.
Barak, M. (2020b) Teaching problem solving in the digital era. In J. Williams & D. Barlex (eds), *Pedagogy for technology education in secondary schools – research informed perspectives for classroom teachers*. The Netherlands: Springer.
Barak, M. &, Albert, D. (2017). Fostering Systematic Inventive Thinking (SIT) and Self-Regulated Learning (SRL) in problem-solving and troubleshooting processes among engineering experts in industry. *Australasian Journal of Technology Education*, 4, 1–14.
Barlex, D. (1987) Technology project work, ET887/897, Units 5–6, Module 4, *Teaching and Learning Technology in Schools*. Milton Keynes: Open University Press.
Barlex, D. (2007) Assessing capability in design & technology: The case for a minimally invasive approach. *Design and Technology Education: An International Journal*, 12(2), 9–56. Wellesbourne: Design & Technology Association ISSN 1360-1431.
Dewey, John (1916). *Democracy and education: An introduction to the philosophy of education*. New York: Macmillan.

del-eliciting activities for primary mathematics classroom. *The* ...7–66.

rchitectural design?: A STEM project based learning unit suitable for ...llesbourne: Design and Technology Association.

ould be. Princeton, USA: Princeton University Press.

h, S. J. (2019) Promoting student creative problem-solving ...nal leadership and teacher creative practices matter?. In T. ...*and career construction in education* (pp. 78–99). Pennsylvania:

Gunderson, S., Jones, R., & Scanland, K. (2004). *The jobs revolution: Changing how America Works.* Pennsylvania: Copywriters Inc.

Kapture 8 (2020) www.kapture8.com (accessed 4 June 2020).

Kay, K.. (2020) 21st century skills. www.educatored.com/instructors/ken_kay.html (accessed September 28 2020).

Kimbell, R. (2007) e-assessment in project e-scape. *Design and Technology Education: An International Journal*, 12(2), 66–76, Wellesbourne: Design & Technology Association.

Moore, T. & Roehrig, G. (2019) Beyond the Textbook: MEAs in Action. University of Minnesota Vision 2020 blog. http://cehdvision2020.umn.edu/cehd-blog/stem-meas-in-action/ (accessed 4 June 2020).

Seery, N. (2020) Pedagogy involving social and cognitive interaction between teachers and pupils. In Williams, J. & Barlex, D. (eds), *Pedagogy for technology education in secondary schools – research informed perspectives for classroom teachers*. The Netherlands: Springer.

T Levels (2020) www.tlevels.gov.uk/subjects?gclid=CIH5ss64uuUCFQccGwodb7oOTw# (accessed 4 June 2020).

Wagner, A. (2008) *The global achievement gap*. New York: Basic Books.

CHAPTER

7

Enabling the 'E' in STEM

Introduction

It is important here to clarify the difference between the T and the E in the acronym STEM. Sometimes the T is interpreted as indicating that learners will study computers and computing. This is understandable as the use of computers does figure largely in our society and in some aspects of technology but not all. Hence in the original National Curriculum in England a clear distinction was made between the school subjects design & technology and information communication technology, each being seen as a separate component of the overall learning area 'technology'. And it is important to tease out the distinction between technology and engineering. Our view is that technology is a socio-technical phenomenon as opposed to an activity. It has characteristics that can be used to describe its intrinsic nature and these are to some extent in dispute among philosophers of technology (de Vries et al., 2019). Is it inside or outside our control; is it overall beneficial or harmful to humans; does it detract from what it means to be human and the possibilities of being human or does it enhance human potential and possibilities?

Its relationship with capitalism is seen by some as beginning to marginalise human activity and the availability of paid work for humans (O'Reilly, 2017) – or perhaps it may provide as yet unthought-of work activities. Engineering is a goal-orientated activity and as such utilises technology to achieve its goals. In doing this it is subservient to the nature of technology and will inevitably have undesirable consequences for humanity in the long run if the views of technology as being outside our control and dehumanising etc. are valid.

The rest of this chapter is in three parts. The first part deals with the situation in the USA where the national engineering standards are incorporated into the US Framework for K-12 Science Education (National Research Council (NRC), 2012) as opposed to the national technology standards. It includes comments by Philip A Reed, Professor in the Department of STEM Education and Professional Studies at Old Dominion University in Norfolk, Virginia. The second part deals with the situation in England where the uptake of engineering as a school subject is extremely low and describes the research funded by the Royal Academy of Engineering to support school-based education for engineering. The third part speculates about future possibilities for the way the E in STEM might evolve in the light of the experiences in the USA and England.

Engineering in USA secondary schools

In the USA, engineering is being subsumed into the science curriculum. The US Framework for K-12 Science Education (NRC, 2012) indicates that engineering should be taught as part of the science curriculum. Of the over 160 pages that form the second part of the document (i.e. the majority) Scientific and Engineering Practices counts for 42 pages, and Disciplinary Core Ideas – Engineering, Technology, and Applications of Science counts for 15 pages. Hence this is not a tokenistic approach. Christine Cunningham and William Carlsen (2013) have reviewed this approach and acknowledge that the broad rationale is to teach engineering before teaching science. Such a teaching and curriculum strategy avoids teaching science first, which students often find abstract, unappealing and difficult to understand. Instead, science can be embedded in an engineering experience in a way that will predispose students to learn science later, when they have been motivated by the engineering experience. Cunningham and Carlsen comment at length on the way the document considers scientific and engineering practices as having parallel features although with a different intent. This epistemic similarity is shown diagrammatically in Figure 7.1. We think it is worth considering some of the features of practice and question to what extent they will achieve their espoused aims of enhancing science education.

First let us consider asking questions/defining problems and link this to designing solutions. Expert designers have considerable substantive and disciplinary knowledge. They know about materials, their properties and applications, manufacturing methods, ways to achieve functionality of many sorts and they know how to deploy this knowledge through designing. In addition, their disciplinary knowledge equips them to explore the contexts in which their proposed designs will be used such that their suggestions meet the requirements of stakeholders. They often suspend judgement and avoid becoming definitive too early in the process and intuitively use case-based reasoning from their considerable experience to develop unexpected and provocative solutions. Lawson (2004: 20) describes this as follows:

> Designing then, in terms of chess, is rather like playing with a board that has no divisions into cells, has pieces that can be invented and redefined as the game proceeds and rules that change their effects as moves are made. Even the object of the game is not defined at the outset and may change as the game wears on. Put like this it seems a ridiculous enterprise to contemplate the design process at all!

The extensive range of uncertainties embedded in the exercise seems to militate against the analogy that Cunningham and Carlsen are drawing. These uncertainties require those designing to be solution focused, as opposed to problem focused. It is not that one activity is better than the other; it is that they are fundamentally different in their intention and hence require significantly different approaches within themselves. It is also worth noting that, in seeking solutions in engineering, there will be no

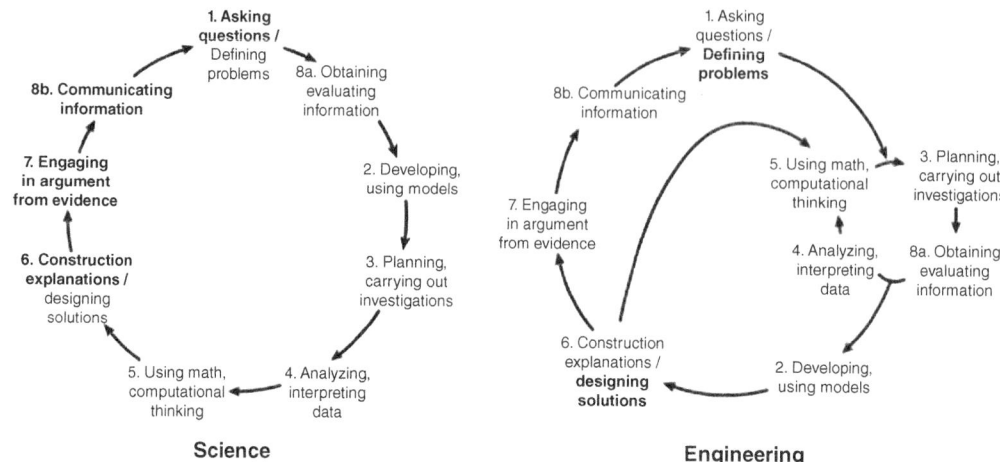

FIGURE 7.1 Cycles of epistemic practice in science and engineering
Source: Cunningham and Carlsen, Pre-college engineering education in handbook of research in science education.

single correct solution but many possible solutions of varying degrees of acceptability. This is not the case with regard to developing explanations in science.

Second, let us consider the practice of developing and using models. Cunningham and Carlsen argue that science deals with conceptual models, whereas engineering deals with concrete models that are more accessible. However, in many science courses these are physical models that students use to explore concepts, such as:

■ the bubble raft for explaining the properties of metals;
■ ball and spoke models for exploring mechanisms in organic chemistry;
■ different coloured beads representing dominant and recessive genes.

And, more recently,

■ computer-based models have been developed for exploring the rules governing the behaviour of objects under various forces.

Such models can then inform the scientific imagining that takes place in the 'mind's eye' when pupils are constructing and reconstructing their science understanding. Students can be asked to construct physical models to develop science understanding. In contrast, engineering models often start with sketches as opposed to 3D models, and these can present students with a considerable conceptual challenge: lines on a 2D surface representing a 3D item. Such modelling skills require teaching, as indeed do 3D modelling skills. We have to ask the question, will the science educator know about different sketching/drawing/modelling techniques and be able to teach them?

Third, let us consider the practice of analysing and interpreting data. Here, the thrust seems to be that giving students design/construction problems that they can solve (as opposed to engaging them with ideas they find difficult to understand) will develop students' self-efficacy. There is also the assumption that this will somehow spill over into disciplinary agency in other disciplines. What seems to be missing here is any sense of just how wonderful a really good explanation based on ideas derived from interpreting data can be. The argument is almost anti-intellectual with regard to scientific thinking.

Fourth, let us look at the practice of engaging in argument from evidence. This feature is clearly important for both science and engineering. With pressure to cover a large amount of content, science teachers often teach the prevailing paradigm as something to be memorised as opposed to something to be developed through reflection and arguing from evidence. This is understandable, if regrettable. However, if engaging in argument from evidence is important for developing science understanding, will skills in arguing about aspects of engineering transfer to science understanding when it is required? It must also be remembered that the item under scrutiny in science will not be a physical object but an explanation.

Fifth, we must consider the practical question of who will teach the engineering practices that are expected to enhance science learning. Engineering, as Cunningham and Carlsen rightly state, leads to the investigation and creation of products in which to some extent science understanding is embedded. If such products are to be other than construction-kit based then the quality of manufacture becomes a serious issue. The generally applicable scientific principles underpinning the site-specific design of a bridge amount to naught if the bridge is so poorly constructed that it fails. In the USA, those teachers who have themselves a wide range of appropriate construction skills, and are able to teach these to students, are likely to be technology teachers but will not necessarily have the science understanding required to engage with teaching science by design through engineering type projects. By the same token, the science teacher who has the science understanding to help pupils use their developing science knowledge in designing products is unlikely to be able to teach construction skills. She may well lead pupils to develop designs that are well beyond their construction capabilities.

Hence the issue here seems to us to be one of achieving suitable collaboration as opposed to poorly prepared science teachers invading and acquiring the 'construction' territory of technology teachers. Tanner Huffman (2019) writing on the Advancing Excellence in P-12 Engineering Education (AEEE) website disputes this and is optimistic about the way science teachers are responding to the engineering standards within the science standards. He writes, 'With the implementation efforts of the Next Generation Science Standards (NGSS), science teachers throughout the country are teaching engineering design as a "complement to" and "a vehicle for" science learning'.

Since the inclusion of engineering in the US Framework for K-12 Science Education (NRC, 2012), matters have moved on with the development and publication of the Next Generation Science Standards (2019) An organising principle throughout the document is the use of three categories to illustrate grade related content from Kindergarten to the end of middle school: science and engineering practice, disciplinary core ideas and cross cutting concepts in each of the items within the subjects

Philip Reed, Professor in the Department of STEM Education and Professional Studies at Old Dominion University in Norfolk, Virginia, has reservations about the way engineering education is developing at school level (Reed, 2018). He is concerned that the T in STEM in the USA is nebulous and unless it defines itself clearly then the positioning of engineering with the science standards will further fragment its identity as technology teachers are inevitably drawn into teaching the E and spend less time on the core business of T. He makes a compelling plea for a laser-like focus on the nature of the T in STEM to clarify it so that it is worthy of a place in the general education of all young people. He cites the work of Stephen Petrina (2007) in explaining, in his opinion, the misguided view of many practitioners who are beguiled by the 'technoenthusiam' apparent in many enhancement and enrichment activities. Hence they are guilty of 'technonaïvete' with regard to the extent to which the technology education learners receive are sustained and substantial. So, Philip is particularly pleased that the Standards for Technological Literacy are currently being revised (ITEEA, 2019) with the expectation that the revised standards will be presented at the 2020 ITEEA Conference in Baltimore.

However the curriculum politics of the USA play out, and there is little doubt that the inclusion of engineering in the Next Generation Science Standards came as a shock to the board members of the ITEEA, we think it is well worth exploring how science and technology teachers might collaborate. The science teachers will be teaching science in the light of the learning that students are achieving in technology lessons and technology teachers will be teaching technology in the light of the learning that students are achieving in science lessons. This is a mantra that we have already chanted in this book!

Engineering in secondary schools in England

Attempts to introduce engineering into the secondary school in England have not been successful in terms of uptake or survival. The percentage of the national cohort aged 16/17 years studying the Engineering General Certificate for Secondary Education (GCSE) for 2016, 2017 and 2018 were 0.53 per cent, 0.49 per cent and 0.36 per cent respectively compared with 12.74 per cent, 11.69 per cent and 9.18 per cent for the design & technology GCSE (JCQ, 2017 and 2018). The Engineering Diploma introduced in 2009 was available only until 2013 (Joint Council for Qualifications, 2019). Despite this low uptake it is worth exploring the nature of some of the engineering courses available to learners. In Wales there is the Welsh Joint Education Council (WJEC) Level 1/2 Vocational Award in Engineering (WJEC, 2019). According to the specification, this qualification offers a learning experience that focuses learning for 14–16-year-olds through applied learning (i.e. acquiring and applying knowledge, skills and understanding through purposeful tasks set in sector or subject contexts that have many of the characteristics of real work). This course has three components as shown in Table 7.1 requiring a total of 120 guided learning hours (GLH).

TABLE 7.1 Unit structure of the WJEC Level 1/2 Vocational Award in Engineering

WJEC Level 1/2 Vocational Award in Engineering			
Unit number	Unit title	Assessment	GLH
9791	Engineering Design	Internal	30
9792	Producing Engineering Products	Internal	60
9793	Solving Engineering Problems	External	30

In *Engineering Design* students learn about the design process. They learn how to analyse a product so they can see what features make it work and how it meets certain requirements. They learn how to take ideas from different products in order to produce a design specification for a product.

In *Producing Engineering Products* students learn to interpret different types of engineering information in order to plan how to make engineered products. Students develop the skills needed to work safely with a range of engineering processes, equipment and tools. With these skills, they learn to make a range of engineered processes that are fit for purpose.

In *Solving Engineering Problems* students learn about how engineers in the past have found solutions to problems and how other engineers use their ideas to solve problems today. Students learn about materials, processes and mathematics that engineers use and how they are used to solve problems. In solving problems, students learn to follow a process and develop drawing skills to communicate their solutions.

It is worth noting that nowhere in this course are students required to design AND make an engineering product. In Engineering Design, they are required to develop a specification. In Producing Engineering Products, they are required to make an engineering product to someone else's design. In Solving Engineering Problems, students have to tackle short and extended answer questions based around applied situations in a written examination. They are required to use stimulus material to respond to questions that could relate to mechanical engineering, electronic engineering, structural engineering or any combination of the three types of engineering. Inevitably, in this paper-based assessment the student is not required to take what we might term 'practical engineering action'. Be that as it may, the specification suggests that successful completion of this qualification can provide students with opportunities to access a range of higher level qualifications including General Certificate of Education, apprenticeships and vocationally related qualifications.

In England, the examination board known as the Assessment and Qualifications Alliance (AQA) offers a GCSE engineering qualification involving candidates studying for two years usually between the ages of 14–16 years (AQA, 2019a). The core content is divided into six areas:

- engineering materials;
- engineering manufacturing processes;
- systems;
- testing and investigation;
- the impact of modern technologies;
- practical engineering skills.

Assessment is in two forms; an externally assessed two-hour written paper accounting for 60 per cent of the marks and a internally assessed non-exam assessment (NEA) accounting for 40 per cent of the marks. The questions in the written paper consist of multiple-choice questions assessing breadth of knowledge, short answer questions assessing in depth knowledge, including calculations, multiple-choice questions related to the application of practical engineering skills and extended response questions drawing together elements of the specification. Sample papers are available (AQA, 2019b). In response to the NEA, candidates produce engineering drawings or schematics to communicate a solution to a brief supplied by AQA towards the end of the first year of study and an engineering product that solves a problem. Inspection of the sample written paper reveals the range of question types required and a thorough coverage of the specification content. The sample NEA presented candidates with this problem:

> It is often necessary to sort through waste products such as litter. Due to the fact that this can be unhygienic, it is helpful if this can be done mechanically. Your task is to engineer a solution to this problem that will reduce the negative impact on the people who are involved in collecting or sorting litter.

The candidates are informed that their solutions must include both mechanical and electronic components to provide an integrated product. In addition, they are provided with three examples of how the problem could be solved and told that they can choose a solution from this list or can create their own. The examples given were:

- Engineer a product that someone can use outdoors to pick litter without bending over.
- Engineer a product or system that reduces the bulk of litter.
- Engineer a product or system that sorts three objects of different sizes.

There are, of course, many stereotypical and unimaginative possibilities for each of these examples and it would be easy to suggest that the bar for such responses has been set low. However, this would not be fair as a good teacher will inspire her students to use the content they have learned to develop imaginative responses and perhaps encourage them to move outside the given examples. It is worth noting that, in responding to the NEA, candidates are not required to submit a design portfolio describing the process they went through to arrive at their product. They have to produce engineering drawings or schematics to communicate a solution to the brief supplied and the engineered product.

The Royal Academy of Engineering gave considerable backing to the Engineering Diploma and were disappointed by its failure, but you would not expect such an astute organisation to put all its 'eggs in one basket'. The Education for Engineering group (E4E) is the body through which the engineering profession offers coordinated and clear advice on education to UK government and the devolved assemblies. It deals with all aspects of learning that underpin engineering. It is hosted by The Royal Academy of Engineering with a wide membership drawn from the professional engineering community including all of the professional engineering institutions. It is significant that E4E spent some considerable effort in developing a re-conceptualisation of the school subject design & technology. The key proposals were launched in March 2013 with the publication of the document New Principles for Design & Technology in the National Curriculum (E4E, 2013). The re-conceptualisation configured design & technology as a toolbox of key concepts. These are shown in summary in Figure 7.2.

Interestingly, there is an explicit acknowledgement that the subject is not a vocational subject and the way the subject has been redefined enables design & technology courses to offer learning experiences that mirror engineering activities to a considerable extent. Given the large uptake of design & technology compared with

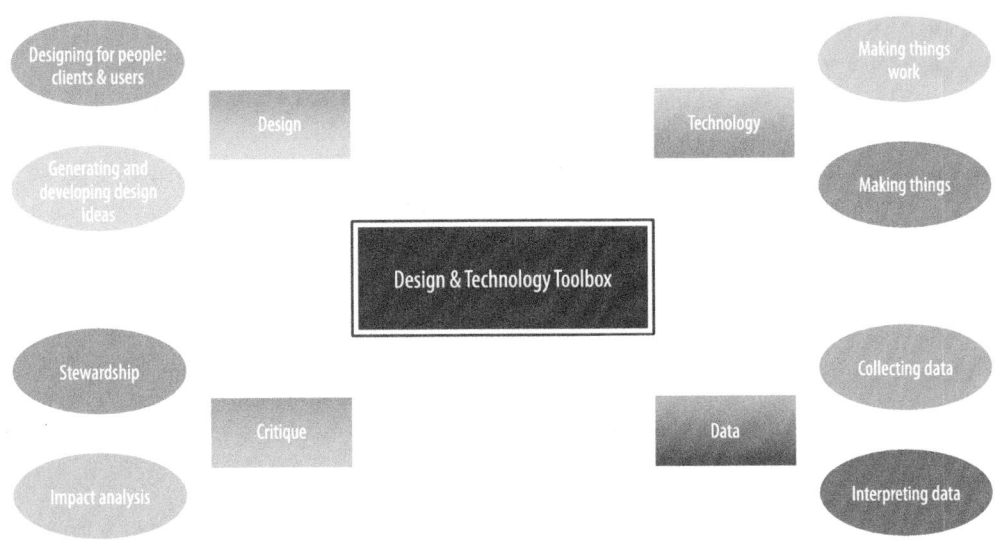

FIGURE 7.2 The design & technology toolbox as envisaged by E4E

the Engineering Diploma, it was hoped that in its redefined form design & technology would enable many aspects of engineering to find their place as a legitimate component of a general education for a majority (if not all) students. The design & technology education community welcomed this approach, but it has to be acknowledged that while it had significant influence in the reshaping of the government's proposals for national curriculum design & technology, this has not prevented the continued decline in the numbers of young people studying the subject to 16+ years. Disappointment over the Diploma and the continuing decline in the popularity of design & technology gave the Royal Academy of Engineering pause for thought and led to new direction in their thinking. The findings of the Aspires Project (Archer et al., 2012) raised awareness of the idea of 'habitus', which embraces family values, practices and a sense of 'who we are' and 'what we do'. In some cases, this leads young people to see possible scientific or technical careers as unthinkable. The Royal Academy of Engineering in consultation with Bill Lucas and Janet Hansen from the Centre for Real World Learning, University of Winchester, reasoned that one way of tackling the habitus problem might be to deliberately develop habits of mind in young people that overcame the 'not for someone like me' mindset. Disciplinary habits of mind are well established in a variety of fields (e.g. mathematics (Cuoco et al., 1996) and science (Çalik & Coll, 2012)) and are seen as contributing to academic achievement.

The first step in enacting this new strategy was for the Royal Academy of Engineering to commission Bill Lucas, Janet Hanson and Guy Claxton to carry out research that would lead to understanding more about how successful engineers think and act and then to consider how best these might be cultivated at school, college and university. This work was deemed necessary because of the following contextual factors:

- Demand for engineers to meet the needs of the UK economy at both graduate and non-graduate levels currently exceeding supply.

- Too few women opting for engineering careers and courses.
- Perceptions of engineering among young people not being positive enough and the public perception of the value of engineering not capturing the full variety and value of engineering.
- Engineering education being patchy in quality and quantity virtually absent at primary, under-represented at secondary and with variable take up at college and university level.
- A lack of theoretical discussion about the pedagogy of engineering teaching and learning that crosses education sectors.

There were two research questions:

1   How do engineers think, especially when they are working to solve challenging problems? (What are the engineering habits of mind (EHOM) which engineers and society value?)

2   How can schools, colleges and universities select learning methods that are more likely to cultivate EHOM?

The research was divided into four phases:

a   Reviewing literature and developing draft EHOM.

b   Refining EHOM with engineers and engineering educators.

c   Exploring the most effective ways of cultivating EHOM via appreciative inquiry with engineers/engineer educators, supported by engineering pedagogy literature review.

d   Synthesising (a)–(c) and producing a report that maps pedagogy (learning and teaching methods) on to desired EHOM along with clear proposals for action.

The report (Lucas et al., 2014) in response to this work was *Thinking like an Engineer: Implications for the Education System*. It identified and described EHOM and placed these within a general framework of learning habits of mind as shown in Figure 7.3.
    The report made three broad recommendations:
    The Royal Academy of Engineering to disseminate its findings to ensure:

1   Wide engagement in the conversation about how engineering is taught.

2   The engineering teaching and learning community to seize the opportunity of the National Curriculum and the report's new thinking to bring about a mindset shift in schools and redesign engineering education, especially at primary level.

3   For employers, politicians and others to engage in a dialogue with schools and colleges about the EHOM they think are most important, suggesting practical ways in which they can help.

The report was welcomed by the Royal Academy of Engineering and although it established proof of concept questions remained: would teachers take to it, would schools adopt the ideas, would there be a difference in response at primary and secondary levels. In response to these questions three more research projects were commissioned. These in their turn led to the following reports.

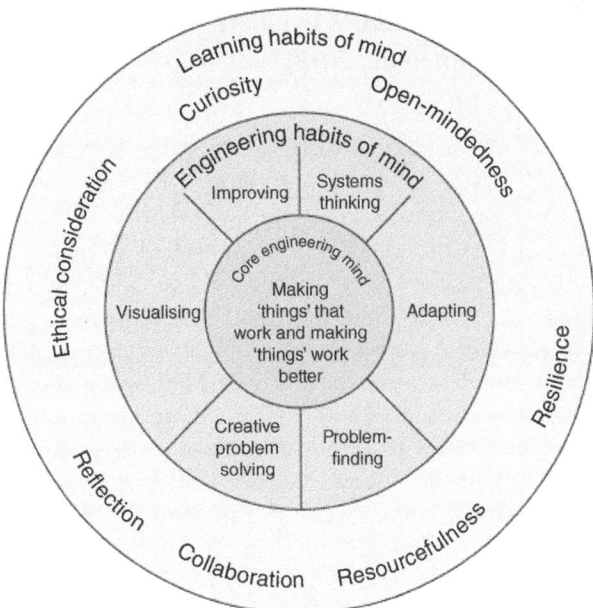

FIGURE 7.3 Engineering habits of mind as identified by Lucas, Hansen and Claxton

Learning to be an engineer implications for schools (Lucas et al., 2017c)

This report identified four principles that underpin the kinds of teaching that are most likely to encourage young people to develop a passion for engineering in today's busy schools and colleges:

1   Clear understanding of engineering habits of mind by teachers and learners.

2   The creation of a culture in which these habits flourish.

3   Selection of the best teaching and learning methods, the 'signature pedagogy' of engineering.

4   An active engagement with learners as young engineers.

Tinkering for learning: Learning to teach engineering in the primary and KS3 classroom (Bianci & Chippindall, 2018)

This report identified seven principles for engineering in primary schools as follows

1   Pupils are engaged in purposeful practical problem solving.

2   Pupils take ownership of the design and make process.

3   Pupils embrace and learn from failure.

4   Pupils' curiosity and creativity is responded to.

5   Pupils demonstrate mastery from other curriculum areas.

6   Pupils draw on a range of thinking skills and personal capabilities.

7   Pupils' learning experiences are guided by a whole-school approach.

Engineering the future: Training today's teachers to develop tomorrow's engineers (Hanson et al., 2018)

This built on the report published in 2017 and made recommendations for taking action to develop an understanding of EHOM in trainee teachers and those in post such that they could devise lessons that developed EHOM and invited engineering employers and professional bodies to engage with schools in supporting such lessons.

The Institution of Mechanical Engineers published a report entitled *'We Think It Is Important but We Don't Quite Know What It Is' The Culture of Engineering in Schools* (Finegold, 2017b). The report was the culmination of two research studies that explored perceptions and experience of engineering in secondary school education. The first study sought to understand how 11–14-year-old pupils, their parents, teachers, school governors and school leaders, frame engineering. The second presented a deeper engagement with engineering through the experience of post-16 students, participating in bespoke engineering debating competitions run jointly by the Institution of Mechanical Engineers and the Institute of Ideas. The report called on government, education practitioners and the engineering community to act together to ensure that more young people discover what engineering is, both as a creative intellectual process and a rich source of future career opportunity.

The report made four recommendations:

1   As part of its industrial strategy, government should situate engineering at the heart of schools education by:

   a   Setting up a working group of leading educationalists and other stakeholders to review and report on innovative ways to integrate engineering into young people's education

   b   Appointing a nationally respected Schools Engineering Champion to provide a channel of communication between schools, government and industry, and to advocate the wider cultural value of greater technological literacy alongside the economic rationale for investing in skills to prepare for the Fourth Industrial Revolution

2   National Education Departments should begin this process by ensuring that engineering is integral to classroom learning by:

   a   Advocating curricula that better reflect the importance of the made world to modern society, and make explicit reference to the engineering applications of science, mathematics, and design and technology

   b   Promoting approaches to teaching that emphasise and value engineering 'thinking skills' and problem-based learning

3   Individual schools should adopt an engineering vision and strategy, with support from local employers and national governors' associations, which would include:

   a   Appointing a member of the school senior leadership team as an Engineering & Industry Leader to establish and communicate a vision for the school and to drive change

   b   Appointing a dedicated Industry School Governor to work alongside and advise the Engineering & Industry Leader, and to embed employer relationships in school governance

    c   Implementing a robust careers strategy such as the benchmarks set out in The Gatsby Foundation's Good Career Guidance report, with special emphasis on embedding careers awareness in the curriculum

4   The engineering community should present a unified narrative around engineering that will be attractive and relevant to a wider range of students by:

    a   Stressing the creative problem-solving nature of engineering, its social benefits and relevance to individuals

    b   Providing opportunities for students to take part in activities that explore the political, societal and ethical aspects of technology.

For those who advocated a prominent place for engineering in the school curriculum, these recommendations seemed eminently sensible, but taking these recommendations forward required agent provocateurs to ensure that this report like many others before it did not fall by the wayside. Such actors were not forthcoming, and the impact of the report was negligible. In terms of classroom practice Recommendation 2 argues for raising the importance of the made world to modern society in science, mathematics and design & technology curricula. One would expect that the school subject design & technology would already be giving the made world a significant role in its curriculum (and this is indeed the case) but would mathematics and science curricula respond in a similar way? Teachers of those subjects would surely ask, 'What is the benefit to my subject?' If they could see none, then they would simply ignore the request – and this has been the case.

So, we have a situation that over some four years in which there has been cogent, well researched investigations into how education might best support engineering, making sensible recommendations and providing useful guidance, and yet it seems likely that all of this has had little effect on practice outside those schools and institutions involved in the research; definitely an example of seed falling on stony ground.

The Royal Academy of Engineering has changed tack and has asked Bill Lucas and Janet Hansen to explore the meaning and place of practical learning in the school curriculum. This is a much broader approach and given that practical activity can take place in a variety of schools subjects across the arts, technology, science, mathematics and the humanities, this may give rise to an endorsement of particular sorts of learning activities that support STEM education in general and education for engineering in particular. It will be interesting to see how this project develops particularly with regard to the way practical learning is developed and validated in different subjects within the curriculum.

Possible futures for E in STEM

So, how might we bring developments in both the USA and England to bear on possible futures for the E in STEM? One way is to attempt some scenario development. The initial task is to identify the critical uncertainties for the axes needed to create the four scenarios. Whether engineering habits of mind (EHOM) is developed just within specific engineering courses or across various subjects' courses seems a crucial factor. Whether these subjects are as important for general education for all young people or just a vocational option for a much fewer number of young people will

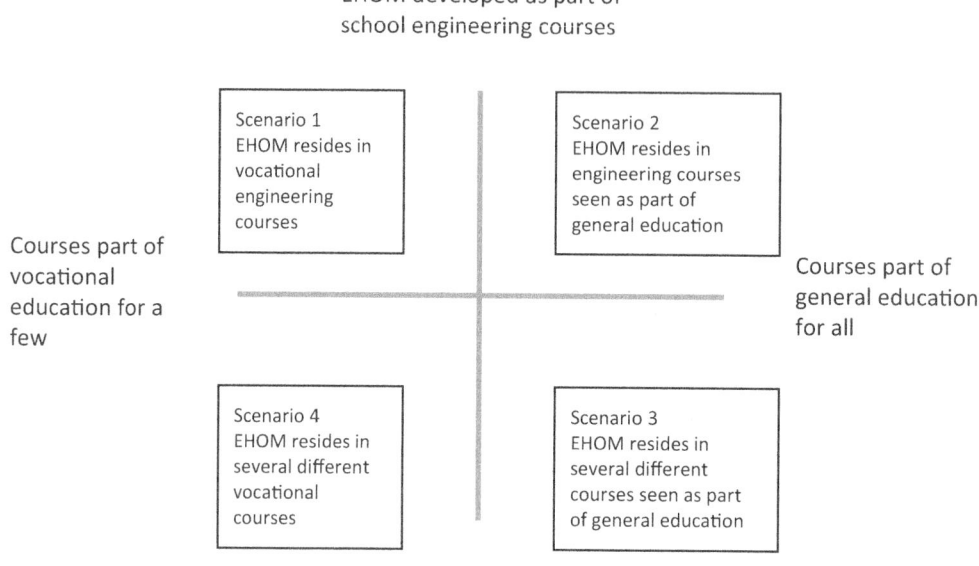

FIGURE 7.4 Scenarios for embedding EHOM in the school curriculum

also be important. This gives rise to the axes shown in Figure 7.4 and four different scenarios.

Before considering each of these scenarios, it is worth noting that engineering can be seen as a subject with both substantive and disciplinary knowledge. The substantive knowledge provides engineers with relevant technical knowledge while the disciplinary knowledge provides engineers with the means to deploy their substantive knowledge. EHOM is concerned primarily with disciplinary knowledge.

Considering scenario 1

In this scenario such school courses will support the acquisition of disciplinary engineering disciplinary knowledge, but such courses may have only limited uptake due to the vocational label.

Considering scenario 2

In this scenario such school courses will support engineering disciplinary knowledge but will require engineering courses to be seen as part of general education

Considering scenario 3

In this scenario each of such school courses are likely to give only limited support for the acquisition of engineering disciplinary knowledge and there will need to be co-ordination across such courses to ensure breadth in developing engineering disciplinary knowledge

Considering scenario 4

In this scenario such school courses will only be able to support the acquisition of engineering disciplinary knowledge if course developers explicitly embed EHOM in the courses they develop.

It is possible that in a large school several of these scenarios might operate side by side, but for small schools it is likely that only one or two will be possible. Each scenario has its pros and cons.

- Scenario 1 enables a focus on engineering as a discipline and enables the development of engineering specific courses that contribute to both substantive and disciplinary knowledge but runs the risk of low uptake because of the vocational label.

- Scenario 2 situates engineering courses within a general education remit, but it is arguable that this is unlikely to happen as the rhetoric surrounding the introduction of engineering into the school curriculum is often couched in terms of its economic utility (see, for example, Institute of Engineering and Technology, 2016), which inevitably lessens its appeal as a 'general education' subject.

- Scenario 3 situates EHOM across several different general education subjects and coordination of teaching will be required to achieve breadth. The advantage of this approach is that many learners will be involved in general education courses. The disadvantage may be that the required coordination may not be forthcoming. Scenario 4 requires that the different vocational courses that incorporate EHOM as part of their disciplinary knowledge coordinate their teaching such that a combination of such courses achieves breadth as is the case in Scenario 3. Such coordination takes time and effort, which might not be forthcoming. If a learner takes only one such course, which does not have a particularly strong engineering focus, then their exposure to EHOM will be limited.

So where do you stand with regard to the E in STEM in secondary schools? The research that has identified EHOM has clarified the disciplinary knowledge of engineering and it is possible to embed this into a range of different education for engineering endeavours ranging across the vocational – general education spectrum. Some will argue that situating this in general education will pay the greatest dividends, but this will not be easy. Would you align yourself with this position? The reality is that when engineering is available as a school subject, schools that opt to teach it do often see it as vocational. Would you prefer to see it as part of the science curriculum as is being promoted in the USA? If you are a science teacher, you might feel that this was asking too much of you and that you wouldn't be able to meet the practical requirements of such a programme. However, you might feel that by collaborating with technology teacher colleagues it would indeed be possible to meet the requirements. This might take diplomacy as some technology teachers might see engineering within science as 'stealing' their curriculum territory. Or would you reject the option of teaching engineering and instead opt to teach design & technology in such a way that it provided young people with the knowledge and skills needed for creative design, innovation and engineering without being overtly vocational as is being suggested in England and also argued for by Philip Reed in the USA in his appeal that technology education should not lose its focus. Where you stand will depend on your school situation, your professional knowledge and skill and your views on the purpose of secondary school (high school) education. Whatever your position, we would suggest that it will not be tenable if you do not engage with colleagues across the STEM subjects.

Recommended reading

Finegold, P. (2017a) *'We think it is important but we don't quite know what it is': The culture of engineering in schools*. London: Institution of Mechanical Engineers.
Lucas, B., Hanson, J., Bianchi, L. & Chippindale, J. (2017a) *Learning to be an engineer: Implications for schools*. London: Royal Academy of Engineering.
Lucas, B., Hanson, J., & Claxton, B (2017b) *Thinking like an engineer: implications for the education system*. London: Royal Academy of Engineering.

References

AQA (2019a) *GCSE engineering*. www.aqa.org.uk/subjects/engineering/gcse/engineering-8852 (accessed June 12 2020).
AQA (2019b) *Sample papers*. www.aqa.org.uk/find-past-papers-and-mark-schemes (accessed June 12 2020).
Archer, L., DeWitt, J., Osborne, J., Dillon, J., Willis, B. & Wong, B. (2012) Science aspirations and family habitus: How families shape children's engagement and identification with science. *American Educational Research Journal*, 49(5), 881–908.
Bianchi, L. & Chippindale, J. (2018) *Tinkering for learning: Learning to teach engineering in the primary and KS3 classroom*. London: Royal Academy of Engineering.
Çalik, M. & Coll, R. K. (2012) Investigating socioscientific issues via scientific habits of mind: Development and validation of the scientific habits of mind survey. *International Journal of Science Education*, 34(12), 1909–1930.
Cunningham, C. & Carlsen, W. (2013) Pre-college engineering education. In N. G. Lederman (ed.), *Handbook of research in science education*, 2nd edn. Abingdon: Routledge.
Cuoco, A., Goldenberg, E. P. & Mark, J. (1996) Habits of mind: An organizing principle for mathematics curricula. *Journal of Mathematical Behaviour*, 15, 375–402.
de Vries, M. J., Halstrom, J. & Dakers, J. (eds) (2019) *Reflections on technology for educational practitioners*. The Netherlands: Brill Sense O'Reilly.
Engineering for Education (E4E) (2013) *New principles for design & technology in the national curriculum*. London: Royal Academy of Engineering. www.solutions4schools.org.uk/site/module_publications/RAeng_E4E_Final_Report_DT_Curriculum.pdf (accessed June 12 2020).
Finegold, P. (2017b) *'We think it is important but we don't quite know what it is': The culture of engineering in schools*. London: Institution of Mechanical Engineers.
Hanson, J., Hardman, S., Luke, S., Maunders, P. & Lucas, B. (2018) *Engineering the future: Training today's teachers to develop tomorrow's engineers*. London: Royal Academy of Engineering.
Huffman, T. (2019) *Framework first. Let's talk engineering standards*. www.p12engineering.org/post/framework-first-let-s-talk-engineering-standards (accessed June 12 2020).
Institute of Engineering and Technology (2016) *Skills and demand in industry 2016 survey*. London: Institute of Engineering and Technology.
ITEEA (2019) *Revision of the standards of technological literacy*. www.iteea.org/Activities/2142/STL/151449.aspx#tabs (accessed June 12 2020).

Joint Council for Qualifications (2019) *The Engineering Diploma*. www.jcq.org.uk/examination-results/diploma (accessed June 12 2020).

Lawson. B. (2004) *What designers know*. Oxford, UK: Elsevier.

Lucas, B., Hanson, J., Bianchi, L. & Chippindale, J. (2017c) *Learning to be an engineer: Implications for schools*. London: Royal Academy of Engineering.

Lucas, B., Hanson, J. & Claxton, B. (2014) *Thinking like an engineer: Implications for the education system*. London: Royal Academy of Engineering.

National Research Council (2012) *A framework for K-12 science education: Practices, cross cutting concepts, and core ideas*. Washington, DC: The National Research Council.

Next Generation Science Standards (2019) *The three dimensions of science learning*. www.nextgenscience.org (accessed June 12 2020).

O'Reilly, T. (2017) *WTF? What's the future and why it's up to us*. London: Penguin.

Petrina, S. (2007). *Advanced teaching methods for the technology classroom*. Hershey, PA: Information Science Publishing.

Reed, P. (2018) Reflections on STEM, standards and disciplinary focus. *Technology and Engineering Teacher*, 77(7), 16–24.

WJEC Level 1/2 Vocational Award in Engineering (2019) *Vocational award in engineering*. www.wjec.co.uk/qualifications/engineering-level-1-2/#tab_overview (accessed June 12 2020).

8

The role of STEM enhancement and enrichment activities

Introduction

Recently, I reviewed some of the comments I've heard while eavesdropping on pupils' informal 'corridor' conversations.

Student A: Miss said we've got a STEM careers day coming up – all sorts of science and technology stands to visit; all of Year 9 have got to go. And in the evening there's an info session for our parents.

Student B: Are your mum and dad coming?

Student A: I think so; my dad said I needed to think about what I wanted to do. Did you hear about the STEM club the science teachers are setting up? Sounds as if you get to do cool stuff – like what we don't get to do in lessons. Sir said he's organising a STEM Challenge Day for us at the local college. Something about robots and there'd be the chance to build one and talk to some engineers. You have to be picked to go through. Should we ask if we can go together?

These indicate some of the activities that make up enhancement and enrichment activities and the excitement that they can generate. Generally, they are outside the mainstream curriculum that we considered in Chapter 2. In many countries, the main rationale for these activities is an economic one. Their aim is seen as supporting and encouraging a larger number of pupils to consider, and ultimately enter, a STEM-based career – the economic argument that has underpinned a variety of STEM initiatives. This is in contrast to the other main rationale for STEM, which is epistemological in nature and contends that the contributing subjects, although different in nature and intention, have sufficient in common and such reciprocal utility that it makes good educational sense to see them in some sort of curriculum relationship.

A question immediately arises. Why are such activities necessary? Is the mainstream curriculum experience not engaging enough to attract young people into STEM-related careers? We have seen in Chapter 3 that the intrinsic nature of science may render it unattractive to many young people and, as indicated in Chapter 5, a significant proportion of young people in both Europe and the USA become alienated towards mathematics as they move through secondary education. This apparent disenchant-ment with STEM is corroborated by the findings of the ROSE (Relevance of Science Education) project (2010). This is a well-regarded international study investigating young peoples' attitudes towards science and technology. Participating countries range across northern Europe, Africa, India, the Far East and South America. Given disen-chantment with the in-school STEM curriculum, and governments' concerns over growing STEM skills gaps, it is easy to see enhancement and enrichment activities as an important strategy for persuading young people to overcome their resistance to so-called 'hard' subjects such as science and mathematics and gain STEM qualifica-tions and move onto a STEM career track.

In this chapter we explore a variety of STEM enhancement and enrichment activ-ities at different scales of implementation. We begin by considering some initiatives that are global in scale and discuss their nature and intentions. Then we consider two individual countries, America and England, and describe and discuss developments that are taking place. Finally, we consider some of the evaluations of enhancement and enrichment activities in England.

Global STEM enhancement and enrichment activities

We will consider here three enhancement and enrichment activities that are global in scale. The first two are competition based and concerned with developing relatively tra-ditional STEM products and systems: FIRST Lego League, which focuses on robotics; and F1 in Schools, which focuses on utilising CAD/CAM in the context of Formula One (F1, 2020) racing. The third is the iGEM Competition, which operates in the sphere of synthetic biology and from this perspective is completely different from the first two.

FIRST Lego League

FIRST Lego League (FLL, 2020) is a well-established STEM enhancement and enrichment activity. FIRST is an acronym – For Inspiration and Recognition of Science and Technology. It operates in over 61 countries and representatives from these countries can attend the annual FLL World Festival. It is a robotics program for 9–16-year-olds (9–14 in the USA, Canada and Mexico), which the organisers say is designed to get children excited about science and technology and teach them valu-able employment and life skills. The challenge facing the participants is in two parts: the Robot Game and the Challenge, both of which are underpinned by the FLL Core Values. Teams of up to ten young people, with one adult coach, participate in the challenge by programming an autonomous robot to score points on a themed playing field (the Robot Game) and developing a solution to a problem they have identified (the Project). The FLL Core Values are significant. They are listed in Panel 8.1.

The terms 'gracious professionalism' and 'coopertition' are significant. The term 'gracious professionalism' was coined by Dr Woodies Flowers, National Adviser to FLL. He defined this as 'learning and competing like crazy, but treating one another with respect and kindness in the process'. Gracious professionals avoid treating any-one like losers. This is strongly linked to 'coopertition', which requires displaying unqualified kindness and respect in the face of fierce competition. According to the

- **Discovery:** *We explore new skills and ideas.*
- **Innovation:** *We use creativity and persistence to solve problems.*
- **Impact:** *We apply what we learn to improve our world.*
- **Inclusion:** *We respect each other and embrace our differences.*
- **Teamwork:** *We are stronger when we work together.*

PANEL 8.1　First Lego League core values

organisers, coopertition is founded on the concept and a philosophy that teams can and should help and cooperate with each other even as they compete.

The FIRST Lego League presents pupils with a socially relevant task – the Innovation Challenge. Past challenges have been based on topics such as nanotechnology, climate, quality of life for the handicapped population, and transportation.

In 2019, the Innovation Challenge required the teams to identify a problem with a building or public space in their community, design a solution, share their solution with others and then refine it. There is a wide range of information to support the participants available on the FFL website. For the team members there is extensive information on building with Lego and programming the processor controlling the robot. For the adults who act as coaches and mentors, there is also advice and guidance.

Similar Lego resources are available to schools, for example, Mindstorm kits, so it would be possible to build FLL into a school curriculum. But the vast majority of FLL activity occurs as part of afterschool clubs, very often with parental support. Teams who take part in FLL can attend official tournaments organised by so-called Operational Partners such as National Instruments, Rockwell Automation, John Deere and 3M. There are, in fact, a wide range of FIRST Lego programmes, of which FLL is just one.

F1 (Formula One) in Schools

F1 in Schools is a multi-disciplinary challenge in which teams of students aged 9–19 deploy Project Management skills and CAD/CAM software to collaborate, design, analyse, manufacture, test and then race miniature gas-powered F1 cars made from model block. It is the brainchild of Andrew Denford, the Chief Executive of Denford Limited a UK manufacturer of CAD/CAM machines and technology. F1 in Schools is a well-established enhancement and enrichment activity. Since its inception, it has grown from operating in just England to involve students from 51 countries. In the period 2004–2017 student involvement has grown from 200,000 to 1,300,000 with the number of schools taking part growing from just over 5,000 to just over 25,000. The organisers claim that it provides a global platform for the promotion of STEM education in partnership with Formula One and partners to a youth market. In order to compete, teams must raise sponsorship and manage budgets to fund research, travel and accommodation. A criticism that these events will not appeal to girls has been rejected by the Williams Deputy Team Principle, Claire Williams, who is a patron of F1 in Schools. The most recent figure available from the organisers indicate that 40 per cent of the contestants are girls indicating that the criticism is unfounded.

The challenge faced by competing teams is as follows:

- Working in teams of between three and six students, each member is assigned a role. The team prepares a business plan, develops a budget and raises sponsorship. Teams are encouraged to collaborate with industry and forge business links.
- Using 3D CAD (Computer Aided Design) software, the team designs a model Formula One car of the future.
- Aerodynamics are analysed for drag coefficiency in a virtual reality wind tunnel using Computational Fluid Dynamics Software (CFD).
- Using 3D CAM (Computer Aided Manufacture) software, the team evaluates the most efficient machining strategy to make the car.
- Aerodynamics are tested in wind and smoke tunnels.
- In the race the cars travel at more than 70 kph and compete side-by-side along 20-metre straights.
- Teams are judged on car speed, as well as supporting evidence of their design, verbal presentation and marketing display stand in 'the pits'. Teams compete regionally, nationally and internationally for the Formula One F1 in Schools World Championship trophy.

Teams who enter the competition are bound by extensive competition regulations – the manual runs to 31 pages. Teams also have to abide by strict technical regulations defined in a 24-page manual. These include a requirement to use CAD/CAM in the production of the car and the organisers recommend the use of Autodesk Fusion 360 for the CAD and the use of DENFORD QuickCAM PRO software for CAM. It is a requirement that the body is manufactured from model block using a CNC router/milling machine. The organisers recommend the use of a DENFORD CNC Router.

An interesting aspect of the competition is that in the schools' World Finals competing teams can be made up as a collaboration of two teams each from a different country. The organisers believe that this will develop participants' communication and collaboration skills and raise levels of tolerance and understanding. It is possible for schools to build F1 in Schools activities into their curriculum. There are clear possibilities for design & technology with the CAD/CAM development of the cars themselves but this can be linked to strongly to science and mathematics in considering and taking into account the drag on particular designs. However, to compete at regional level and above requires considerably more commitment. The organisers believe that participating in the competition will help change young peoples' perceptions of science, technology, engineering and maths and enable them to develop an informed view about careers in engineering, Formula One, science, marketing and technology.

Comparing FLL and F1 in Schools

While there are similarities between FLL and F1 in Schools – their economic justification, global scale and the fact they are both competition based – there are two significant differences. The main curriculum difference is that FLL introduces a new challenge each year that deals with quite different STEM domains. In recent years, the problems faced by the participants have involved nanotechnology, climate change, transportation and disaster management. The F1 in Schools challenge has remained essentially unchanged since its inception and does not differ significantly year from year. There is a more explicit identification and emphasis throughout FLL on their Core Values of 'gracious professionalism' and 'coopertition' than is apparent in F1 in Schools. At the World Championship level in F1 in Schools, such values do become apparent to some extent by the requirement that each competing team can be is made up from teams from different countries.

The International Genetically Engineered Machine (iGEM) competition

Biology is not usually seen as a science subject that has a significant contribution to STEM where the accent is often on physics and, to a lesser extent, chemistry. It is noteworthy that in the USA the technology curriculum includes agricultural technology and related biotechnologies and in New Zealand biotechnology is an identified area of optional study. Hence it is possible that topics within biology that inform biotechnology might be seen as part of STEM. Any study of biology will, of course, deal with genetics but it is unlikely that that there will be any in-depth consideration of synthetic biology. This is a new area of research in which engineering principles are combined with knowledge of genetics to enable the design and construction of new biological functions and systems not found in nature.

The iGEM challenge

iGEM (2020) is a worldwide synthetic biology competition initially aimed at undergraduate university students but now extended to high school students. Given that the treatment of synthetic biology is at best very limited in school science courses, iGEM can be seen as a global example of STEM enhancement and enrichment. Student teams are given a kit of biological parts at the beginning of the summer from the Registry of Standard Biological Parts. This is a collection of genetic parts that are used in the assembly of systems and devices in synthetic biology. The iGEM competition facilitates this by providing a library of standardised parts (called BioBrick standard biological parts) to students, and asking them to design and build genetic machines with them. Student teams can also submit their own BioBricks. Working at their own schools, the teams use these parts and new parts of their own design to build biological systems and operate them in living cells. Successful projects produce cells that exhibit new and unusual properties by engineering sets of multiple genes together with mechanisms to regulate their expression. Information about BioBrick standard biological parts, and a toolkit to make and manipulate them, is provided by the Registry of Standard Biological Parts. This is a core resource for the iGEM program, and one that has been evolving rapidly to meet the needs of the program.

The organisers of the iGEM competition believe that the competition has goals beyond that of just building biological systems. They identify these as:

- Enabling aspects of biology to be considered as engineering.
- Promoting the open and transparent development of tools for engineering biology.
- Helping to construct a society that can productively apply biological technology.

The organisers argue that requiring the teams to be self-organised and engage with the imaginative manipulation of genetic material provides a new way to arouse student interest in modern biology and to develop their independent learning skills.

In 2012, 41 high school teams registered for iGEM; 31 were from the USA, seven were from Asia and four were from Europe. In 2019 this had grown to 80 high school teams including 46 from China and 11 from the USA. In 2019, as in 2012, all the projects submitted indicated significant sophisticated synthetic biology. The winner was a team from China and involved collaboration between students from 13 different high schools operating under the collective name of Great Bay SZ (2019). The team manufactured recombinant spider silk using E.coli and dyed the silk with microbial natural pigments deoxyviolacein and indigo. To obtain better colour and a more convenient dying process, they fused the repetitive region to chromoproteins and mixed them with silk proteins during spinning.

Comparing global STEM enhancement and enrichment activities

The organisers of all of these global enhancement and enrichment activities use an economic rationale to justify themselves. Their aim is to engage school pupils with STEM-related activities in the expectation that this will lead them to consider a STEM-based career and so ensure an increase in the numbers of young people who can be employed in STEM-based occupations. FIRST Lego League (FLL), F1 in Schools and iGEM are competition based. Whereas FLL and F1 in Schools focus on 'traditional' STEM content, for example, hi-tech making, computing and the physical sciences, the iGEM competition is different in that it focuses on synthetic biology, which is a newly emerging area of technological activity taught to only a limited extent in school science courses. Indeed, for young people who cease to take science courses at 16 years of age, it is likely that they will have learned little about synthetic biology. Given that this technology is likely to have a significant, if not disruptive, impact in the near future, the iGEM competition is particularly important in raising public awareness.

STEM enhancement and enrichment activities in the USA

In 2009 Barak Obama made a plea for STEM education to the National Academy of Sciences as follows:

> Think about new and creative ways to engage young people in science and engineering, like science festivals, robotics competitions and fairs that encourage young people to create, build and invent—to be makers of things, not just consumers of things.

In response to Obama's plea, DARPA and Cognizant became involved in STEM enhancement and enrichment activities. Both of these initiatives related strongly to the Maker Movement, which bases its philosophy on a constructionist view of education. Those in the Maker Movement not only promote making things as a fun and enjoyable activity but insist that making activities develops a wide range of cognitive skills.

The DARPA initiative focused on engaging high school-age students in a series of collaborative design and distributed manufacturing experiments. DARPA envisioned deploying up to 1,000 Computer-Numerically Controlled (CNC) manufacturing machines – such as 3D printers – to high schools across the USA. The involvement of DARPA has been problematic for some members of the Maker Community who expressed concerns with regard to the involvement with the military and intellectual property rights. Mitch Altman, a San Francisco-based hacker and prominent member of the maker community, withdrew from participating in Maker Fairs in 2012 (Altman, 2012). Dale Dougherty, the editor and publisher of *Make*, felt required to publish a justification for involvement with DARPA and dispel misapprehensions (Dougherty, 2012) . However, in the event, despite an apparently auspicious start, the DARPA MENTOR high school program never really got off the ground because it lost its funding in President Obama's big sequestration budget cut of March 2013.

Cognizant identified three troubling trends in the USA that underpinned Obama's plea:

- a relative decline in math and science proficiency;
- a decline in interest in the STEM fields;
- a decline in measured creativity.

Cognizant also noted that these trends threatened the competitiveness of the US economy and quality of life for future generations. Cognizant responded by initiating the Making the Future programme, which has an after-school and summer programme as its flagship. Developed in partnership with the Maker Education Initiative and the New York Hall of Science, the programme provided grants to community organisations to run hands-on, Maker-movement inspired programmes in an after-school or summer camp setting, or within the school day when conditions allow. Making the Future grants may cover costs for tools, materials, instructor fees, and other expenses essential to meeting the needs of the children participating in the programme. Cognizant issued over 20 programme grants in 2013, based on an established pool of funding. Each grant was in the region of $15,000–$30,000.

Since then Cognizant has changed its focus from schools to reskilling those already in the workforce with the skills necessary for digital technology jobs. As a result, the Cognizant US Foundation was launched in 2018 with an initial $100 million investment from Cognizant and the Foundation has since awarded $12 million to organisations working to educate and train the next generation of workers in communities throughout the US (Cognizant US Foundation, 2018). Interestingly Cognizant's Centre for the Future of Work (CCFoW) produced a guide to likely areas of future employment – *21 Jobs of the Future a Guide to Getting and Staying Employed for the Next Ten Years* (CCFoW, 2017) – taking into account the impact of some disruptive technologies in the short term: AI, virtual reality and big data, in particular. Key features of many employment opportunities will be (a) being part of human–machine collaborations and (b) facilitating human–machine collaborations that are likely to require both technical understanding and 'soft' skills such as communication, leadership and team work. Some nine years after Obama's initial plea, the National Science and Technology Council produced *Charting a Course for Success: America's Strategy for STEM Education* (National Science and Technology Council, 2018b). One of the four pathways identified in the report as supporting this strategy is to develop and enrich strategic partnerships which the report describes as 'bringing together schools, colleges and universities, libraries, museums, and other community resources to build STEM ecosystems that broaden and enrich each learner's educational and career journey'. This clearly embraces enhancement and enrichment activities and some of the considerable funding to support STEM education (some $345 million) referred to be Betsy DeVos (DeVos, 2019) in her speech about STEM investment priorities will find its way to supporting such activities.

STEM enhancement and enrichment activities in England

This section considers seven STEM enhancement and enrichment activities. They are as follows: (1) British Science Week, (2) The Big Bang, (3) engaging with STEM Ambassadors, (4) TeenTech, (5) the CREST Award scheme, (6) Nuffield Research Placements and (7) the work of a single teacher who has developed a wide range of such activities.

British Science Week

British Science Week, organised and run by the British Science Association (British Science Association, 2020ba, 2020b) is an annual ten-day celebration of science, technology, engineering and maths, featuring entertaining and engaging events and activities across the UK for people of all ages. It provides a platform to stimulate and support teachers, STEM professionals, science communicators and the general public to produce and participate in. The aim is to encourage schools to use the time as an opportunity to link STEM to other curriculum subjects and to student's own backgrounds, lives and interests.

The British Science website provides a wide range of information and free downloadable resources to support schools who want to use British Science Week as an enhancement and enrichment activity. These include advice and guidance of using volunteers, activities at home, gathering resources, follow-up activities and involving students in deciding what to do. Importantly, schools can decide for themselves what to do during the week so can adjust what they do to their own circumstances and the needs of their students. In 2019 the organisers estimated that 180,000 children and young people were involved in the week.

The Big Bang

The Big Bang consists of the Big Bang Fair, the Big Bang Competition and Big Bang Near Me. It is organised by Engineering UK, a not-for-profit organisation, which works in partnership with the engineering community to inspire tomorrow's engineers and increase the talent pipeline into engineering (Engineering UK, 2020a).

The Big Bang UK Young Scientists & Engineers Fair is an annual celebration of science, technology, engineering and maths (STEM) for 7–19-year-olds made possible thanks to the collaborative efforts and support of over 100 organisations (Engineering UK, 2020b).

It provides a combination of exciting theatre shows, interactive exhibits and careers information and takes place at the NEC, Birmingham, each March. It allows young people to meet hundreds of scientists and engineers, to get hands-on with science and engineering and shows them where their STEM studies might lead in the future.

The fair is free for schools and families to attend and regularly attracts 80,000 visitors and significant national media coverage.

The Big Bang Competition

The Fair hosts the finals of *The Big Bang Competition*, which recognises and rewards young people's science and engineering project work and identifies the GSK UK Young Engineer and GSK UK Young Scientist of the Year. Projects presented have included a low-cost, flat-pack 3D printer, an app that helps people escape safely from fires and an investigation into new ways of separating 'Mirror Image Molecules'.

Big Bang Near Me

With events taking place from as far north as the Isle of Skye, to Plymouth in the south, the *Big Bang Near Me* is a complementary programme involving some 900 organisations, that extends the reach of The Big Bang across the UK.

Using a framework that's flexible to suit local needs, the Big Bang Near Me comprises regional events attracting up to 10,000 visitors and local fairs welcoming around 2,000. The Big Bang Near Me programme reaches over 170,000 young people each year.

STEM Ambassadors

The STEM Ambassador scheme is run by STEM Learning, a national organisation that also provides significant professional development for primary and secondary school teachers (STEM, 2020). STEM Ambassadors are volunteers from a wide range of science, technology, engineering and mathematics (STEM) related jobs and disciplines across the UK. They offer their time and enthusiasm to help bring STEM subjects to life and demonstrate the value of them in life and careers. Their services are free of charge. They bring a wealth of experience from their varied STEM roles and diverse backgrounds, working for over 5,000 different employers. More than half (59 per cent) are under 35 years of age, 43 per cent are female and 13 per cent are from black, Asian and minority ethnic backgrounds. STEM learning argues that it is not just teachers and pupils that benefit from the STEM Ambassador scheme. There are significant benefits for employers in that being a STEM Ambassador increases staff engagement and boosts their confidence, communication and presentation skills. Volunteering increases ambassadors' job satisfaction and knowledge, as well as opportunities to develop their own professional network within and beyond their own organisation or sector

TeenTech

TeenTech was founded in 2008 by Maggie Philbin and Chris Dodson to help students see the wide range of career possibilities within science, technology and engineering

(TeenTech, 2020). The programmes are structured to guide students through key academic choices with a coherent journey through age appropriate interventions and continued opportunities even beyond its initiatives to gain experience, knowledge and skills. The rationale given by TeenTech for its activities is definitely economic utility in developing ICT-based competencies on the grounds that:

- By 2020, the EU (not withstanding Brexit) could face a shortage of up to 900,000 ICT professionals.
- 47 per cent of jobs will disappear over the next 20 years due to technology but for every one lost, two will be created.
- 132,000 job opportunities possible in big data over the next five years.
- Only 12.8 per cent of the total STEM workforce in the UK are women.
- The UK has the lowest number of female engineering professional in Europe

To achieve their goals, the organisation established TeenTech Awards in 2012 in which students are required to develop their own ideas for making life better, simpler, safer or more fun. Participating schools are provided with a suggested structure and industry contacts. There are 20 categories including 'Teacher of the Year'. All submitted projects receive feedback and a bronze, silver or gold award. Every year the best projects go forward to the TeenTech Awards Final at the IET London for judging and the winners receive £1,000. In addition, there is a City of the Future competition, which offers students the opportunity to apply their knowledge of science, technology and engineering to create their City of Tomorrow where no ideas are too big and everything is smarter, kinder and safer than before. The activity highlights careers pathways in construction, engineering and technology and promotes good citizenship.

CREST Awards and Nuffield Research Placements

Two other national schemes are worthy of mention. These are the CREST Awards (2020) and Nuffield Research Placements (2020).

The CREST Awards scheme requires young people at school to undertake projects of their own choice in the STEM subjects and, depending on the demand of the project, pupils can achieve bronze, silver or gold awards. Bronze awards are typically completed by 11–14-year-olds. Over the course of ten hours, teams of students design their own investigation and record their findings, giving them a taste of what it is like to be a scientist or engineer in the real-world. The Bronze level works well in a STEM club setting or as something that is completed across a term, although there is no deadline for completion. The projects are assessed by the teacher against the CREST Bronze award criteria, which are available on line. Upon completion, teachers upload a sample of your students' work on to the CREST online platform for moderation purposes. Students' personalised CREST certificates are sent to the schools within four weeks of the project's submission. The Silver and Gold levels are designed to stretch students. They are long-term, in-depth projects that are run by the students themselves. Students choose the topic and type of project they want to run from four options: a practical investigation, a design and make project, a research project or a communication project they want to run. Silver projects are typically completed by students aged 14+ years and Gold by students aged 16+. At this level, students are encouraged to collaborate with a CREST mentor – an academic or person from industry with expertise in their project's theme. All Silver and Gold level projects are assessed externally via the online platform by assessors who are experts across various STEM areas and have received assessment training. The costs for entering students for a CREST award in England are modest; Bronze award £5.00 per student, Silver award £10.00 per student and Gold award £20.00 per student. Importantly, UCAS (the organisation responsible for managing applications to higher education courses in the UK) have endorsed CREST Awards for inclusion in young people's personal statements in their application for admission into university.

Nuffield Research Placements provide over 1,000 students each year with the opportunity to work alongside professional scientists, technologists, engineers and mathematicians. Students in the first year of a post-16 science, technology, engineering or maths (STEM) courses are eligible to apply. Placements are available across the UK, in universities, commercial companies, voluntary organisations and research institutions. The organisers are particularly keen to encourage students who don't have a family history of going to university or who attend schools in less well-off areas. To this end students must meet one of the following the eligibility criteria:

- Be living, or have lived in, local authority care.
- Come from a family with a combined household income of below £30,000 a year, or be entitled to free school meals, either now or at any time in the last six years.
- Be the first in their immediate family to participate in higher education (not including any older siblings). This means neither of their parents/carers have participated in higher education, either in the UK or another country. If their siblings have attended higher education, but their parents/carers have not, they are still eligible.

To ensure that no one is excluded on a financial basis, students' travel costs are covered. Some students may also be eligible for a weekly bursary in addition to travel expenses.

An exceptional teacher

The majority of teachers who engage in enhancement and enrichment activities do so through events like British Science Week, the Big Bang or through working with STEM Ambassadors. However, it is possible for an individual teacher to develop their own brand of enhancement and enrichment activities. David Baker has taken the position that although the initial 'D' is missing from STEM it is through design activities that young people can be engaged in STEM. While teaching at Latymer School in Hammersmith, London, he organised a whole range of extra-curricular design-based activities including design days at weekends and design camps during the summer holidays. Pupils from neighbouring schools were invited to attend. Recently, he has organised some STEM activities under the banner of the STEM Academy, which were supported by funding from the charity 'Shine'. The programme Scrape, Rattle and Blow was concerned with the science, design and mathematics of music and musical instruments and ran for five consecutive Saturday mornings. During this time pupils aged 14 years learned about sound and how it is produced and how it can be altered according to the size, shape and materials used in the instrument. They explored how a sound can be measured in terms of waveforms and frequencies, and how this links to the pitch of a note. They used mathematics to work out different formulae to create tuning systems. They used a variety of acoustic and electronic devices to amplify sound. They built their own design of musical instruments and produced a CD recording of their performance. The work of David Baker shows that an individual teacher with energy can make a significant contribution to enhancement and enrichment activities. It is noteworthy that his approach was not predicated on the economic argument but rather on using a 'designerly' approach to show how science, mathematics and technology could come together under a single context, in this case musical instruments, and each contribute considerably to pupil's knowledge understanding and skill.

Evaluation of STEM enhancement and enrichment activities in England

In 2016 the Nuffield Foundation commissioned an independent evaluation of the Nuffield Research Placement Scheme led by Dr Gillian Paull at Frontier Economics. This six-year evaluation is tracking the education and career destinations of three cohorts in comparison to similar students that have not undertaken a placement. In addition to providing a robust assessment of impact, the evaluation makes recommendations for how the programme might be improved. An interim report was published in 2019 (Cilauro & Paull, 2019), which highlights the following:

- The Nuffield Research Placement Scheme increases the likelihood of enrolling in a STEM course at a 'Russell Group' higher education institution (similar to Ivy League in the USA), with similar-sized impacts for all pupils and for more disadvantaged (free school meals (FSM)) pupils.
- The placement experience is perceived to enhance transferable skills, including study motivation, overall confidence in abilities and specific skills in presenting, writing and time management.
- Students report that placements improve their understanding of what STEM researchers do on a day-to-day basis but do not appear to influence their attitudes on how interesting a STEM career is or how much they enjoy the study of STEM subjects. Placements also do not appear to influence students' aspirations and plans, mainly because most placement students aspired to study for a STEM degree even prior to application for a Nuffield Research Placement.
- The Nuffield Research Placement Scheme increases the number and quality of STEM A levels achieved, which may support the impacts on HE enrolment. Undertaking a placement is associated with a higher average number of STEM Advanced Levels of 0.2 for all pupils and 0.3 for FSM pupils, and a higher average point score for STEM A levels of 7 for all pupils and of 11 for FSM pupils (which roughly corresponds to one grade for one A level).

STEM Learning produced an Impact Report in 2019 and provided the data as to the impact of the STEM Ambassador scheme as follows:

Concerning the impact on young people

- ■ 76 per cent of young people surveyed rated their experience with STEM Ambassadors good or very good.
- ■ 91 per cent of schools, colleges and community groups report increased enjoyment, interest and engagement in STEM subjects as a result of STEM Ambassadors.
- ■ 74 per cent report that STEM Ambassadors have significantly increased young peoples' pursuit of STEM subjects and aspirations for STEM-related study and careers.

Concerning the impact on teachers and ambassadors:

- ■ 81 per cent of teachers say that working with STEM Ambassadors has increased their understanding of STEM business and industry. Positive impacts include increased confidence, motivation and enthusiasm for STEM.
- ■ 90 per cent of STEM Ambassadors say that volunteering as an Ambassador has increased their job satisfaction.
- ■ Employers report improved presentation, communication, organisation, leadership and mentoring skills among staff who volunteer as STEM Ambassadors.

The impact as reported does appear very positive but it is perhaps worth a note of caution. Clare Gartland of University Campus, Suffolk (Gartland, 2014), looked at the way ambassadors from higher education interact with school students. Her work questioned the prevailing wisdom that young people of similar ethnicity and gender, as school students, will necessarily provide role models to emulate. School students can sometimes be suspicious of the 'marketing approach' and can feel alienated because they are seen as lacking appropriate ambition. Gartland's work shows that a much more nuanced approach may be required with ambassadors working more closely with teachers in subject-specific contexts as opposed to simply providing 'look what I've done – you can do it too' sessions. This is to some extent echoed by the following statement from Baroness Brown, the chair of STEM Learning in her introduction to the Impact Report (2019).

> As you will see in this report, our work is having a positive impact, but there is still more to do. Challenges include tackling 'cold spots' around the country that are underachieving in **STEM**, as well as social mobility, diversity and inclusion. We want every young person to feel confident in **STEM**, regardless of where they live, their family background, gender or ethnic identity. **STEM** can be, and should be, for all.

In 2016 Pallavi Banerjee published a paper summarising a research plan for a longitudinal evaluation project conducted on the population of secondary schools and pupils in England indicating that the impact of these activities would be evaluated in terms of school and pupil educational outcomes (Banerjee, 2016). It was hoped that the research findings from the study would form an evidence base for policy and practice and recommendations would be useful for academic and non-academic beneficiaries. In 2017, Banerjee published the results from the longitudinal study of the impact of science, technology, engineering and mathematics (STEM) 'enrichment and enhancement activities' that included science practical lessons, supported by ambassador visits, trips to laboratories, STEM centres and higher education institutions (Banerjee, 2017).

The common theme for these activities was their aim to improve understanding and enjoyment of science in the short term and encourage STEM participation in the long term. The 2007 cohort across all state maintained secondary schools in England was followed up from the beginning of Key Stage 3 to the end of Key Stage 5 making use of school and pupil level datasets from the national pupil database. The study investigated whether engaging in these STEM programmes, run for 11–16-year-olds, in secondary school is likely to affect subject choices during post-compulsory education. It asked whether young people sparsely represented in STEM courses such as those from a lower socio-economic class and black ethnic minority engage better with STEM subjects because of actively participating in these activities. A direct noticeable impact of these activities was not seen on STEM take-up. Banerjee's analysis concluded that there is no evidence to suggest continued engagement in these activities is manifested in terms of increasing or widening STEM participation. Clearly, this makes sad and worrying reading for those who see the enhancement and enrichment activities that Banerjee considered as worthwhile and effective.

The Education Endowment Foundation (EEF) carried out a randomised controlled trial of the CREST Programme (EEF, 2019) involving 2,810 pupils from 180 schools. The study found no evidence that Year 9 pupils offered participation in the CREST Silver Award made any additional progress in science attainment, compared to similar pupils that were not offered the programme. This finding had moderate to low security due to the high number of students that dropped out of the trial, and the risk that remaining students may not be representative of the overall student population. Also, there was no evidence to suggest that the CREST Silver Award improved self-efficacy in science or increased the proportion of students aspiring to a STEM career; however, small positive impacts were found for pupil confidence and attitudes to school. Again, this makes worrying reading for those who are advocates of enrichment and enhancement activities. However, intriguingly the results from the EEF trial are at odds with that revealed by Pro Bono Economics for the British Science Association (Stock Jones et al., 2016). This focused on students in English state schools, the majority of whom were aged 14–16 and who took part in CREST Silver Awards between 2010 and 2013. This study found that students who took CREST achieved half a grade higher on their best science GCSE result, compared to a statistically matched control group. In addition, the study found that those CREST students eligible for free school meals saw a larger increase in their best GCSE science score (two thirds of a grade) compared to a matched control group of other students who were also eligible for free school meals. These results were statistically significant. The study also found that 82 per cent of CREST students took a STEM qualification (an AS level) post 16 compared to 68 per cent of a statistically matched control group. CREST students were therefore 21 per cent (or 14 percentage points) more likely to take a STEM AS level than students in the control group. For students who had been eligible for free school meals this difference was larger (38 per cent or 21 percentage points). These results were statistically significant. The report does acknowledge that the possibility of other unobserved variables is affecting GCSE results and AS level subject choice cannot be ruled out. The report made several recommendations for further work, including replicating this analysis through a Randomised Control Trial (mirroring the EEF work), broadening it to cover Discovery, Bronze and Gold CREST Award types and conducting a cost-benefit analysis for schools. Additionally, it made three broader recommendations: that charities ensure accurate and usable data collection, that young people consider taking part in project/inquiry-based learning such as CREST and finally that the British Science Association consider targeting CREST at students from low income families (mirroring the approach being taken by the Nuffield Foundation).

Given the contradictory nature of these evaluations, it is best to remember that the contexts of enrichment and enhancement activities vary considerably and that there are many influences at play. Hence the work of Louise Archer in the ASPIRES Project is particularly relevant (Archer et al., 2020b). It is an extensive piece of research which tracked a cohort of young people in England from age 10 to 19 (2009–2018), through over 40,000 surveys and 660 in-depth interviews with young people and parents/carers. Data were collected from the cohort at five time points (when the young people were in school years 6, 8, 9, 11 and 13, at ages 10/11, 12/13, 13/14, 15/16 and 17/18, respectively). Follow-up interviews were also conducted at age 19. Some the key findings were that young people's career aspirations are relatively stable over time, with no evidence of a poverty of aspiration and the proportion of young people specifically aspiring to be a scientist is around 16 per cent), established fairly early and remains stable from age 10 to 18. Persistent, low science aspirations are not due to lack of interest in science. The Project identified three features that contribute to science identity and aspiration shown diagrammatically in Figure 8.1.

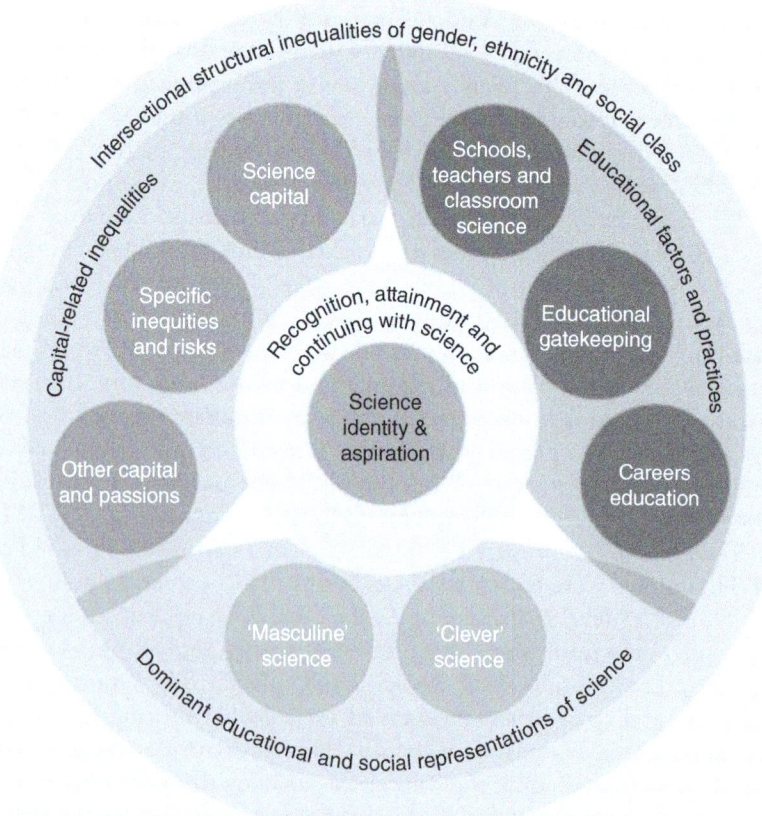

FIGURE 8.1 The Aspires Project model of factors shaping young people's science identities and aspirations age 10–19

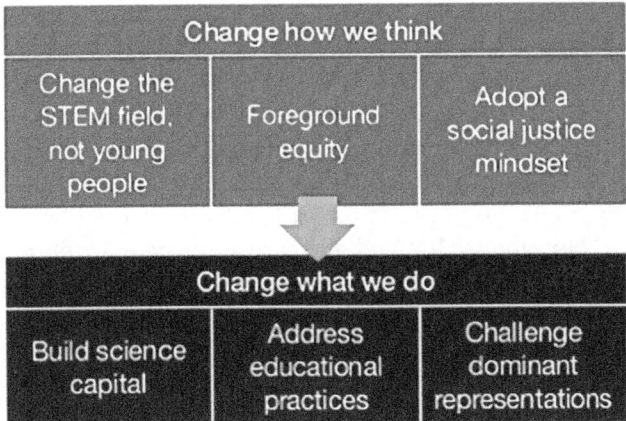

FIGURE 8.2 Aspire Project overview of recommendations for policy and practice

Given the range and complexity of the influences revealed it is not surprising that any single enhancement and enrichment activity is unlikely to pay particularly strong dividends, however laudable its intentions. The Project makes important recommendations as changes to how we think about STEM engagement and inspiration work and changes to what we do in practice as summarised in Figure 8.2.

Those engaged in developing enhancement and enrichment activities and those teachers who use such activities would do well to take these recommendations into account.

In conclusion

Several questions remain concerning the provision of enhancement and enrichment activities for all STEM subjects. Why is the school experience of such subjects perceived as being so impoverished that stakeholders feel that there is a need to initiate enrichment activities outside the mainstream school provision? We must acknowledge that some enhancement and enrichment activities (e.g. F1 in Schools and FIRST Lego League) do have a place in the mainstream curriculum but this is not seen as a key feature, although the involvement of school teachers in these activities through extracurricular activities is crucial to their success. We asked the same question in Chapter 2: should it not be possible to develop a 'business as usual' curriculum that does not require such enhancement or enrichment? Might some of the activities initially envisaged as sitting within enhancement and enrichment migrate into the mainstream provision? One way of looking at the activities in enhancement and enrichment activities could be to see them as a means of curriculum development in which activities could be devised and piloted with pupils before transfer into the mainstream curriculum. A particular feature of some enhancement and enrichment activities which makes them attractive is the extent to which they allow those taking part to choose what they do. This can create problems when a syllabus requires certain features to be taught and pupils choose to do things that do not meet these requirements. However, it should be possible to run a mixed economy and provide significant choice at times and limited choice at others. Migration into the mainstream would in no way detract from the work of those currently engaged in supporting enhancement and enrichment. On the contrary, it could be argued that it would see their contribution to the curriculum having a more pervasive effect, concentrating on developing a curriculum with both appeal and intellectual coherence for all pupils, as opposed to a minority. Indeed, a useful intention for some enhancement and enrichment programmes would be to develop activities that could migrate into the mainstream and the evaluation criteria for such activities would be the extent to which this occurred. This approach could also take into account the recommendations of the Aspires Project.

Recommended reading

Archer, L., Moote, J., MacLeod, E., Francis, B., & DeWitt, J. (2020a) *ASPIRES 2: Young people's science and career aspirations, age 10–19*. London: UCL Institute of Education.

National Science and Technology Council (2018a) *Charting a course for success: America's strategy for STEM education*. www.whitehouse.gov/wp-content/uploads/2018/12/STEM-Education-Strategic-Plan-2018.pdf (accessed June 12 2020).

Nuffield Research Placements (2020a) www.nuffieldfoundation.org/students-teachers/nuffield-research-placements (accessed June 12 2020).

STEM Learning Impact Report (2019) http://magazines.stem.org.uk/stem-learning-impact-report-2019.html (accessed June 12 2020).

References

Altman, M. (2012) *Google+ post regarding his objections to MENTOR*. No longer available.

Archer, L., Moote, J., MacLeod, E., Francis, B., & DeWitt, J. (2020b) *ASPIRES 2: Young people's science and career aspirations, age 10–19*. London: UCL Institute of Education.

Banerjee, P. A. (2016) A longitudinal evaluation of the impact of STEM enrichment and enhancement activities in improving educational outcomes. *International Journal of Educational Research*, 76(1), 1–11.

Banerjee, P. A. (2017) Is informal education the answer to increasing and widening participation in STEM education? *Review of Education*, 5(2) https://onlinelibrary.wiley.com/doi/full/10.1002/rev3.3093 (accessed June 12 2020).

British Science Association (2020b2020a) *CREST Awards and Science Week*. www.britishscience-week.org (accessed June 12 2020).

British Science Association (2020b) *CREST Bronze Awards criteria* https://help.crestawards.org/portal/kb/articles/crest-bronze-criteria-guidance (accessed June 12 2020).

Cilauro, F., & Paull, G (2019) *Evaluation of Nuffield Research Placements: Interim report*. London: Nuffield Foundation.

Cognizant Centre for the Future of Work (2017) *21 Jobs of the future*. www.cognizant.com/white-papers/21-jobs-of-the-future-a-guide-to-getting-and-staying-employed-over-the-next-10-years-codex3049.pdf (accessed June 12 2020).

Cognizant US Foundation (2018) *Reskilling the US workforce*. www.cognizantusfoundation.org (accessed June 12 2020).

DeVos, B. (2019) *U.S. Department of Education advances Trump Administration's STEM Investment Priorities*. www.ed.gov/news/press-releases/us-department-education-advances-trump-administrations-stem-investment-priorities (accessed June 12 2020).

Dougherty, D. (2012) *Makerspaces in education and DARPA.* Makerspace.com. http://blog.makez-ine.com/2012/04/04/makerspaces-in-education-and-darpa/ (accessed June 12 2020).

EEF (2019) CREST report. https://educationendowmentfoundation.org.uk/public/files/Projects/Evaluation_Reports/CREST_Silver.pdf (accessed June 12 2020).

Engineering UK (2020a) www.engineeringuk.com/about-us/overview/ (accessed June 12 2020).

Engineering UK (2020b) *Information about the Big Bang.* www.engineeringuk.com/our-programmes/the-big-bang/ (accessed June 12 2020).

FIRST Lego League (2020) FIRST lego. www.first-lego-league.org/en/ (accessed June 12 2020).

F1 in Schools (2020) Formula 1 www.f1inschools.com/ (accessed June 12 2020).

Gartland, C. (2014) *STEM ambassadors and social justice in HE.* London: Trentham Books.

Great Bay SZ (2019) https://2019.igem.org/Team:GreatBay_SZ (accessed June 12 2020).

iGem (2020) *Synthetic biology iGem.* https://igem.org/Main_Page (accessed June 12 2020).

National Science and Technology Council (2018b) *Charting a course for success: America's strategy for STEM education.* www.whitehouse.gov/wp-content/uploads/2018/12/STEM-Education-Strategic-Plan-2018.pdf (accessed June 12 2020).

Nuffield Research Placements (2020b) *Research placements.* www.nuffieldfoundation.org/students-teachers/nuffield-research-placements (accessed June 12 2020).

Obama, B. (2009, April) *'Educate to innovate' Remarks made at the National Academy of Sciences Annual Meeting.* https://obamawhitehouse.archives.gov/issues/education/k-12/educate-innovate (accessed 28 September 2020).

Stock Jones, R., Annable, T., Billingham, Z., & MacDonald, C. (2016), *Quantifying CREST: What impact does the Silver CREST Award have on science scores and STEM subject selection?* A Pro Bono Economics research report for the British Science Association.

ROSE (2010b2010a) *Relevance of Science Education Project.* http://roseproject.no (accessed June 12 2020).STEM (2019) *Learning impact report.* http://magazines.stem.org.uk/stem-learning-impact-report-2019.html (accessed June 12 2020).

STEM (2020) *Information about the STEM Ambassador Scheme.* www.stem.org.uk/stem-ambassadors (accessed June 12 2020).

TeenTech (2020) *TeenTech Awards.* www.teentech.com (accessed June 12 2020).

Computing, digital competence, computer science, TEL and STEM

Introduction

This chapter is out of date. I am writing it in June 2020 in rural mid-Wales and I know that by the time you are reading this, many aspects of our lives and that of our pupils will have been affected, even changed, by further developments in 'new technology'. I will therefore set out here some cutting edge examples that I think will affect the teaching of STEM over the next few years, but also tackle some current teaching concerns in the use of digital media that are likely to be difficult to overcome.

First, however, to demonstrate the rate of development in this area it is worth setting out some general changes that have come about in the five years since the first edition of this book was written:

- Online shopping in the UK has risen from 13 per cent to 22 per cent of retail sales (ONS, 2019).
- Online banking in the UK has risen from 53 per cent to 73 per cent (Statistica, 2019). More than a third of UK bank branches have closed since 2015.
- The chief executives of the UK's four mobile phone networks have agreed to form a new company to help boost phone coverage in rural areas such as mid-Wales.
- In 2017, contactless cards accounted for 15 per cent of all payments, but UK Finance – which represents the major banks – has predicted this proportion to rise to 36 per cent by 2027. An estimated 3.4 million people hardly used cash at all during 2017.
- In 2018, 24 per cent of UK phone users pay by using a payment app. By 2022, it's estimated that the transaction value of mobile payment apps worldwide will reach nearly \$14 trillion.
- In 2018, 81 per cent of people booked their holiday online – in October 2019 Thomas Cook the world's oldest holiday company (founded 1841) went into liquidation.
- The data analytics firm Cambridge Analytica harvested millions of Facebook profiles of US voters in one of the tech giant's biggest ever data breaches, and

used them to build a powerful software program to predict and influence choices at the ballot box.

- The 2020 'lock-down' across the world due to the COVID-19 pandemic saw a massive increase in the use of social media for keeping families in touch with their friends and relations; and in the use of new technologies by teachers supporting their pupils when the schools were closed. Despite the increase in use, ISPs maintained internet speeds across the EU & UK and the connections stable.

- There is growing concern about the potential associations between social media use and mental health and wellbeing in young people (Viner et al., 2019).

- In 2018, pupils up to the age of 15 in France were banned from using their mobile phones during school hours after a law was passed prohibiting their use. In 2019, the Australian State of Victoria was the first state to ban the school use of cell phones from the first school term of 2020.

I was shopping in my local supermarket the other day and a mum was pushing a screaming small child in the seat at the front of the shopping trolley. The tantrum quickly faded, and I looked over to see how she had placated the child. He was holding her mobile (cell) phone and was mesmerised by a cartoon playing; after a short while, the two-year-old swept the icon to select another cartoon. We all shopped in relative calm, me going through the self-service check-out. As I scanned the barcodes, I pondered two things. The first was how almost everyone, even a very small child, uses a mobile phone every day. That led me on to the way they are rarely exploited in schools. The first time I used a computer in school was in 1969, the year of the first Moon landing, as a teleprinter terminal had been installed in the school where I was a senior pupil, connected to the 'main frame' computer at County Hall. I never saw that computer; it was the size of a room and guarded by the 'high priests', the computer technicians. As I typed at the terminal, it produced a roll of punched tape that listed the computer commands – as I had to programme the computer for what I wanted, which was mainly simple calculations, I learned the computer language FORTRAN to do so. When ready, the punched tape then ran through a 'reader' to give the commands to the 'main-frame' computer 25 miles away. That was then the norm – what was unusual was the Apollo 11 computer, which in contrast was a mere 1 foot (30 cm) cubed, using the first integrated circuits to shrink the size.

In 1976, my brother-in-law bought himself a Tandy TSR-80 personal computer and I remember laughing out loud about it. Why would anyone want their *own* computer? As we now know, he was a personal-computing pioneer, and just a decade after I sat at that teleprinter I, too, was using personal computers as part of my teaching – Commodore PETs with far less memory capacity at 8 Kb than the cheapest mobile phone in 2020. Programmes for the machine could be saved onto cassette tapes – and that is the significant point. All these early personal computers and the cheap home hobby computers such as the Sinclair ZX 80 through to the BBC Micro of the 1980s had to be programmed, although now in the more user-friendly BASIC computer language. Everyone who wanted to use a computer was learning programming in order to be 'computer literate'. It seems there really was a 'golden age' of personal computing where users could not only 'drive' the computer, they also knew in some detail 'what goes on under the bonnet'. The students then entering university computer science courses already knew something about computer architecture and programming, and all who had used a computer had some understanding of the basics of computer science.

Now that phone apps and Microsoft Office are so easy to download and use, some have questioned where the next generation of computer scientists will come from. However, we need to discuss how computers can be used by *everyone* in teaching and learning STEM, as well as how computing science itself can best be taught.

In this chapter we will consider:

- apps and the use of computers in teaching and learning the STEM subjects;
- digital competence and computer science in school;
- technology enhanced learning (TEL), what we might be using and where we might be going both in home and school learning through the almost ubiquitous use of new devices;
- social media and pupil wellbeing.

SOME DEFINITIONS

Computing
The broad subject area; roughly equivalent to what is often called IT in industry, as the term is generally used.

Computer science
The rigorous academic discipline, encompassing programming languages, data structures, algorithms, computer architecture etc.

Information technology (IT)
The use of computers, schools, in industry, commerce, the arts and elsewhere, including aspects of IT systems architecture, human factors, project management, etc. (Note that this is narrower than the use in industry, which generally encompasses computer science as well.)

Digital competence
The general awareness of and ability to use computers; a set of skills and understandings rather than a subject in its own right.

TEL (technology enhanced learning)
TEL is the support of any learning through the use of technology, so breaking down barriers of when and where one can learn and setting one's own level of the pace of learning. This is often done through the provision of a virtual learning environment (VLE).

(Adapted from Royal Society, 2012: 5)

Computing in teaching and learning the STEM subjects

Before we consider for what we might use computers – smartphones, iPads, laptops and notebooks – in our teaching, we need to pause and think through our beliefs about the relationship between the pupil and the teacher. Who is in control of the learning process? (See Chapters 6 and 10.) What are our attitudes to 'hands-on' skills and mathematical processes rather than computer simulation, calculator use and computer aided activities? Can augmented reality (AR) or virtual reality (VR) blur the distinction between 'hands on' and 'computer simulation'? What do we think of the use of smartphones being available during lessons?

TABLE 9.1 Models of learning with information technology

Instructional	Revelatory	Conjectural	Emancipatory
Drill and practise type programs.	Playing a simulation game or adjusting the conditions on a simulation experiment or using AR or VR.	Looking at a set of data and drawing conclusions.	Using the computer or mobile phone as a tool to do calculations or other labour-saving activities.
Using YouTube or other 'how-to' video sites to follow a technique or process. Use of QR codes to access information.	Amending a given design. Using an AR 'cube' to manipulate a simulated 'environment'.	Modelling and testing a hypothesis. Trying out a possible new design – CAD.	Data capture, using STEM apps, word processing, constructing graphs. CAM linked to 3D printing.
Computer leads the learning.	⟵————————————————⟶		Pupil leads the computer.

Source: Kemmis, Atkin and Wright, 1977, adapted

One of the most useful ways of thinking about these matters, although suggested so long ago, was by Kemmis, Atkin and Wright in 1977! Their ideas applied to information technology is set out in Table 9.1

In most schools when IT is used we can list several teaching modes that are currently in operation:

- Mode 1: as a tool for demonstrating and illustrating, e.g. using an electronic white board or data projector and screen.
- Mode 2: a computer/smartphone as part of a circus of activities or as a when-needed support to class activities.
- Mode 3: with half a class sharing and discussing around a few tablet computers.
- Mode 4: with a whole class using a set of tablet computers or smartphones in the classroom.
- Mode 5: independent use (e.g. at home, in the library, or the learning resource centre) which might link to AI developments.

Although mobile phones are so ubiquitous and access to information so easy that even a 'pub quiz' is prone to the teams illicitly searching for answers, we need to remember to consider the intended learning objectives of any lesson and the way that IT can support or detract from that learning. We also need to face up to the way that smartphones can be an enormous benefit in STEM activities but, if their use is not managed properly, they can also be a severe distraction. As we noted above, some countries and states have completely banned mobile phones from schools. Other places, such as the UK have left it to each individual school to decide on its own policy and consequently the pattern of use is mixed. Many UK schools impose a France-like total ban, some allow use of phones during break-times, but some schools such as Portsmouth High School in England and Ysgol Uwchradd Caergybi, Holyhead, in Wales, have now relaxed the ban and use phones as a 'powerful educational tool'. Head teacher Adam Williams at Ysgol Caergybi said he had not seen any deterioration in pupil behaviour or wellbeing, but he recognises that the education of pupils in the appropriate use of social media is a responsibility of schools. There are times when we would wish pupils to work unaided such as when searching for patterns, doing mental arithmetic or practicing some manual skills. But with blanket bans on the use of calculators or mobile phones at school, state or country level, we seem to be in danger of cutting off school learning from everyday life. A camera on a mobile phone can keep a record of project work or record ideas; or capture group work using the audio record function as well as connecting to the internet to look up information and receive stimuli for new ideas. Everyone can now have a 'library in their pocket' but also hold a device that can interact with and make measurements on their environment too.

It is not only the *way* the computer or smartphone is used as a learning tool that needs to be thought through, but also *how* the use of an app or computer programme is taught. For example, taking pupils through all the different possible commands of a design package in a lock-step manner is very different from allowing the pupil, working in pairs, to explore different possibilities using supporting tutorial videos as and when needed. However, when even in a restaurant it is not unusual to see two diners looking at their phones rather than talking to each other, the ubiquitous 'library in the pocket' could be a classroom distraction rather than a support. We will consider this later, but first let us consider some specific uses of apps in contributory STEM subjects, although the apps described could be used across STEM.

Using phone apps: Science

As we have already discussed in earlier chapters, process skills are very important in science and many science teachers cling to a 'seeing is believing' principle of practical work. I am a firm believer in 'doing' science too, and new techniques can change the ways that experiments are carried out to vastly improve learning. The mobile phone is indeed a 'powerful education tool' and it is worth pointing out the different sensor inputs that are available for experiments – some rather surprising! All smartphones have the first six main sensors listed below as they are what enable it to give you the usual expected touch-enabled functionality, but nearly all smartphones have the other sensors too:

1 **Microphone:** As it is a smartphone, there is obviously a microphone available.

2 **Accelerometer:** Used by apps to detect the orientation of the device and its movements, as well as allow features like shaking the phone to change music.

3 **Gyroscope:** Works with the accelerometer to detect the rotation of your phone, for features like tilting phone to play racing games or to watch a movie.

4 **Digital compass/magnetometer:** Helps the phone to find which way is north for use with maps.

5 **Ambient light sensor:** This sensor alters the brightness of the screen automatically depending on the ambient light level, which helps save battery life and when surrounding light level is low it helps to reduce eye strain.

6 **Proximity sensor:** During a call, when the phone is near your ear, it automatically locks the screen to prevent unwanted touch commands.

7 **GPS:** Global Positioning System (GPS) units in smartphones communicate with the satellites to determine our precise location on Earth.

8 **Touchscreen sensors:** The screen sensors in a touchscreen have an electrical current passing through them at all times and touching the screen causes a change in the signals to give an input.

9 **Fingerprint sensor:** This is like the screen sensor where the ridges in your fingerprints touch the surface whereas the hollows between the ridges have a slight separation. When stored in the phone, only you can access it. This is useful in apps that require authentication such as mobile payment apps.

10 **Pedometer:** The pedometer is used for counting steps, and generally use the values generated by the accelerometer to monitor your movements like running or walking.

11 **Barcode/QR code sensors:** Most smartphones have barcode sensors that can read a barcode by detecting the reflected light from the code. Barcode sensors are useful in scanning the barcodes products or QR codes.

12 **Barometer:** The barometer measures the air pressure, so it is quite useful in detecting weather changes and in calculating altitude.

13 **Heart rate sensor:** The heart rate sensor that measures heartbeat. An LED emits light towards the skin, and this smartphone sensor looks for the light waves reflected by it. There is a difference in the light intensity when there is a pulse and the heartbeat is measured by counting the changes in light intensity between the minute pulsations of the blood vessels.

14 **Thermometer:** Every smartphone comes with an inbuilt thermometer. If the phone overheats the system shuts down itself to prevent any damage.

15 **Air humidity sensor:** This sensor measures the humidity in the air, and the data collected by it would tell the user whether the given air temperature and humidity are optimum.

16 **Geiger counter:** This rather specialised app was first introduced to smartphones in Japan. The Geiger counter can measure the current radiation level in the area.

(Adapted from Fossbytes, 2018)

CAMEO 1: FREE FALL AND THE USE OF THE PHYPHOX APP

The free 'phyphox' app, developed by RWTH Aachen University in Germany, gives access to the sensors of your phone either directly or through ready-to-play experiments that analyse your data and let you export raw data to a laptop along with the results for further analysis. It is possible to measure the duration of free fall using your smartphone and an acoustic stopwatch available in phyphox. The acoustic stopwatch starts with a loud noise and stops at a second loud noise.

In this experiment arrangement, a balloon is set up at a height S to carry a weight that can fall onto a metal plate (see Figure 9.1). A pin pops the balloon to give the first sound that starts the acoustic stopwatch and the weight falls freely onto a metal plate which both protects the floor and gives the second loud sound to stop the watch. The time t between the two sounds is how long the weight took to fall the distance S. $S = ut + 1/2\ gt^2$ where g is the acceleration due to gravity. As it falls from rest $u = 0$ so $S = 1/2\ gt^2$. Re-arranging $2S/t^2 = g$ and g can be therefore be calculated and averaged for a range of heights.

FIGURE 9.1 Free-fall experiment

The phyphox app allows access to the following sensors:

- accelerometer;
- magnetometer;
- gyroscope;
- light intensity;
- pressure;
- microphone;
- proximity;
- GPS.

The barometer sensor can measure air pressure and is so sensitive it can follow the change in air pressure as a lift (elevator) ascends. A word of caution is necessary when using sensors and data-loggers. Sound experimental technique is, obviously, still necessary – so stirring a solution after adding reagents before (say) measuring the temperature is important whether by a mercury-in-glass thermometer or a digital probe.

Computing in science lessons

Considering putting the pupil at the centre of their science learning, Table 9.2 illustrates science activities and possible IT tools to support that learning:

TABLE 9.2 Science activities and possible IT tools

Pupils' science activity	What IT tools will help?
Planning an investigation	Flow charting software; word processing.
Researching/learning about a topic	Internet, e.g. You Tube, Wikipedia, online tutorial, databases.
Taking measurements	An app such as phyphox to enable the use of smartphone sensors plus data-logging software.
Making results tables	Spreadsheets.
Drawing graphs	Data-logging software, spreadsheets, databases.
Doing calculations	Spreadsheets, data-logging software.
Searching for patterns	Spreadsheets, databases, simulations, modelling programs.
Asking 'what if…?' questions	Simulations, databases, modelling programs, augmented reality apps.
Comparing pupils' results with other people's (reviewing a topic)	Social media sites.
Presenting information in a report	Word processing, desk-top publishing, spreadsheets.

Using phone apps: Design & technology

The Merge Cube (2020) as shown in Figure 9.2 allows users to physically hold and interact with 3D objects using augmented reality (AR) technology. It is made out Styrofoam and is about 12 cm on each edge. A smartphone app (some are free, some bought) uses the phone's camera to enable a pupil to look through the screen at the patterned cube, which triggers a range of 3D images such as the human body or the solar system. As they hold and manipulate the cube in the palm of their hand, they can look at different body parts – or different planets.

In design & technology, a pupil can design an artefact using a 3D design programme such as Google Poly or SketchFab and upload them to the Merge Cube web portal. It supports a range of file formats (.fbx, .obj, .stl, .dae, .blend, and .gLTF) but there is a 100 Mb file size limit. An 'Object Viewer' app will then enable the pupil to

FIGURE 9.2 The Merge Cube

'hold' and manipulate what they have designed as they view it in three dimensions by physically turning and inspecting their own design, which is 'duplicated' by the Merge Cube. Pupils can make amendments to their design before creating it on a 3D printer.

Computing in design & technology and engineering lessons

Looking at Table 9.1, using IT can be used in design & technology and engineering for:

- Context exploration
  - o Use of word processing packages and presentation software to create questionnaires
  - o Use of digital photography to capture contexts
  - o Use data-logging equipment to carry out preliminary investigations
- Idea generation
  - o Use of scanners to capture 3D form
  - o Use of software to support development of brainstorms, mind maps and spider diagrams
- Idea development
  - o Use of software to develop surface decoration
  - o Use of CAD software to develop ever more detailed digital representations of design ideas providing accurate descriptions of both form and performance
- Idea communication
  - o Use of social media to enable communication with others
  - o Use of Photoshop software to develop detailed realistically rendered digital presentations of design proposals
  - o Use of spreadsheet data to provide performance data in both table and graphical form
- Planning
  - o Use of flow chart software and GANNT chart software
- Manufacture
  - o Use of CAM software to drive dye sublimation printers, vinyl cutters, engravers, laser cutters, CNC lathes, milling machines and routers, and 3D printers
- Control
  - o Use of programming software to embed instructions in programmable products and systems (see Table 9.3).

These activities span the spectrum from instructional to emancipatory. A food probe can record the temperature profile when making bread or melting chocolate or producing jam to prevent scorching – an IT version of what could be done by traditional means but easier – or a control programme can be written to automatically control the windows of a greenhouse, for example.

As CAD/CAM programs and equipment become more and more affordable for schools, how should we balance the new skills of using computer support for design and manufacture with the development of pupils' psychomotor skills that are promoted through basic hand and machine tools? It is now possible to define a design that

TABLE 9.3 Some members of the online 3D printing community – weblinks available in the references

Name	Some features – taken from website
Shapeways	*Tutorials* *Printing* in up to 25 materials including metals, ceramics and glass *Uploading and ordering designs*
Ponoko	*Choose and buy a product.* The designer will order the parts from Ponoko and send it to you direct.
Making outlets in five countries	*Choose and buy a design* and download the files to your PC. You can then customise it if the Creative Commons copyright licence allows. When you're ready to make it, upload the files to your My Ponoko account, select the materials, and get an instant price for us to make it and deliver it to your door.
i.materialise Headquarters in Leuven Belgium plus branches worldwide	*3D print your designs* Upload your 3D design and instantly see the price for your models. ● no login necessary; ● choose from a large selection of materials and colours; ● scale your model to the ideal size; ● order as many copies as you want. *Sell your designs* Want to show off your design talent and make some money at the same time? Offer your designs for sale in our gallery and once a month, we will pay you a fee for every one of your items sold.
Thingiverse	*Digital designs for a wide range of objects* Examples of a wide range of things made via 3D printing Examples of tools in the following categories: automated, clamping, crafting, cutting, electrical, hand, measuring and power
Creativity Essentially a trading site for a variety of digital media – pictures, models, music and software	A suitable platform for trading of digitalised 3D objects

is then produced by a computer-controlled machine, just as in industry, to an accuracy a pupil could rarely achieve themselves. And as the software improves in its usability, the time invested in becoming competent shortens and outcomes can become progressively more sophisticated. However, it will always be a balance between investing in what is appropriate to fabricate in school and what can be better made using the global network of FabLabs. The concept of the FabLab emerged from MIT's 'Center for Bits and Atoms'. There, the goal was to 'provide access to the tools, knowledge and the financial means to educate, innovate and invent using technology and digital fabrication to allow anyone to make (almost) anything, and thereby create opportunities to improve lives and livelihoods around the world' (Johns, 2018: 3). There are well over 1,500 FabLabs in over 100 countries which give access to a 'maker' environment and manufacturing equipment that is at the cutting edge of what is available.

In addition to FabLabs, an online 3D printing community has emerged and some members of this community are shown in Table 9.3. Barlex and Stevens (2011) suggest that it is possible that pupils could use such organisations in two ways:

■ developing simple digital designs for products, which the organisation prints and then sends to them by post – the school and/or pupils would need to pay for this service;

■ developing simple digital designs for products that the organisation prints and sends by post to others who are prepared to pay for the design – the school and/or pupils would make money from this service.

Although many schools possess a 3D printer, in neither of these cases is it necessary for the schools to do so. The pupils could be involved in designing but not making, although it would be possible for them to acquire made versions of their designs. The making function would have been devolved to one of any number of online 3D print bureaux. If such a 'designed by me but made by somebody else, somewhere else' practice became widespread in England, for example, it would be a major disruptive departure from the prevailing school practice where pupils are invariably required to make what they have designed.

Therefore, pupils can now use CAD to develop designs that they could not make using traditional 'school making skills' but which they can realise using the latest manufacturing techniques either in school or accessed externally in places such as Fablabs or 3D print bureaux. That pupils can now design and make artefacts that would be difficult to achieve in a school environment is no doubt a considerable step forward in design & technology learning.

However, it is worth noting that CAD is not a substitute for 'designerly imagination' and intuition. There is the problem that the nature of the software and what it can do easily may overly influence the nature of any resulting design and limit the creativity of the designer. Experienced designers have reported that they leave the use of CAD until as late as possible in their designing to avoid this problem (Carr, 2015). In addition, the continually changing software for CAD presents challenges to pupils and teachers alike. Dr Debi Winn (2012: 6), head of faculty at a school in Cambridgeshire, UK notes:

> Teachers often struggle to learn the programmes themselves and as teaching the programme is only a small part of the curriculum a limited time is allowed for training. This often restricts the teachers' knowledge to the basic commands and so when trying to teach a class of students and problems occur, the teacher is often unable to solve them. This is especially so if a length of time has passed between the teacher last using the program or the program has been updated. This is frustrating for both the students and the teacher, and because of this teachers can sometimes avoid teaching the more difficult CAD software. This problem is further compounded when one considers the way CAD is often taught to students. The 'traditional' method of teaching involves the entire class following either a written set of instructions or a video clip in order to make identical products at the same pace in a 'lock-step' manner. Those students that pick up the commands quicker become bored whilst they wait for the others to catch up and those that experience problems are waiting for help, which in a large class can be a several minutes. This restricts progress for both of these groups of students. This style of teaching is demotivating for both the teacher and the students and does not encourage either to take risks.

To tackle this problem, Debi worked with the pupils to design a computer adventure game based on 'wizards' and requiring the pupils to use CAD to make their own items for the game such as 'keys', 'drinking cups' and, finally, 'a castle'. Working in pairs, this more 'strategic' approach was shown to be much more successful than the traditional 'lock-step' learning of programme commands in producing different and more creative idea.

CAMEO 2: MY SCHOOL HAS BEEN COLLABORATING WITH PUPILS ON THE NAVAHO RESERVATION IN ARIZONA, USA.

Using video-conferencing, we exchanged ideas about designing and manufacturing products and discussed different preferences that brought out the importance of considering the values of the client and the maker. They had access to a laser cutters and 3D printers, as did we, and we exchanged files of our ideas as well as producing some products jointly. Most interestingly, the Americans had the idea of including as a motif a good luck spirit image of a 'mustang'.

(Adapted from Learning Schools Open Educational Resources)

Physical computing in design & technology lessons

In Table 9.4 below, Torben Steeg attempts to capture the various strands of progression for school pupils working in design & technology with systems that enable the designing of artefacts that include embedded intelligence or control. Only hardware and software is listed that interfaces with real hardware (simulations are excluded) and that can programme external hardware (so systems where a PC is the controller are excluded). In other words, Torben sees the tables as centred on systems that enable the designing of real-world artefacts that include embedded intelligence, which is the domain of modern digital design & technology. He gives a word of caution, although some aspects are easier to use than others it really is 'horses for courses' and some 'easy to use' software may be entirely appropriate for both a primary classroom and an undergraduate project.

The progression in difficulty of software and in program concepts is from left to right.

Some possible 'physical control' projects suggested are:

- Design and make a device that can explore the environment in a small stream.
- Design and make a small weather station that can collect data concerning temperature, pressure, light levels and rain fall.
- Design and make a plaything to engage and amuse young children on a long car journey.
- Design and make an electronic dice to be used in a snakes and ladders game to be played by children aged between four and six years.
- Design and make a device that will keep small valuable items at home safe from theft.
- Design an anti-theft system to be installed in a small jewellery box.
- Design and make a communication device that utilises the ability to receive and transmit infrared signals.

TABLE 9.4 Progression in aspects of physical computing – controlling artefacts

Software progression	Blocks: Crumble, Blockly for PICAXE, EduBlocks for micro:bit, ArdBlock for Arduino, MakeCode for Mindstorms	Flowcharts: Genie, Flowol, PICAXE, Flowcode Arduino	Other graphical: Kodu for micro:bit, LabVIEW and Robolab for Mindstorms, Minibloq for Arduino	High level language: Python, Circuit Python, C++, C, BASIC	Assembler/machine code (specific to each microcontroller)
	These are broadly equivalent in difficulty				

Device Family progression	Crumble	Micro:bit	Mindstorms	Genie	PICAXE	Arduino	Raspberry Pi*	mbed
Programming concepts	Sequence with waits (e.g. controlling a digital output); Unconditional loops	Branching (e.g. If … then … else … for responding to a digital input); Boolean branching (AND, OR, NOT)	Integer variables (e.g. responding to an analogue input); Subroutines	Repeat using a variable	Arithmetic operators	Interrupts; Variable types	List processing	Indexing and table lookup (Arrays)
Hardware	Simple digital outputs (LEDs, buzzers); Music actuator (Piezo sounder); Intelligent output (e.g. Neopixel/Sparkle)	Digital sensors (switches, analogue sensors read digitally); DC Motor control; Analogue sensors (ADC, e.g. temperature, light, hall effect)	Analogue output control (e.g. PWM)	Data providing sensors (e.g. ultrasonic distance, accelerometer, compass)	Bluetooth communication	Multiplexed inputs (e.g. keypad)	Multiplexed outputs (e.g. LED matrix)	Communication protocols (e.g. internet communication)

* The Raspberry Pi is really a single-board-computer rather than a microcontroller system (see below).

- Design and make a device that enables parents to listen in on a sleeping child to ensure they are breathing normally and not in distress (Barlex, Gardiner, & Steeg, 2011: 3).

Using phone apps: Mathematics

Quick response codes (always referred to as QR codes) are the two-dimensional barcodes that are often seen, for example, on exhibit cards in a gallery or a museum. The code contains data that points to a website or an application. In the museum, therefore, scanning the QR code using a smartphone with a QR reader could open a website with more information or perhaps a YouTube video showing the archaeological dig where it was discovered. Obviously, the code can be linked to any website, and it is very easy to generate QR codes for free using a QR Code Generator (2020). Wyn Owen in the mathematics department at Ysgol Uwchradd Caergybi uses QR codes that link to support websites to help 11-year-old pupils with homework and to YouTube videos to support 16-year-old pupils with revision. Although, like in other subjects, pupils are allowed to use or share their smartphones in class, the maths department also use laptops/iPads too (QR Codes in Maths, 2020).

Computing in mathematics lessons

Just as we have seen in our consideration of the examples in other STEM subjects, IT can be used to help us teach more efficiently but doing activities that have been part of the subject for many years; it can expand the possibilities of what can be taught; and it can transform what and how we teach. For example, teaching the relationship between the equation $y = mx + c$ and its graph can be done with a pencil and ruler and a pile of graph paper, but graph plotting software or a graphic calculator could allow many more possibilities to be investigated in the time available. In teaching statistics, a 'revelatory' opportunity is possible (see Table 9.1 above). Is a dice loaded? If one is thrown 100 times and there are 25 sixes, is the dice fair – could that just be chance? Using a simulation programme, a pupil could re-run the number of sixes in 100 throws many times and produce a frequency chart. From that she could consider how often she might see as many as 25 sixes in a fair dice. This process would be very tedious if done manually. Table 9.5 shows mathematics activities and possible IT tools to support that learning.

You may have noticed how many times the word 'explore' was used in the table suggesting a teaching of mathematics that is focused on investigating and experimenting with numbers. It suggests a 'trial and error' approach where a pupil can try something and the software will provide feedback that reflects what they have done – non-judgmental and impartial. I think some of the dislike pupils sometimes express about mathematics comes from what is perceived as a wholly right or wholly wrong outcome. Using IT takes away the fear of public failure – if it does not work, just try again and there is no external humiliation. The mathematics teacher associations have for many years suggested that pupils have an entitlement to learn using IT by:

TABLE 9.5 Mathematics activities and possible IT tools

Pupils' mathematics activity	What IT tools will help?
Explore the shape of families of graphs such as $y = a(x-b)^2 + c$ either on a tablet or graphical calculator.	Graph-plotting software
Explore number patters; find optimum solutions; solve equations numerically and graphically; investigate sequences and iteration; display statistical information on charts.	Spreadsheets
Explore geometric transformations; construct geometric figures; study relationships through measuring co-ordinates, lengths, angles and areas; construct loci; develop ideas of invariance and dependency.	Dynamic geometry software
Taking out the reparative calculations often associated with statics to focus on the important statistical ideas. Specialist statistical software is often more powerful than needed at school level, but spreadsheets can be used for a range of statistical manipulations such as cross tabulation.	Statistics software
Manipulate algebraic functions, arithmetic, data handling and matrices and 3D plotting.	Algebra software
On screen or using a physical floor rover or 'turtle' or AR Cube to explore shape and position; develop the ideas of a function and a variable; learn about algorithms.	NetLogo or Scratch or Merge Cube
Focusing on the mathematics rather than the calculation.	Computation software
Independent study of topics and revision using online tutorials.	E-mail and the WWW

Source: Adapted from Richardson and Johnson-Wilder (1999) updated

- learning from feedback;
- observing patterns;
- seeing connections;
- working with dynamic images;
- exploring and generating data;
- sequencing logical steps.

I was in a mathematics classroom in Uttar Pradesh in India watching the teacher drawing on a rather old blackboard as his class of about 20 15-year-old boys sat on the floor. The topic was about calculating angles within a circle. I marvelled as he drew freehand, perfectly, circle after circle without any aids – just using the rather crumbly chalk. Many teachers in the West would now use a whiteboard and software like Cabri to demonstrate a range of geometric shapes in two or three dimensions. As is set out in Table 9.5, what was once done using paper, ruler and compasses can now been done much quicker and with many more iterations using such dynamic geometry software.

The same issues return. Should the use of free-hand sketches be preserved rather than use of CAD in engineering; what is the place of 'hands-on' science over simulations; what is the role of calculation – including mental arithmetic in mathematics education? In all cases there are those who argue for using IT much more in schools and those who regret the passing of some of the traditional hand skills. We return to this topic as we look to the future of STEM in Chapter 12. However, here we turn to the place of computer science in schools as a replacement for what has become in much of the UK the discredited subject of 'ICT'.

Digital competence and computer science in schools

In the early twenty-first Century, the lead that the UK had in the 1980s in home computing and computer coding was in danger of slipping away. In schools, 'ICT' – information and communication technology, which concentrated on the use of office software – had become dominant, including what one pupil described as 'the boring bits of my Mum's job'. Moves by The Royal Society and the UK government changed this focus on just Microsoft Office and also highlighted a number of problems in introducing computing in school that are also true in countries across the world:

■   teaching by unqualified staff;
■   the problem with keeping school computers up to date;
■   a national (or state) curriculum that due to its inevitably 'fixed' syllabus finds it difficult to be 'future proof'.

The Royal Society's main argument for curriculum change was similar to that of the teaching of science – it needs to serve the needs of the citizen as well as the future scientist. The school IT curriculum needs to be appropriate for all citizens too so that they have appropriate digital competence, but it should also include an entitlement for all pupils to be able to study computer science.

Digital competence

The 'Curriculum 2022' in Wales sets out a Digital Competence Framework. This is a useful cross-curriculum model that all countries might consider for what pupils should study to gain what some call 'digital literacy' and covers much of the general aspects of computing in the STEM subjects mentioned above. The example below is taken from the suggested curriculum for those aged 11–14 years.

Citizenship – Through the elements shown in Table 9.6 learners, will engage with what it means to be a conscientious digital citizen who contributes positively to the digital world around them and who critically evaluates their place within this digital world. They will be prepared for and ready to encounter the positive and negative aspects of being a digital citizen and will develop strategies and tools to aid them as they become independent consumers and producers.

Interacting and collaborating – Through these elements learners will look at methods of electronic communication and know which are the most effective. As shown in Table 9.7, Learners will also store data and use collaboration techniques successfully.

TABLE 9.6 'Citizenship' in Digital Competence Curriculum 2022 Wales

Elements of 'citizenship'	Selected examples from curriculum document
Identity, image and reputation	I can understand how to protect myself from online identity theft, *e.g. identifying secure sites, phishing, scam websites.* I can understand that I have a digital footprint and that this information can be searched, copied and passed on. I can recognise the risks and the uses of data/services on personal devices, within the terms and conditions of a range of software and web services…
Health and wellbeing	I can demonstrate healthy online behaviours and identify unacceptable behaviour. I can identify ways of reporting unacceptable online behaviour. I can take reasonable steps to avoid health problems caused by the use of technology and suggest strategies to prevent or reduce the problems, both physical and psychological.
Digital rights, licensing and ownership	I can understand copyright and can explain the legal and ethical dimensions of respecting creative work, *e.g. exploring the ethical and legal ramifications of piracy and plagiarism and know that they are irresponsible and disrespectful,* and I can apply my understanding of the rules and regulations to different scenarios.
Online behaviour and cyberbullying	I can understand the implications of online actions, including my digital footprint and the legal implications of sharing inappropriate material. I can apply appropriate strategies to protect the rights, identity, privacy and emotional safety of both myself and others in online communities.

TABLE 9.7 'Interacting and collaborating' in Digital Competence Curriculum 2022 Wales

Elements of 'interacting and collaborating'	Selected examples from curriculum document
Communication	I can select and use different online communication tools for specific purposes with higher levels of competence, *e.g. set up and manage an address book, organise contacts, use advanced features of e-mail provider (signature, auto reply, read receipt, widgets).*
Collaboration	I can independently select and use a range of online collaboration tools to create a project with others in one or more languages, *e.g. making use of online technology to share and present ideas to others.*
Storing and sharing	I can independently select and use a range of online collaboration tools to create a project with others in one or more languages, *e.g. making use of online technology to share and present ideas to others.*

Producing – These elements cover the cyclical process of planning (including searching for and sourcing information), creating, evaluating and refining digital content. Although this process may apply to other areas of the framework, what is illustrated in Table 9.8 is of particular importance when creating and producing digital content. It is also essential to recognise, however, that producing digital content can be a very creative process and this creativity is not intended to be inhibited. Digital content includes the production of text, graphics, audio, video and any combination of these for a variety of purposes. As such, this will cover multiple activities across a range of different contexts.

TABLE 9.8 'Producing' in Digital Competence Curriculum 2022 Wales

Elements of 'producing'	Selected examples from curriculum document
Sourcing, searching and planning digital content	I can independently use a range of complex searches, *e.g. and/or/+/-/not.* I can search efficiently for information for my digital work and evaluate the reliability of sources of information, justifying opinions and reasons for choices, and I can reference work using appropriate methods.
Creating digital content	I can select and use a variety of appropriate software, tools and techniques to create, modify and combine multimedia components for a range of audiences and purposes such as: • text and images, *e.g. explore and use effectively image manipulation techniques; explore and use appropriately the many aspects of document layout; use animation, video and audio effects such as echo, tempo, envelope, layering, frame rate, key frames.* • presentation, *e.g. use design tools; adapt themes and colours to suit the purpose; create master templates.*
Evaluating and improving digital content	I can suggest and make improvements that are relevant for audience and purpose, based on feedback and self-evaluation of my digital work.

Data and computational thinking – Computational thinking is a combination of scientific enquiry, problem-solving and thinking skills. Before learners can use computers to solve problems they must first understand the problem and the methods of solving them. Through these elements shown in Table 9.9, learners will understand the importance of data and information literacy; they will explore aspects of collection, representation and analysis. Learners will look at how data and information.

Many countries are looking again at their school IT curriculum. For example, 'digital technologies' was added to the New Zealand curriculum in 2018 but it is also, rather confusingly, referred to as computing, computational thinking, computer science or coding. The 2019 'State of Computer Science Education' report in the USA showed that 45% of states teach computer science, although as each state can define the subject, again it is not always clear what is included under the computer

TABLE 9.9 'Data and computational thinking' in Digital Competence Curriculum 2022 Wales

Element of 'Data and computational thinking'	Selected examples from curriculum document
Problem solving and modelling	I can identify the different parts of an algorithm to determine their purpose. I can develop logical solutions to determine the input, outputs and processes of a program, *e.g. following pseudocode or a flowchart to come to an outcome, developing a written sequence of steps that could be followed.*
Data and Information literacy	I can create a data capture form, capture data, search data and create a database and spreadsheet with appropriate data input method. I can use my data to explain and add validity to conclusions and, where possible, modify conclusions and/or hypothesis

science curriculum banner. Israel undertook a review in the 1990s and around 20,000 students there now study computer science. In Lithuania, Finland, Korea and Japan, initiatives and policies were made to introduce the development of computational thinking skills and programming in their schools. The UK has had a lead in computing since the pioneering code-breaking work of Turing and others during the World War II; and the periodic investment in school computing has led to an expertise in video games and cinema visual effects and a wider exploitation of IT for industry. Following the Royal Society report there was a re-launch of computer science in schools in England.

Computer science in school

Taking the Royal Society Report as a starting point, a computer science school curriculum was devised for England with the following topics:

Pupils aged 11–14 years should be taught to:

- design, use and evaluate computational abstractions that model the state and behaviour of real-world problems and physical systems;
- understand several key algorithms that reflect computational thinking [for example, ones for sorting and searching]; use logical reasoning to compare the utility of alternative algorithms for the same problem;
- use two or more programming languages, at least one of which is textual, to solve a variety of computational problems; make appropriate use of data structures [for example, lists, tables or arrays]; design and develop modular programs that use procedures or functions;
- understand simple Boolean logic [for example, AND, OR and NOT] and some of its uses in circuits and programming; understand how numbers can be represented in binary, and be able to carry out simple operations on binary numbers [for example, binary addition, and conversion between binary and decimal];
- understand the hardware and software components that make up computer systems, and how they communicate with one another and with other systems;
- understand how instructions are stored and executed within a computer system; understand how data of various types (including text, sounds and pictures) can be represented and manipulated digitally, in the form of binary digits;
- undertake creative projects that involve selecting, using, and combining multiple applications, preferably across a range of devices, to achieve challenging goals, including collecting and analysing data and meeting the needs of known users;
- create, re-use, revise and re-purpose digital artefacts for a given audience, with attention to trustworthiness, design and usability;
- understand a range of ways to use technology safely, respectfully, responsibly and securely, including protecting their online identity and privacy; recognise inappropriate content, contact and conduct and know how to report concerns (DfE, 2013).

Interestingly, the Royal Society Working Group consider that computer science with its mathematical foundations, its scientific approach to experimentation, its design, construction and testing of artefacts and its use of a range of technologies is 'a quintessential STEM discipline, sharing attributes with Engineering, Mathematics, Science, and Technology' (CSWG, 2012: 4).

Raspberry Pi and BBC micro:bit

In the third decade of the twenty-first century, what is the equivalent of the cheap hobby computers that were such a stimulus to budding computer programmers 40 years ago? In 2006, Ebden Upton and his colleagues at the University of Cambridge decided there was a need for a small and cheap computer for young people. Having left the university, and working in his spare time, Ebden took three years to create the Raspberry Pi. The Raspberry Pi as illustrated in Figure 9.3 is a credit-card sized computer that plugs into a TV and keyboard and costs about £25 ($35). In 2020, it is in its fourth generation. The speed and performance of the Raspberry Pi 4 is a step up from earlier models and now a complete desktop-style experience. A user can edit documents, browse the web with a bunch of tabs open, juggle spreadsheets or draft a presentation. But the Raspberry Pi can be programmed using Scratch or Python and pupils are able to make cartoons and games and use computing concepts in practice. It has two USB 2 ports and two USB 3 ports for fast data streaming, a gigabit Ethernet port for network connection, wireless working and Bluetooth and is capable of Blu-Ray quality playback. It is booted up from a SD card and can be powered from the mains or four AA batteries.

Figure 9.4 shows a similar 'credit card sized' computer, the BBC micro:bit. It was launched in 2015 as a response for a cheap £12 ($15) computer that can respond to the need for computer coding in UK schools. It has also been taken up in schools in Finland, Iceland, Singapore and Sri Lanka. The device can be used for teaching robotics and games. It has LED outputs on the board, and it can be coded from any web browser using Blocks, Javascript, Python and Scratch. The board has accelerometer and magnetometer sensors, and can be connected by Bluetooth or by use of a USB port.

The launch of these small and cheap computers combined with the new push for computer science in schools around the world has created enormous interest. The impetus for the development was to see cheap, accessible computers back in the hands of young people everywhere. With free open source software available, a new wave of computer programmers may start to enter higher education. Free software and training in the use of Python and computer languages are available from the Raspberry Pi Foundation (2020) and from Micro:bit (2020).

FIGURE 9.3 Raspberry Pi 4

FIGURE 9.4 BBC micro:bit

Technology enhanced learning (TEL)

In many schools in Europe, the UK and the USA, the chalk-board has given way to the interactive whiteboards connected to the internet, which can bring audio–visual materials into the classroom on a daily basis. But in many ways the pedagogy has not fundamentally altered. In fact, despite the name it has sometimes *reduced* classroom interaction and active learning opportunities, and in some cases locked teachers into nineteenth-century exposition pedagogy. However, the speed and use of audio inter-action such as Echo, Assistant or Alexa will change the interaction between teachers and pupils and their computers. We have reached a 'tipping point' recently in what we can expect students to have in terms of access to computing technology and we need to revisit the issue of 'mobile phones in school'.

It seems that there are the first signs of a relaxation of the prohibitions of pupils using their own ICT devices such as smartphones, tablets or netbooks in school. In the past, schools have provided the necessary hardware and controlled IT use in all aspects of pupils' learning. If pupils do 'bring their own device', what are the implications? Teachers considering this future have suggested:

- schools may need to cope with diverse student-owned devices, develop strategies for this and employ staff who can help;
- a possible shift to less interventionist pedagogies or 'minimally invasive education' (a term linked to self-organised learning environments, discussed below);
- all teaching staff need to develop knowledge of a range of common devices and understand their capabilities and limitations;
- a shift away from the external agency or local government authority model of provision;
- a requirement for social based networks for teachers in all disciplines.

(Berry et al., 2012: 7)

If the use of smartphones and other devices are encouraged in school, rather than banned, there is a need for some changes in attitudes to authority in the classroom. Although they are proficient at 'pressing the buttons', pupils will need to be taught to understand much better ideas of information reliability and the concept of plagiarism as suggested by the Digital Competence Framework described above. Project work can be set that expects the learner to generate their own learning content when the carry a 'library in their pocket'. Pupils who are able to use social network sites to link with other pupils and schools with other schools can create a learning community of pupils and of teachers too. While being realistic about not wanting to overload the pupils, and being careful about the need to monitor and protect young people when given access to the internet, the technology enhanced learning (TEL) report (2012: 4) notes the following benefits:

- helping children to learn in and out of school, through activities that start in the classroom and then continue in the home or outside, enhanced by technology that reinforces, extends and relates formal and non-formal learning;
- putting children in touch with the expertise and alternative perspectives of people other than their teachers, as well as increasing their awareness of places outside the classroom, strengthening the relevance of classroom learning;
- collecting data 'in the wild' to take back into the classroom, enabling authentic and original investigations that ground the development of abstract knowledge in observation and experimentation in the real world;
- unobtrusively capturing individual children's interests and learning strategies;
- making use of communities and social interactions that happen outside the classroom.

When we considered the teaching of Debbi Winn in her work on design & technology CAD we saw that she constructed a game involving 'Wizards', with keys, goblets and castles that the students had to draw using 3D CAD. In a similar way, games can be used to experience different phenomena using augmented reality through devices such as the Merge Cube discussed above:

Computer-based simulations, games and 'augmented reality' – where the real world is overlaid with information from the digital world – hugely expand the variety of problems students can study, and their ability to use this new knowledge. Simulation authoring tools such as SimQuest, enable them to explore, for example, the physics of motion with skaters on ice, trains on railways and lorries on roads.

(TEL report, 2012: 25)

The closure of schools during the 2020 COVID-19 pandemic saw teachers supporting their pupils, and their parents, who were home-schooling. Resources were sent via schools' websites, YouTube was used to give examples, suggest safe home experiments, and broadcasters such as the BBC gave 'bite-sized' topics, which were extended to include 20-minute shows, each designed to target a specific age group, from ages five to 14, and for pupils throughout the UK covering what should be learned that day in maths and science for the different year groups. For older students, BBC 4 presented evening programmes to support the GCSE and A level curriculum in England and Wales, and BBC Scotland presented content specific to the NQs and Highers.

These examples show what can be done and what some schools are already doing to link school to home to everyday 'real life' and raise the possibility that, as the school/home divide has so necessarily become very slim, new technologies will be used even more to support the education of young people in a range of settings. As we move to ensure a more robust infrastructure to make society resilient for future pandemics, teachers providing technology enhanced learning for those studying at home and collaboration between teachers supporting pupils attending schools in shifts might become more prevalent.

Using AI in the classroom

The Open University report *Innovating Pedagogy 2020* (Kukulska-Hulme et al., 2020: 10–12) noted:

> The term 'artificial intelligence' (AI) is used to describe computer systems that interact with people and with the world in ways that imitate human capabilities and behaviours. AI-powered learning systems are increasingly being deployed in schools, colleges and universities, as well as in corporate training around the world. While many people fear that AI in education means robot teachers, the reality is less dramatic but potentially still transformative. Student-facing applications of AI include intelligent tutoring systems, dialogue-based tutoring systems, exploratory learning environments, automatic writing evaluation, and conversational agents.
>
> Intelligent tutoring systems (ITS) present students with some information, a related learning exercise and often a quiz or test. Having closely monitored the student's interactions and responses, the ITS then adapts the next set of information, exercise and quiz to the student's individual strengths and weaknesses. In short, ITS adopt an instructionist pedagogy. Each student proceeds step-by-step along a learning pathway that is automatically personalised for them. This personalised approach is promoted as being more effective than standard classroom practices (in which students progress through the same materials together and at broadly the same pace), although there remains insufficient evidence to support this view. Furthermore, a typical ITS personalises only the learning pathways and not the learning outcomes. The aim is still for everyone to learn the same materials, often to pass examinations, while little attempt is made to enable students to develop their personal aims or individual interests. Meanwhile, ITS also reduce human contact between students and with teachers. In short, typical ITS make various pedagogical choices with important ethical implications.
>
> Two alternatives to ITS are dialogue-based tutoring systems and AI-enabled exploratory learning environments: Dialogue-based tutoring systems (DBTS) adopt a Socratic pedagogy, which means they are designed to engage the student in a conversation, written or sometimes spoken, using questions to guide them towards an understanding of the topic being studied. However, DBTS also only personalise the learning journey not the outcomes.

AI-enabled exploratory learning environments (ELEs), on the other hand, adopt constructivist pedagogy, which is to say that they provide more open opportunities for the student to explore a topic and to construct their own understandings. However, although exploratory learning can be very powerful, it usually does not work well without guidance. In an AI-enabled ELE, it is the role of the AI to provide appropriate guidance in the form of automated feedback.

As is mentioned above, the use of AI has ethical implications concerning the role of the teacher and strikes at the heart of why children go to school. If authentic learning can take place anywhere, what is the function of a school and where is the social interaction taking place?

Social media and pupil wellbeing

The fact that France, the state of Victoria and many schools around the world have banned the use of mobile phones in schools is indicative about worries that many parents have that a 'library in the hand' and a powerful personal 'computer in the pocket' that can be used for STEM and other learning is overshadowed by the possible negative aspects of personal phones. While no one would belittle the harm done by sites encouraging young people to question their body image or offer advice on self-harm and even questioning their personal worth, how and where pupils should be taught to react to such information needs to be addressed. It is a fact that in any mainstream high school class in 2020 at *least* 60 per cent of the pupils will have their own phone, and that cuts across all income and social groups, and ownership can only increase as the power of phones soars and competition forces costs down.

The teachers at Ysgol Uwchradd Caergybi in Holyhead, Wales, are clear that pupils need to be educated in the use of their smartphone – from how to keep safe on social media through to using it for more than the latest online gaming craze – and that using phones appropriately in the classroom quickly becomes normalised. The school moved away from a phone ban as both parents and pupils complained that confiscating phones caused domestic difficulties as to where and when pupils were to be picked up; and any changes of plans are now done by phone. The school staff are adamant that now that the 'confrontation has gone' a phone can readily be used in a lesson when needed or put away when requested. Charlie Wilson coordinates the Digital Competence Framework (DCF) and Personal and Social Education (PSE) at the school. She is clear that the explicit teaching of keeping safe online is both a necessary part of the curriculum and a responsibility of a school that encourages the mobile phone in the day-to-day curriculum – such as taking a picture of the homework from the board and conducting surveys around the school. In PSE time, age-appropriate lessons – such as the showing of 'Kayleigh's Love Story' (Leicestershire Police, 2020) – make clear the dangers of engaging in online relationships with strangers.

In Chapter 1 we suggested that 'the STEM subjects cannot be divorced from other dimensions of human thinking and behaviour since the beliefs and values of individuals and communities are influenced by, and exert pressure on both science and technology themselves'. The use of mobile phones is ubiquitous. It is important that in STEM subjects pupils not only use the technology available to them but also consider the implications of their use.

Conclusion

As long ago as 1993, Larry Cuban wrote an article titled, 'Computers Meet Classroom: Classroom Wins' (Cuban, 1993). He wondered why schools lagged behind other organisations in the use of technology and suggested that technological innovations have never been central to national school improvement movements and the dominant cultural belief about teaching, learning and proper knowledge; and about the way schools are organised for instruction inhibits computer use. Given that this chapter, over three decades on, is describing schools that are so concerned about use of mobile phones that they were banned, maybe in some schools Cuban's views still hold. However, we have explored in this chapter, I believe, a new and profound change in attitude that is coming much more widespread. The computing power of even a cheap mobile phone puts video, sounds and information in the hands of every learner; the ready availability of sophisticated computing power can no longer be ignored. Rather than computers only being available in a special room or through a few tablet computers that can be borrowed just for special and infrequent use, the personal 'library in your pocket' is now accessible to all – and not only in resource rich countries. A tablet computer is available in 2020 for $64 and it will only become lighter, faster and more powerful over the next decade and as 5G is rolled out very much faster.

But we should also be cautious and at the start of this chapter I said, 'Before we consider for what we might use computers – smartphones, iPads, laptops and notebooks – in our teaching, we need to pause and think through our beliefs about the relationship between the pupil and the teacher'. The role of the teacher is crucial especially in developing understanding through high quality thinking. You can teach a class about the Newton's laws of motion or ionic bonds in solids, but engaging students so that they think hard enough to gain some understanding requires much more than enabling recall. As we have seen, teachers using computers and computing can do this, but it requires pedagogy with a distinctly human touch, lots of listening, appropriate questioning and discussions as the learners construct their understanding. Research by Adrian O'Connor (2019) in the use of a virtual learning environment (VLE) to support the teaching of eight pre-service teachers and the learning of 104 pupils, showed that the learning process became much more explicit and the design ideas developed were much more robust than in previous situations in which the VLE was not available. This indicates strongly that the successful use of IT in teaching and learning requires that the pedagogy being used is sound in the first place and the IT *supports* and enhances this pedagogy. In this case the IT was able to facilitate enhanced discourse between teachers and pupils.

The computer, in all its forms, is a teaching tool that is integral to STEM activities and indeed essential for teaching and learning in all subjects. It is in computing, digital competence, IT, and computer science, that STEM teachers not only need to look sideways at what other colleagues are doing, but be the advocates of technology enhanced learning to help and support the students and their teaching colleagues across the whole curriculum.

Recommended reading

Computers in teaching and learning the STEM subjects

Atherton, P. (2018) *50 ways to use technology enhanced learning in the classroom*. London: Sage.
Luckin, R. (ed.) (2018) *Enhancing learning and teaching with technology*. London: UCL/IoE Press.

Computer science in school

Halfacree, G. (2020) *The official Raspberry Pi beginner's guide*. Cambridge: Raspberry Pi Press.
Matthes, E. (2019) *Python crash course*. San Francisco: No Such Press Inc.
Simmons, C., & Hawkins, C. (2015) *Teaching computing*. London: Sage.

What we might be using both in home and school learning in the future

Baker, T., Smith, L., & Anissa, N. (2019) *Educ-AI-tion rebooted? Exploring the future of artificial intelligence in schools and colleges*. London: NESTA.
Bernhardt, C. (2019) *Quantum computing for everyone*. Cambridge, MA: MIT Press.
Wilkins, N. (2019) *Internet of things* (Independently published ISBN: 978-1799092216).

Social media and pupil wellbeing

House of Commons (2019) *Impact of social media and screen-use on young people's health* (HC 822), Report of Science and Technology Committee. London. https://publications.parliament.uk/pa/cm201719/cmselect/cmsctech/822/822.pdf (accessed June 5 2020).
Twigg, L., Duncan, C., & Weich, S. (2020) Is social media use associated with children's wellbeing? Results from the UK Household Longitudinal Study. *Journal of Adolescence*, 80, 73–83.

Useful websites in teaching using IT

Creatity (2020) Teaching and AI www.creatity.com (accessed June 5 2020).
i.materialise (2020) Using online 3D printing http://i.materialise.com/ (accessed June 5 2020).
Ponoko (2020)Laser Cutting using IT www.ponoko.com (accessed June 5 2020).
QR Code Generator (2020) www.qr-code-generator.com/ (accessed June 5 2020).
Shapeways (2020) 3D printing on demand www.shapeways.com/ (accessed June 5 2020).
Thingiverse (2020) Virtual prototyping www.thingiverse.com (accessed June 5 2020).

References

Barlex, D., Gardiner, P., & Steeg, T. (2011) *Learning journeys for computing in D&T: Embedded control/intelligence*, Unpublished (written for the Design and Technology Association to inform the consultation about Computing in Schools carried out by the Royal Society).
Barlex, D., & Stevens, M. (2011) Making by printing – Disruption inside and outside school? In T. Ginner, J. Helstrom, & M. Hulten (eds), *Technology education in the 21st century*. Proceedings of the PATT 26 Conference 2012, pp. 64–73, Stockholm, Linkoping University, ISBN Proceedings: 978-91-7519-849-1.
Berry, M., Brooks, B., Coombs, S., Deepwell, M., Jennings, D., Schmoller, S., Slater, J., Twining, P., & Webb, J. (2012) *Better learning through technology – A report from the SchoolsTech conversation*. London: Naace and ALT.
Carr, N. (2015) *The glass cage*. London: Penguin.

CSWG (Computing at School Working Group) (2012) *Computer science: A curriculum for School.* www.computingatschool.org.uk. Written March 2012 (accessed June 5 2020).

Cuban, L. (1993) Computers meet classroom: Classroom wins. *Teachers College Record,* 95(2), 185–210.

DfE (2013) *Computing programmes of study: Key stages 3 and 4.* London: Department for Education.

Fossbytes (2018) *Mobile phone sensors.* https://fossbytes.com/?s=mobile+phone+sensors (accessed June 5 2020).

Johns, J. (2018) *FabLab guide.* Bristol, UK: Bristol University.

Kemmis, S., Atkin, R., & Wright, E. (1977) *How do students learn?* Occasional paper No. 5. CARE: University of East Anglia UK.

Kukulska-Hulme, A., Beirne, E., Conole, G., Costello, E., Coughlan, T., Ferguson, R., FitzGerald, E., Gaved, M., Herodotou, C., Holmes, W., Mac Lochlainn, C., Nic Giolla Mhichíl, M., Rienties, B., Sargent, J., Scanlon, E., Sharples, M., & Whitelock, D. (2020). *Innovating Pedagogy 2020: Open University Innovation Report 8.* Milton Keynes: The Open University.

Leicestershire Police (2020) *Kayleigh's love story.* www.youtube.com/watch?v=WsbYHI-rZOE (accessed June 5 2020).

Merge Cube (2020) *Augmented reality in the classroom using MERGE Cube.* https://mergeedu.com/ (accessed June 5 2020).

Micro:bit (2020) *Support for the BBC micro:bit.* https://microbit.org/ (accessed June 5 2020).

O'Connor, A. (2019) Using technology to support discussion in design and technology. In P.J. Williams & D. Barlex (eds), *Explorations in technology education research.* Singapore: Springer.

ONS (2019) *Office of National Statistics: Online shopping, by age group, 2018 to 2019,* Great Britain. www.ons.gov.uk/peoplepopulationandcommunity/householdcharacteristics/homeinternetandsocialmediausage/adhocs/10347onlineshoppingbyagegroup2018to2019greatbritain (accessed June 5 2020).

QR Codes in Maths (2020) *Examples of QR codes in mathematics lessons.* http://new-to-teaching. blogspot.com/p/math-qr-codes.html (accessed June 5 2020)

Raspberry Pi Foundation (2020) *Support for using the raspberry Pi in the classroom and beyond.* www.futurelearn.com/partners/raspberry-pi (accessed June 5 2020)

Richardson, J., & Johnston-Wilder, S. (1999) *Teaching in mathematics: Learning schools programme.* Milton Keynes: Open University/Research Machines.

Royal Society (2012) *Shut down or restart? The way forward for computing in UK schools.* London: The Royal Society.

Statistica (2019) *Data on development of IT use in society.* www.statista.com/search/?q=online%20 banking%20users (accessed June 5 2020).

Technology Enhanced Learning (2012) *System upgrade: Realising the vision for UK Education.* London: Tel.ac.uk, Institute of Education (see http://tel.ioe.ac.uk/).

Viner, R.M., Gireesh, A., Stiglic, N., Hudson, L.D., Goddings, A.-L., Ward, J.L., & Nicholls, D.E. (2019) Roles of cyberbullying, sleep, and physical activity in mediating the effects of social media use on mental health and wellbeing among young people in England: A secondary analysis of longitudinal data. *The Lancet Child & Adolescent Health,* 3(10), 685–696.

Winn, D. (2012) *CAD and creativity at Key Stage 3: Towards a new pedagogy,* Unpublished PhD Thesis. Milton Keynes: The Open University.

10

Creating an environment for sustaining STEM

When you visit a school, I think it is possible to gain quite a good impression of it from its environment. I don't mean by this where the school is located or the nature of the buildings but rather what one 'feels' on entering. I'll give you a couple of examples. As an external examiner for an initial teacher education course offered in the North of England, I went to a small school in the Yorkshire Dales, a particularly beautiful spot; so much so that I stopped my car to look along the valley to take in the view before driving on towards the school buildings built out of an attractive weathered stone. I reported to the office and said who I was and who I wanted to see. There was much confusion, but I was asked to wait. Looking around, I noticed that the interior had been extensively refurbished within the older building shell but in the reception area there were couple of reproduction paintings, a cabinet of trophies for different sports and a rather incongruous list of previous head boys and girls, which seemed to have stopped over ten years ago. The corridors were bright and cheerful and in a 'pristine' condition, although the refurbishment had taken place three years ago. I went to see quite a good design & technology lesson in a large space with some safety notices on the walls. During my visit I saw no examples of pupils' work on display and no posters other than those concerning safety.

The following week, I want to a school in Birmingham in the industrial heart of England that had been due to be refurbished under a 'Building Schools for the Future' initiative but the plans and funds had been dropped with a change in government. It was a flat-roofed building of the 1970s and quite hard to find as it was hidden at the back of a rather run-down housing estate. In contrast to the welcome in Yorkshire, here I was greeted by a smiling secretary who had a visitor's badge already made out with my name on it and again I was asked to wait as she made a call to the head of department. Looking around, the reception area was full of artwork that the pupils had done, and on the wall the TV scrolled through a series of photos of recent school trips, action sporting shots and messages about forthcoming events. Walking along the corridors on the way to the department, the escorting teacher pointed out the departments we went through by the pupils' different work on display including some tessellations in the mathematics department and some photos of measuring the speed of sound in science – all done in the previous two terms. We entered the design &

technology department through an arch with 'Technology' across the top. Again, I saw some safety notices, but also displayed on the wall was a series of posters of working engineers and scientists of both genders and different ethnicities, some pupil graphical communication examples of ideas for a 'chocolate wrapper' and some small examples of artefacts from Year 7 to Year 10 in a display cabinet.

From the way I have described these two schools you can guess which 'environment' I thought was best for teaching and learning. In Birmingham, without overtly evangelising, the environment gave the pupils messages that contributing STEM subjects were not only interesting and applied to life outside the school; STEM was something that people like them could study or might use in their everyday life too. Spending money on new school buildings is a very good idea, of course, and the school staff in Birmingham were understandably very bitter that they missed out, but they showed that how the building is *used* is much more important in creating an exciting and respectful environment than pretty stonework.

In this chapter, we are going to think about learning in the STEM subjects as more than one-off events such as competitions and career initiatives, fun and interesting though they are. We consider creating an environment that will sustain the contributory subjects of STEM, where their place in the school is explicit and valued, where teachers are supporting each other and working together, and where pupils see the links between the subjects to help their learning.

We will look at:

- the physical environment;
- the pupils' learning environment – for both girls and boys;
- the teachers' professional environment.

The physical environment

'I hate this place – it smells!' However long one has been teaching, certain pupils will always be remembered. That frequent refrain of Cathy's sulky entrance to the lab still echoes in my memory many years on. She had a point. The science and technology area of that school certainly had some interesting aromas caused by the close proximity of laboratories to workshops and the mingling smells of gas, glue and gunk. The workroom environment, I admit, left much to be desired and I expect many other less vocal pupils were also affected by it. I quickly came to realise that the physical environment is very important. Not only does it affect our attitude to the subject, it also has a profound effect on learning. The layout of the room itself says much about the way that the teacher relates to pupils and how pupils can relate to each other and the physical environment is intimately linked to what I call the 'pupil environment', which we will look at later. The safe distribution, use and collection of apparatus, tools and resources in an efficient and controlled manner contribute significantly to appropriate pupil behaviour. If the physical environment of the classroom is set up so that pupils can take responsibility for their work and make informed choices of tools and components as they progress through a task, they are better able to take control of their own learning. But if they have to wait for equipment, materials or attention, pupils become bored and frustrated and sometimes disruptive. In creating an effective learning environment there is a very close link, therefore, between the class layout, resource management, behaviour management and the safety of all who are in the laboratory or workshop.

Primary schools in England have been world leaders in the creation of a visual environment for learning. It probably goes back to an exhibition of children's art in the 1930s that was visited by John Blackie who later became Senior Chief Primary Inspector. Over several generations of primary teachers, it has now become the rule that entering a primary school is often a kaleidoscope of colour and images. But this is far more than mere decoration and secondary schools, although far better in recent years in celebrating pupils' work and using commercial and home-grown resources to support learning, have much to learn from primary colleagues.

A web search will quickly find a range of ideas, images and resources to improve the learning environment and quickly find links to the topic being taught (see Caviglioli, 2019b). Some science and maths classroom ideas have been around for many years – such as the one shown in Figure 10.1 – a word wall of difficult concepts – which rather than being hackneyed, illustrates a clear link between language and learning that is extremely useful.

This example of a 'learning wall' is only one part of making an appropriate learning environment. There is the display of pupil work as was evident in the Birmingham school mentioned above, which explicitly says 'what you do in STEM is important and we want to celebrate your work'. But also important are the messages given by posters and other images that say the STEM subjects are for all pupils and for people of every ethnicity, and are not just studied by old white men! Here is something you might like to try on a professional development day with other STEM colleagues or even over coffee with your immediate department friends: list as many well-known women scientists, mathematicians or engineers as you can. What about well-known scientists, mathematicians or engineers – say Nobel Prize winners – who have a

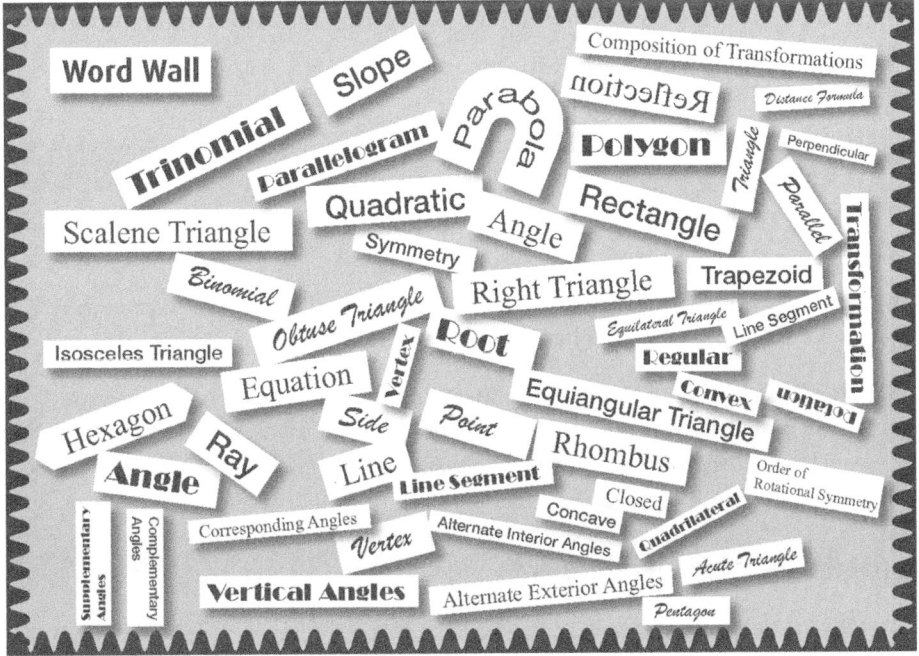

FIGURE 10.1 Learning wall

South Asian heritage, or South American heritage? Websites that list such notable STEM practitioners are given at the end of this chapter and many are available for school display (see Pamona College, 2020; Society of Canadian Women in Science and Technology, 2020; STEM Role Models, n.d.; Women in Mathematics, 2020) .

Creating a successful physical environment for learning is not only at the classroom, corridor and department levels; it extends to consideration of the whole way pupils interact with the school.

I once taught in a school in Wales that had a range of separate buildings. My laboratory was on the top floor of a Victorian building that still had the remains of the pipes on the ceiling for the original gas-lighting. The pupils sat at long teak tables and all the services were around the rim of the room, out of the way until they were needed. Although well over a century old, it was the most adaptable room I have ever worked in, and when I had the opportunity to re-design the lab when the school was rebuilt, I followed the same overall plan as the Victorian design but with a matrix of electrical sockets across the floor. School physics labs *never* have enough electrical sockets. However, that is not the main point of my story. The different buildings of the old school ranged in design as they were built across the decades, and as pupils moved from lesson to lesson they had to go outside to walk to the next building. It rains a lot in Wales, and pupils were always dashing with coats and hoods from place to place. The new school was built as a 'shirt sleeve' environment and the difference in behaviour and attitude of the same pupils was remarkable. Carpeted areas and noise-reduction tiles brought down the acoustic 'temperature' drastically. Clean areas in science and design & technology can be improved by stimulating and informative visual displays but also can cultivate a calm and purposeful acoustic environment. A senior management team interesting in *sustaining* STEM could ask the following:

1   Who is responsible for display in the department and who else is involved?

2   Do pupils have some responsibility in selecting the display in communal areas?

3   How do new STEM staff gain training in display as part of their induction?

4   What are the walls used for – children's work? Puzzles in maths? Showing commercial posters of STEM careers and the practicality of the subjects in everyday life? Is there some unfinished work to debate – 'our first ideas about forces'?

5   In the teacher professional areas what is the tone of the notices – humorous, cynical? Are there articles of interest from magazines or photocopied from journals?

6   Are all areas – including corridors – of a sensible 'acoustic temperature'? If not, what is the strategy for changing that?

The book *Dual Coding with Teachers* (Caviglioli, 2019b) presents a multitude of ideas both for designing general display, and also for ways of using diagrams to support explanations while teaching using exposition.

The pupils' learning environment

What the research shows consistently is that if you face children with intellectual challenges and then help them talk through the problems towards a solution, then you almost literally stretch their minds. They become cleverer, not only in the particular topic, but across the curriculum.

(Adey, 2001: 17)

How do pupils learn? That seems a straightforward question, but you will already know from your day-to-day teaching that an answer is far from obvious. Are your views about how pupils learn the same as those of your colleagues both in your subject and other teachers of STEM subjects? You could ask them, perhaps informally at break-time, the following questions: How do you think pupils learn? What should we do as teachers to help that happen?

Asking these questions in such a blunt way is likely to elicit either a flippant response or maybe a cautious one along the lines of 'Everyone learns in different ways'; 'It depends who they are. I teach depending on the needs of the pupil'. And so on. It is almost certain that your straightforward questions will not get straightforward answers!

All teachers, and parents for that matter, have a 'theory' of learning. It may link to formal ideas but is more often not something grand or grounded in careful research, but rather is a collection of day-by-day assumptions about what we, as teachers, should do to help those we are teaching to learn. New ideas about learning are developing, particularly related to our new understandings about the brain and how it develops, and we need to test them out against our knowledge of pupil behaviour and the views we currently hold.

The following are some views that people, including teachers, hold about how pupils learn:

a   Knowledge and skills can be broken down into component parts and it is the teacher's job to do this for the learner. The teacher then teaches each element and gives the pupil sufficient repetition until the learner can give a 'positive response'. The pupil will generally receive the same instruction as everyone in the class, but if assessment shows that the pupil requires further help, then an additional programme with smaller steps over a longer time scale will be provided.

b   A child constructs meanings by getting to grips with the particular problems in hand. Private problem solving is very important and a teacher should provide the necessary stimulus material and opportunities for the individual pupil to learn something new. A pupil will not progress without plenty of practice in the activities that have already mastered. A child will only be able to 'get' an idea when she has reached a certain stage of maturity and the teacher's job is to be aware of that and to decide when the pupil is 'ready' to move on. Some pupils are never able to 'get' certain ideas.

c   All pupils are educable and are helped in their learning by discussion and other social interaction, including with a more experienced learner or teacher. There is no fundamental difference between the learning of children and that of adults. Rather than waiting for a pupil to be 'ready' to learn, a teacher is finding out what the pupil thinks in order to guide and support what the pupil is trying to do next. By talking with the teacher, and obtaining other support, a pupil is able to grasp ideas and new understandings that they could never arrive at on their own.

These very brief summaries relate to the three main traditions of learning theory; behaviourism, Piagetianism and social constructivism. How do these well-known ideas relate to what you actually do in your 'STEM' classroom? Are you able to 'sign up' to any one of the theories wholeheartedly? As you read these descriptions you may have felt that each of them separately described some aspects of your ideas about learning and those of your colleagues, yet none was wholly satisfactory in its own right. For example, in teaching certain practical skills, a regime of practice and reinforcement in the 'behaviourist' tradition may be appropriate. An individual project will provide problem-solving opportunities and will be successful if the pupil is working largely within his or her capabilities, a Piagetian standpoint. Group practical work in science, and discussion of an idea with others before answering questions posed to the class would reflect a social constructive perspective. That teaching methods should be selected in terms of 'fitness for purpose', rather than adherence to a particular dogma of 'good practice', is clear. Teachers tend to have their preferred way of working, which reflects a personal 'theory' but, nevertheless, are not hidebound by particular ideologies, and will adopt a different teaching strategy if they think it will be helpful. Sometimes it is called a 'folk theory' of learning.

Some people think that good teaching means the same thing as good explaining – keep it clear and simple and all will understand. In fact, some teachers, particularly those in pre-service education get very upset when, despite their greatest efforts, the pupils just don't grasp what they have explained. When pupils just don't 'get it' they take it as a personal failure, or maybe blame the pupils. It is certainly true that a key teaching skill is the ability to explain and describe things clearly. But a belief that clearly transmitting information is *all* that is required for a 'good' teacher is insufficient. However, such a 'folk' theory of how minds work is very common across the world, and also explains the position some parents take to learning and teaching. These common beliefs were investigated by Bereiter and Scardamalia who characterised a folk theory of mind as follows:

1   Knowledge is 'stuff'.

2   Mind is a container.

3   Learning involves putting stuff in the container.

This tends to be reinforced by national curricula and examination syllabuses that emphasise content knowledge above all else. Bereiter and Scardamalia (1996) suggest that the corollaries of such a view of the mind is:

1   Pedagogy: a craft for stocking minds.

2   Educational testing: a process for inventorying mental contents.

Desforges (2001: 25) indicates that the corresponding 'folk pedagogy' to such a view of learning has had some remarkable success in teaching through 'show and tell'.

> But where the 'stuff' metaphor breaks down – as it does with wisdom, creativity, knowledge creation, appreciation, a 'feel' for a subject, we are left floundering.

Folk theories are indeed robust, yet the alternative ideas about teaching and learning outlined above have been considered for a least the last 60 years and linked to a growing understanding about the biology of the brain.

In the last fifteen years 80% of our knowledge about the brain and how it learns has been accumulated. Understanding about the different functions of specific parts of the brain has led to a more sophisticated appreciation of what happens to the brain in learning situations. However, this new knowledge is, for the moment, playing little or no part in influencing the design of the experiences we provide for students in our classrooms. Indeed, much of what happens in classrooms throughout the country conflicts with what is known about the brain and its design.

(Smith, 1996: 13)

Smith wrote that decades ago and his pessimism about 'what happens in the classroom' is at last beginning to change, especially in relation to children with specific learning difficulties. In 2011, the Royal Society published a report on *Neuroscience: Implications for Education and Lifelong Learning* where they pointed out that:

The rapid progress in research in neuroscience is producing new insights that have the potential to help us understand teaching and learning in new ways. [...] Neuroscience is shedding light on the influence of our genetic make-up on learning over our life span, in addition to environmental factors. This enables us to identify key indicators for educational outcomes, and provides a scientific basis for evaluating different teaching approaches.

(Royal Society, 2011: 3)

In summary the Royal Society report suggests:

- Education is about enhancing learning, and neuroscience is about understanding the mental processes involved in learning. This common ground suggests a future in which educational practice can be transformed by science, just as medical practice was transformed by science about a century ago.
- Neuroscience research suggests that learning outcomes are not solely determined by the environment. Biological factors play an important role in accounting for differences in learning ability between individuals.
- By considering biological factors, research has advanced the understanding of specific learning difficulties, such as dyslexia and dyscalculia. Likewise, neuroscience is uncovering why certain types of learning are more rewarding than others.
- The brain changes constantly as a result of learning, and remains 'plastic' throughout life. Neuroscience has shown that learning a skill changes the brain and that these changes revert when practice of the skill ceases. Hence 'use it or lose it' is an important principle for lifelong learning.
- Resilience, our adaptive response to stress and adversity, can be built up through education with lifelong effects into old age.
- Both acquisition of knowledge and mastery of self-control benefit future learning. Thus, neuroscience has a key role in investigating means of boosting brain power.
- Some insights from neuroscience are relevant for the development and use of adaptive digital technologies. These technologies have the potential to create more learning opportunities inside and outside the classroom, and throughout life.

This is exciting given the knock-on effects this could have on wellbeing, health, employment and the economy.

■ We urge caution in the rush to apply so-called brain-based methods, many of which do not yet have a sound basis in science. There are inspiring developments in basic science although practical applications are still some way off.

■ The emerging field of educational neuroscience presents opportunities as well as challenges for education. It provides means to develop a common language and bridge the gulf between educators, psychologists and neuroscientists.

There is no doubt that as we learn more about the way the brain works it will revolutionise how teachers teach and the technologies that they use in the classroom.

But the Royal Society warns about the rush to apply 'so-called brain based methods' and a web-search for brain-based learning methods does indeed produce a large number of companies offering such methods as part of work-based training. But are there some teaching strategies that can be used in classroom today? Daniela Kaufer, a professor at Berkeley, University of California, has drawn on her understanding of neuroscience to set out some key ideas (Kaufer, 2011) as follows:

Key learning principles

■ From the point of view of neurobiology, learning involves changing the brain.
■ *Moderate stress* is beneficial for learning, while mild and extreme stress is detrimental to learning.
■ Adequate sleep, nutrition, and exercise encourage robust learning.
■ Active learning takes advantage of processes that stimulate multiple neural connections in the brain and promote memory.

Neuroscience fundamentals

■ The most effective learning involves recruiting multiple regions of the brain for the learning task.
■ **Moderate stress:** Stress and performance are related in an 'inverted U curve' (see Figure 10.2). Stimulation to learn requires a moderate amount of stress (measured in the level of cortisol). A low degree of stress is associated with low performance, as is high stress, which can set the system into fight-or-flight mode so there is less brain activity in the cortical areas where higher-level learning happens. Moderate levels of cortisol tend to correlate with the highest performance on tasks of any type.
■ We can therefore conclude that moderate stress is beneficial for learning, while mild and extreme stress are both detrimental to learning.
■ Moderate stress can be introduced in many ways: by playing unfamiliar music before class, for example, or changing the format of discussion, or introducing any learning activity that requires individual participation or movement.

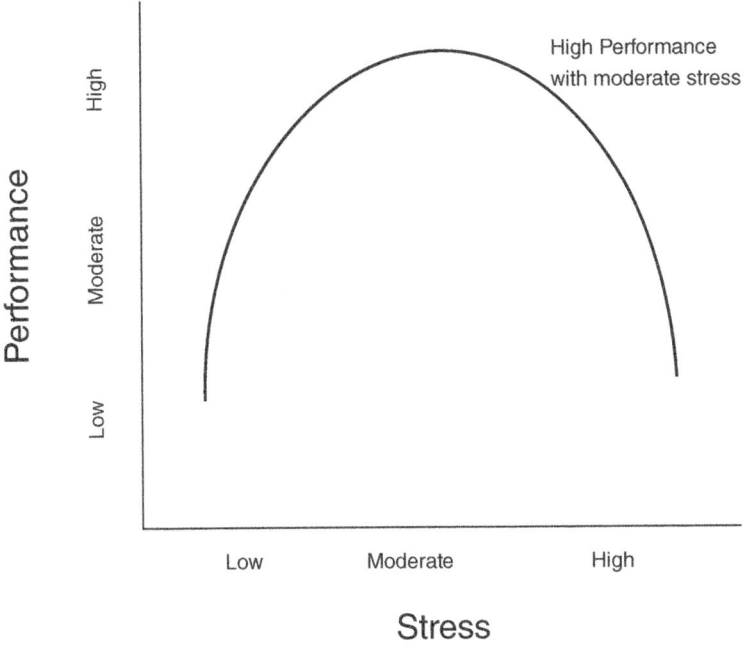

FIGURE 10.2 Performance related to stress level

- People do not all react the same way to an event. The production of cortisol in response to an event varies significantly between individuals; what constitutes 'moderate stress' for one person might constitute mild or extreme stress for another. So, for example, cold-calling on individual students in a large-group setting might introduce just the right amount of stress to increase some students' performance, but it might produce excessive stress and anxiety for other students, so their performance is below the level you know they are capable of.
- Any group dynamic that tends to stereotype or exclude some students also adds stress for them.

So, what does this understanding of how pupils learn impact on the teaching and learning environment for *all* pupils in STEM subjects? We know that there continues to be a disparity in girls and boys uptake of the physical sciences, but it is not inevitable. Work conducted by Microsoft found that 'Most girls become interested in STEM at the age of 11-and-a-half but this starts to wane by the age of 15' (Microsoft, 2017). What can be done to tackle this tail-off in interest?

The pupil environment for both girls and boys

I am sure every teacher wants to provide an appropriate STEM environment for all pupils – girls and boys. But what about a gender bias that we are unconscious of? When teachers are observed they are sometimes surprised to discover that the girls are more likely to be praised for being well-behaved while boys are more likely to be praised for their ideas and understanding. A disruptive girl may be admonished more than a boy who exhibits similar behaviour. Quiet boys are often overlooked. Consequently, girls and boys learn the 'rules of the classroom' – girls do not take risks and boys 'opt out' if they do not 'get it' easily. The Institute of Physics (IoP, 2020) has made a list of ten classroom practices that supports both girls and boys across STEM subjects:

1   **Use everyday language.** Technical jargon can be intimidating for many learners. Avoid it and make sure that you only introduce technical language or equations once the context is understood

2   **Avoid asking for volunteers.** Boys may be more likely to raise their hands, call out answers and volunteer to take part in activities. Other techniques, such as individual whiteboards or selecting students at random, can broaden the range of students participating

3   **Assign roles for practical work.** Boys often dominate the equipment while girls hang back and write down the results. To avoid this you can assign roles, or use single-sex groups for practicals.

4   **Use examples that show how STEM links to their experience.** This is useful for all students, but research shows that girls in particular tend to appreciate context and seeing the bigger picture (we discussed this in Chapter 1).

5   **Use gender-neutral contexts whenever possible.** Try to avoid using examples that focus on stereotypically male or female hobbies or interests.

6   **Allow time for pair or group discussions.** Give time for students to discuss answers to challenging questions before asking them to share ideas with the class.

7   **Challenge discriminatory language.** STEM is for everyone. Always treat sexist language as unacceptable, and tackle the attitudes behind it.

8   **Monitor your interaction with different genders.** You might be surprised at the ratio of different genders asking or answering questions in your class. Keep a note yourself or ask a colleague or student to observe one of your lessons and keep count.

9   **Regularly refer to a range of careers that use STEM-based skills.** Girls are more likely to consider their future career when choosing their options. Emphasise the transferable skills that studying science helps to develop.

10 **Ensure that your students are exposed to a diverse range of STEM practitioners.** Be wary of giving your students the impression that science is only for high achievers. Emphasise that STEM is for everyone, irrespective of their background.

Three-point communication

Caviglioli (2019b: 41) considers the 'fall-out that happens as a result of face-to-face difficult conversations. Negative comments – however professionally framed – hurt the recipient'. As illustrated in Figure 10.3, he suggests breaking down the face-off dynamic by teacher and pupil both discussing a shared third point – such as a diagram, so that 'both parties sit side-by-side and share the same view of the third point – the visual'.

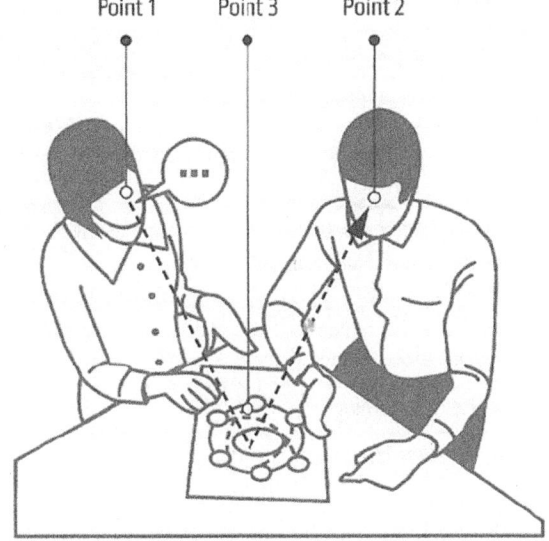

Both eyes and message are directed towards the added third point.

FIGURE 10.3 Shared third point

The pupil learning room environment

You might like to think about the teaching room that you use most frequently. What is the position of the tables or benches, the services such as electricity and water, and the position of the places where tools, equipment and materials can be accessed? You might like to compare your room with that of a colleague who teaches in a different STEM subject and see what the differences are and what they think about your room layout. Science labs and a design & technology workshop might have some similarities but be different to a graphics studio or a textiles room. But what does the location of the tables or benches, and where they are in relation to where you often stand to teach (if you do stand), say about the expectations you have for pupil interaction?

My guess is that it is very easy to think of the times in design & technology and maths lessons when pupils are working on their own, but that very rarely happens during practical work in science. The manufacture of an individual artefact that a pupil can take home is often a key part of design & technology schemes of work.

But in all the STEM subjects, as in other areas of the curriculum, discussion work in pairs or small groups is vital, too, if pupils are to address, for example, ideas about investigating patterns in mathematics, values implicit in science in society and a consideration of the impact that a design & technology product and materials have on consumers and society (who wins, who loses) and on the environment. Discussion is important to enable all to articulate their thinking and clarify their understanding. Of course, as well as group or pair work, the pupils will work as a whole class for presentations and evaluations and perhaps with other groups if there is a guest speaker or a whole-school STEM 'challenge'. In some schools, the timetable is collapsed at certain times of the year to allow for a concentrated period of work with even larger groupings.

The extent to which pupils interact in your lessons has much to do with how you think pupils learn and how you wish to be viewed as a teacher. For example, within each design & technology classroom there is a network of social interactions that 'grows' the knowledge available to the students enabling them to make difficult decisions about the details of their emerging, but as yet unresolved, design proposals. This knowledge is not evenly distributed among the students; serendipity plays a part in who knows what, but a skilful teacher orchestrates the social interaction to ensure that the classroom is a place in which communication between students is the norm, invariably on task and beneficial (Seery, 2020).

Environmental psychology is a discipline that draws on areas of knowledge such as geography, architecture, sociology, anthropology, design and ergonomics and suggests that everyday objects are not only physical but have an impact on how we relate to the world. The ability of a pupil to sit to work, move around the room (or not), to have control over what 'tools' (physical and cognitive) they choose when problem solving, and the nature and usefulness of display material all have a profound effect on their creativity and ability to work constructively with others. Storage systems, for example, can hinder as well as help in the efficient conduct of tasks. The layout of a teaching room should help pupils to understand the classroom environment and support what is expected of them. In Chapter 1, for example, we considered how an electronics system can be represented by three linked building blocks:

Using pieces of hardboard covered in fabric and systems electronic blocks with 'Velcro' glued onto the back, sets of Inputs, Process and Output modules can be attached to the fabric and set out so that pupils can easily choose, select and combine the modules together. More generally, David Barlex (2000) suggested that there are four conditions that a teacher needs to meet if their teaching of design & technology is to be effective:

1   The teacher should have the expectation that pupils will be capable. This means that it will be perfectly acceptable for pupils to make decisions and take action based on those decisions. In some cases, the actions will require teaching.

2   The teacher needs to facilitate pupil capability by organising and maintaining an appropriate environment. This means that pupils will have open access to materials, components, tools and equipment. In most cases, like the pupil in Figure 10.4, they will be able to collect what they need, as they need it, use it and return it. In some cases, particularly scarce resources may need to be booked in advance.

Choose it... use it... put it away!

FIGURE 10.4 Choose, use, put away
Source: Swails in Caviglioli (2019: 193)

But it is essential that decisions, once taken, can be acted upon if pupils are not to become disenchanted and lose motivation.

3   The teacher will need to provide the resources for capability by teaching the technical knowledge and understanding, aesthetics, design strategies, making and manufacturing skill and values needed for successful designing and making.

4   The teacher should maintain the motivation for capability through insight into pupils' motivations ensuring that activities are relevant.

Chapters 3 to 9 have suggested that pupils learning can be enhanced by being exposed to and appreciating the connections between the STEM subjects. It may be convenient for schools to compartmentalise knowledge and understanding into specialist subjects in specialist rooms, but, obviously, real life is not like that and, at the very least, teaching a subject in the light of another helps pupils connect their thinking. Schools that have embraced such a coordinated approach to STEM have noted the learning benefits:

- STEM learning is fun and therefore motivating; it helps learners to see the relevance of what they are learning, especially in mathematics and science.
- Co-operative learning is effective and develops Personal Learning and Thinking Skills (PLTS).
- Learning in one subject area when reinforced in others aids pupil understanding.
- STEM projects help teachers to understand the work of their colleagues in other departments better, resulting in schemes of work that are prepared in a coordinated and collaborative way, which increases the efficiency of teaching.
- Targeted STEM interventions can affect results […] and offer opportunities to stretch gifted and talented students
- STEM enhancement and enrichment activities with built-in reflective opportunities have an impact on […] attainment.
                    (Specialist Schools and Academies Trust (SSAT), 2009: 6, abridged)

We know that building a positive, supportive learning environment and maintaining positive self-esteem is important, although often in a busy school environment with the fragmented period day that is not always easy to achieve. As we have seen, the physical environment is important as well as the social in order to put learners in the right 'frame of mind'. The displays in the classroom help those youngsters who prefer to use visual stimuli to aid their learning and set the right learning conditions for all. As we see, there is a close inter-relationship between learning, the interconnectedness of the problems that they face, the social environment and the physical space that pupils work in.

Building on the evidence from SSAT, the traditional ideas about learning above, the Royal Society's report and Kaufer's description about how the brain responds when in different learning situations, here is an adaptation of some work by Alistair Smith. It suggests that to help pupils learn we need to set up conditions of moderate stress but high challenge in our classes. Smith and Call summarise how to create a successful pupil environment by adopting nine principles (Smith & Call, 1999: 33–34, adapted here):

1   The brain develops best in environments with high levels of sensory simulation and cognitive challenge.

2   Optimal conditions for learning involve sustained levels of cognitive challenge with moderate threat.

3   Higher order intellectual activity may diminish in environments the learner considers emotionally or physiologically hostile (remember Cathy smelling the lab!).

4   The brain thrives on immediacy of feedback and choice.

5   There are recognised processing centres in the hemispheres of the brain. This suggests structured activities.

6   Each brain has a high degree of plasticity, suggesting developing and integrating classroom with other experiences.

7   Learning takes place at a number of levels. This requires a range of strategies and personal goals.

8   Memory is a series of processes rather than locations. To access long-term memory is an active not a passive process.

9   Humans are 'hard-wired' for a language response. Discussion is a vital part of learning.

The teachers' professional environment

In this final section we turn to the professional environment of the teacher. Here, I don't mean the state of the staff room, the quality of the coffee at break-time (although they certainly have an impact) or even conditions of service. The professional environment that I am considering is how the school at all levels from the newest member of staff to the senior management team create, manage and sustain an environment that addresses staff needs and aspirations, and allow STEM activities and the associated curriculum to become embedded. Creating the right physical environment and nurturing the pupils' environment for both girls and boys are very important, but for ensuring the sustainability of STEM in a school, addressing the professional needs of staff, in particular teaching staff, is vital.

Tim Brighouse, a Professor of Education and a local government Chief Education Officer for over 15 years used to say 'Teachers get exhausted where the rest of us merely tire'. The 'rest of us' include head teachers and other members of the senior staff in school who often have a little 'down time' during the day that classroom teachers rarely have, even in those periods set aside for planning and assessment. Keeping staff motivated and enthusiastic when they have such an intense and often stressful workload is a key function of the leadership of a school. Tim suggested (1991) that staff require four conditions to create the successful professional environment that enables them to teach effectively and I have adapted them here:

- responsibility;
- circumstances that enable things to happen;
- new experiences;
- respect.

Responsibility

There is a difference between work – which is about things to do and which there is often far too much of, and responsibility – which we quite like, and is about having the final say and looking to improving how something might be. So, there is a real difference between jobs to be done and responsibility to do it, and ensuring the right person at the right level has appropriate responsibility is key in embedding STEM and ensuring it is sustainable long term. I don't only mean teaching staff here. Technician and classroom assistants also need to know the nature and extent of their responsibilities.

Responsibility is often formally established through the job description for staff appointments. I often sit on appointment panels and I think that most descriptions of jobs responsibilities are too numerous and too diffuse, and wonder if they are drawn up to give flexibility as the school is not certain what they want the candidate to actually do. Two of three lead responsibilities and three or four secondary ones makes it clear what is required and is much more likely to attract a candidate who has a clear vision of what they could make of the job. This is an example of a job description for a senior subject teacher of science that I think fits this well:

A JOB DESCRIPTION FOR A SENIOR SCIENCE TEACHER

Job purpose
To promote learner enjoyment and achievement through outstanding teaching that creates an irresistible climate for learning for all learners. To share your skills and experience with other teachers

Key responsibilities
- Take a lead role in the continuing improvement of teaching and learning in the Science Faculty.
- Provide high quality personalised professional development for teachers within the school.

- Support curriculum leaders in planning and resourcing high quality differentiated schemes of work.

You will also have these secondary responsibilities:

- To embrace whole school initiatives, including Assessment for Learning, Accelerated Learning and the use of ICT.
- To promote learner self-esteem and a positive academic self-concept.
- To work effectively as a member of the subject team to improve the quality of teaching and learning.
- To have a thorough and up-to-date knowledge of all the national curriculum and examination courses.
- To keep up-to-date with research and development in pedagogy both within the subject and as a teacher/learner.

Comment

- The responsibilities will be reviewed annually as part of our Performance Management process and may be subject to amendment or modification at any time after consultation with the post holder.

Some points, I think, are worth considering. There is a particular emphasis on supporting other staff in science, working as part of a team and keeping up to date in both subject developments and in new teaching strategies. This job description also matches well against the 'Framework of Teacher Professional Knowledge' that we discussed in Chapter 2.

Teachers of STEM subjects are first and foremost teachers of young people, but they are also a teacher of a subject and feel responsible for keeping up to date as knowledge and processes expand exponentially. Debi Winn in Chapter 9, for example, introduced her game method of teaching CAD not only to make the learning of the software package more efficient and more fun, she also realised that some colleagues were reluctant to teach CAD as they themselves felt inadequate in know-how to use it and the new computer-controlled workshop machines. Answering questions in science such as 'Do mobile phones give you brain tumours?' and 'Why are GM crops called "Frankenstein Foods?" in the newspaper' are similarly challenging. It is increasingly easy to access information through the use of new technology and pupils can more easily be coached to access information for themselves (see Chapter 9), but a professional environment that facilitates different teachers sharing their enthusiasm, knowledge, ideas, resources not only formally through schemes of work but through a communal noticeboard, and by making shared coffee time conversations possible and, in terms of sharing STEM information and ideas, professionally permissible. I have worked in both large and small schools and found that large schools shared a cross-subject STEM ethos less well simply due to where staff chose to meet at break-time; big subject departments stuck together, small subject groups went out to seek company. I know that teachers who have a job specification responsibility to 'Provide high quality personalised professional development for teachers within the school' have to work hard to avoid being labelled 'Billy Wizz' and 'Super-Teacher' by some cynical colleagues. The quickest way to gain credibility and change attitudes is through informal cooperative arrangements promoted by a careful consideration of where teachers can congregate at break and meal times. All staff having clear responsibilities and an understanding of how they relate to those they work with is so important in creating a successful supportive professional environment that reduces stress.

Circumstances that enable things to happen

Having one's responsibilities clear is an important first step but for teachers to be effective the leadership of the school needs to create the right circumstances to make things happen. At the basic level, this is an obvious 'give me the tools and I'll do the job' plea. Both authors of this book have spent time in classrooms in rural India and it is encouraging to see what good teaching goes on in some science and mathematics classrooms with extremely limited resources. So much more could have been achieved, however, with more books, materials and equipment. Having adequate resources is necessary for any teacher, anywhere. Teaching the STEM subjects in a way that develops understanding is best done through interacting with tools and materials so that learning can be 'minds on' as well as 'hands on'. As we saw in Chapter 1, the tradition of practical work in STEM (including mathematics) has been established in schools since the 1960s. Through the influence of ideas such as those of Jean Piaget, pupils became a 'scientist for the day' and learnt from discovery. But over the decades the possibility of such a hands-on approach has mirrored the prevalent economic climate and the money spent on schools has sometimes not been adequate to provide new science equipment or the latest CAD/CAM machines in design & technology. Some headteachers have tried to influence the curriculum and pupils' entitlement to engage across STEM when they have felt that the limited resources possible could not be stretched sufficiently; and so it is encouraging when even world leaders stress the importance of STEM education for all.

I think there are four important circumstances that enable things to happen for STEM to be embedded. The first one is being able to work in teams and learning from each other. We are convinced that encouraging teamwork not only shares work and expertise, it provides a richer and purposeful learning environment for pupils. It is worth stressing once again: **'look sideways'**.

Leaders in school need to ask the following questions:

- Is it possible at the department level for team teaching if it is needed?
- How can the head of department and the teacher responsible for subject development (such as in the above job description) have the support to build teamwork?
- Are there notice boards that enable *all* to keep up-to-date with research and development in pedagogy both generally and within their subject?
- How can teachers be given the circumstances to 'look sideways' at the teaching in other STEM subjects?
- How can experts outside the school contribute effectively to the STEM curriculum? What support do such visitors need?

Ensuring that staff, working in a coordinated and collaborative way, know what is happening within and across STEM subjects seem to us to be the one key factor that would improve pupil learning, attainment and, just as important, a positive attitude and open mind. The composition of the team, and so the ideas, need not be solely from school staff. Due to funding from other organisations, it is often possible for external experts to contribute to a team approach; a 'STEM Ambassador' for example, and also to take pupils out of school to engage with other adults by learning aspects of STEM in 'real-world' contexts.

> STEM learning takes place in the real world. Schools work with outside partners from industry, commerce, government services, higher education and other schools.
> Examples include:
> - *working with environmental agencies to develop a more sustainable school*
> - *bringing space craft into the school*
> - *visits to hydro-electric power generation plants*
> - *working with the motor industry to help careers awareness*
> - *bridge building with engineering consultants*
> - *involving STEM ambassadors in school life*
> - *working with a water company and a university to solve a problem on a sewage treatment plant*
> - *companies providing challenges that can be worked on in clubs*
> - *working with primary partners on STEM*
> - *visiting a botanical garden to see how tropical environments are maintained.*
>
> Schools report that learners enjoy being out of school and seeing how science, design & technology, engineering and mathematics are used in the real world. This reinforces and extends what they learn in the classroom. [...]
> Companies get a chance to inform learners about their work. This long term strategy helps them to recruit and demonstrate their commitment to the community.
>
> (SSAT, 2009: 7, abridged)

However, as discussed in Chapter 12, the benefit of such work needs to be firmly embedded in the school curriculum to be effective and it is well to remember the obvious point that external experts are not teachers, and need to be supported so that their contribution can be an effective and positive experience both for the pupils and for them.

The second circumstance that enables things to happen brings together what we have said about teamwork, the conditions for interaction, and links back to our above discussion of the physical environment. We are all influenced by our social and physical environment. The walls, bookshelf and noticeboards of staff areas influence our attitude to our teaching job just as it influences the learning of pupils. Conversations can be dominated by school politics or (and!) they can be informal debate about projects and pupils' progress. Notice boards that have dusty and curling teacher-union posters give one feel to teaching in a school, but a changing series of cuttings from the *Times Educational Supplement* – humorous as well as the 'cutting edge' – give quite another. A department might have access to journals and other hard-copy resources from subject associations whereas other communal areas might have the more general. As shown by Caviglioli (2019b: 110) developing expertise in display to create such an influencing environment need not be a huge drain on resources but the effect on professional development can be marked.

The third circumstance that enables significant change is the huge support that non-teaching colleagues offer. Getting right the technical resource that supports preparation in science and design & technology can make probably the most significant difference between a lesson that is mediocre and one that goes like clockwork and is an exciting and successful learning experience. The professional development of technicians is important and in most schools is now firmly in the staff development plan. Some school science technicians in the UK have enrolled for courses leading to professional recognition such as Registered Science Technician (RSciTech) through the Science Subject Association (ASE), which requires such knowledge and competences as:

A   Application of Knowledge and Understanding;

B   Personal Responsibility;

C   Interpersonal Skills;

D   Professional Practice;

E   Professional Standards.

Other staff that support teachers in providing photocopy and audio-visual resources, or can give support to finding illustrations and materials for the electronic whiteboard for display in class or Open Educational Resources (OER) for free sharing on the virtual learning environment need to be properly supported, trained and adequately resourced too.

There is one fourth and final point to be made. Probably the most significant circumstance that the leadership of a school can do to enable change to occur is not the physical and staffing issues considered above but rather by giving STEM subjects the permission to experiment and try out new ideas. How can a senior management team encourage this and enable those good ideas to be shared? David Hargreaves suggests that one of the principal tasks of senior management is to know how to manage 'knowledge creation' – how to encourage and nurture such new ideas. Hargreaves uses a five-step gardening metaphor to set out what managers need to do, which I have adapted here:

Steps 1 to 5 are often facilitated through appropriate use of information technology. Some schools have their own internal professional development site on social media with links out to the free Open Educational Resources (OER) sites such as ORBIT from Cambridge University (Orbit, 2020), 'OpenLearn' (OpenLearn, 2020) from The Open University and 'FutureLearn' (FutureLearn, 2020) that draws together courses from universities and other organisations worldwide. The school sites, linked to their own virtual learning environment, provide a forum to discuss aspects of professional knowledge (see Chapter 2) and Technology Enhanced Learning (TEL) as we discussed in Chapter 9. In terms of creating an appropriate professional environment

DAVID HARGREAVES – CULTIVATION OF NEW IDEAS

Step 1: Generating the ideas – sowing

Create a professional environment – a school culture that promotes 'tinkering' – so that teachers actively try out new ideas or adapt old ones and take carefully calculated risks. Enabling teachers to try something new is important (see New Experiences below) but often teachers find it difficult to explain why something that they do 'works'. The knowledge is tacit. By enabling teachers to work together or even team-teach creates the *shared experience*, which generates and transmits tacit knowledge. Also, *dialogue* and collective reflection across the STEM team enables externalisation to turn tacit knowledge into explicit knowledge which can be shared with others.

Step 2: Supporting ideas – germinating

In a school that supports new ideas – new ideas will come, and just as likely (more likely?) from the newly qualified teacher as much as the more experienced. Such ideas may need protection from the cold frost of cynicism.

Step 3: Selecting the most promising ideas – thinning

Not all new ideas can be picked up and enacted at the same time but the ones that are selected need to be done so with a clear rationale. The criteria for selection of the best must be clear and those whose ideas are not pursued immediately should not lose face.

Step 4: Developing ideas into knowledge and practice – shaping and pruning

This is difficult – showing that the new idea is worthwhile and really works. Also, if something is not working any more it is the responsibility of the senior staff to move practice forward and so take on the new methodology. This may, for example, be by embedding the new content or teaching strategy in a scheme of work.

Step 5: Disseminating knowledge and practice – showing and exchanging

An effective school management team will create channels of communication in a school so that the outcomes of knowledge creation are shared across all staff. In earlier chapters, we talked about respecting STEM subjects other than your own and appreciating their value and educational intentions as being essential for STEM to flourish. It is clear that creating and disseminating knowledge of teaching should be across the whole school and be considered a two-way street.

(Hargreaves, 2001: 29–33, abridged and adapted)

for staff and pupils, the use of school social media for important messages has seen the once-ubiquitous school Tannoy system mercifully consigned to the dustbin.

When results and high-stakes inspections are so important, it is a brave head teacher that will back their staff when they wish to move away from the orthodox and try something new. It is exciting and motivating when one is allowed to take risks with one's teaching – it is reasonable, however, that the senior management is told about it first.

New experiences

Everyone needs new experiences to be intellectually stimulated. That often happens, of course, in the classroom. I must have taught 'Ohm's law' tens of times but on every occasion, even when I set about teaching in a similar way, the reaction of the class would be different and the experience would be new. As we saw from the job description above 'The responsibilities will be reviewed annually [...] and may be subject to amendment or modification at any time after consultation with the post holder'. It is important that all staff teaching STEM subjects have clear responsibilities that enable them to 'look sideways' at what others are doing but changing those responsibilities for teaching younger and older students helps to keep staff fresh and the work interesting. Being able to contribute to the teaching of electronics and control and systems in design & technology or to computer science is also a stimulating new experience for a physical science teacher.

As a school leader, one knows that the professional environment is healthy when a colleague comes to ask to run a STEM challenge during a lunch hour, part of an after-school club or as part of a project with a particular group of pupils. I went online today and found a range of possible group challenges such as designing a CAD Formula 1 car, a RoboFest robot and a video game challenge; and others run annually by multinational companies such as BP and Toyota. Although we would suggest that STEM is much more than just these extra-curricular peripheral events, it is certainly the case that such activities gives a buzz to STEM teaching in any school and, if carefully selected, appeals to both boys and girls.

New ideas and new experiences can also come along through INSET professional development and if through systematic appraisal procedures a school is able to contribute to formal qualifications, teachers feel valued and a professional environment that recognises and supports such individual need for teacher development, which can be aligned with the collective department and school agenda.

New experiences need not be lonely ones. I have often been told to 'get knotted' by colleagues but in one school 'knotworking' was used as an interesting technique to kick off something new. A member of the science department had heard of quantum technologies and thought it would be an interesting area of new technologies that could interest colleagues teaching older pupils in design & technology too. She found that by linking together inputs from the UK Quantum Technologies Programme (Quantum Technologies, 2020) and the Engineering and Physical Sciences Research Council, the quantum technologies programme aims to help young people understand concepts of quantum physics and technology. The programme:

- Brings Quantum Ambassadors and researchers in the field of quantum physics into the classroom.
- Brings cutting-edge physics and technology to life.
- Aimed at upper high school students of physics or computer science, the programme demonstrates what fundamental quantum science is and how it works with future technologies.

She became enthusiastic about the new experience that would provide a range of support to help her lead quantum physics activities in the classroom:

- curriculum enrichment visits from leading researchers in quantum technology;
- free teaching resources for her students, written by teachers for teachers;
- professional development for teachers showcasing how to create a buzz in the school around this exciting topic.

Yrjö Engeström and his colleagues (see Engeström et al., 2012), use the idea of 'knotworking' to describe how a group of people can come together to do various strands of activity to tackle a particular task or problem. In knotworking, the tying and untying of a knot from separate threads of activity is not linked to any specific individual or fixed organisational entity, such as a department, as centre of control or authority. Rather, the knot brings together interested participants from different communities of practice to solve a particular problem. In this case, the quantum technologies programme is an attempt to ignite new creative thinking. The knot was created not only from teachers of science, but also design & technology and the two Quantum Ambassadors and researchers. Once the quantum technologies initiative was established and up and running, the 'knot' was untied as it had served its purpose. Knotworking is a useful technique for STEM as it recognises that there are a range of stakeholders, which can all contribute to the different strands of activity needed.

Respect

Probably every generation of teachers, and in Africa and Asia as well as the USA and Europe, has felt a certain lack of respect from the society of which they are a part. Never well paid, teachers are often blamed for the ills of society. A vital factor in creating a professional environment where teachers are committed to working to improve the teaching and learning in STEM is for the senior management team of a school to make sure it knows what is being done and ensuring such commitment is recognised. All teachers are good with people; to last any time in the profession they must be. Senior staff must be the best. It is the task of the senior school leaders to ensure that they recognise the importance of interpersonal relationships and are seen around the school by pupils and staff. Quite simply, they should set aside some time each day for thanking people. However, just as important is that respect is not just 'top down', it is also peer to peer. To be able to look sideways to work with colleagues within one's own department and across STEM subjects one needs to be respectful of that privilege.

An environment that creates respect between staff so that the seeds of new ideas can grow, links naturally into rules for the classroom that creates similar respect between pupils. Although formal 'school rules' are important so that all know what is expected of them, far more important is the 'rules' of how certain activities are carried out that encourages respect between pupils. I was in a school in Wales recently and the following was on the wall: Our collaboration rules:

- every suggestion is written down;
- words already on the sheet will spark off other ideas;
- no one's suggestion is discussed [initially];
- no one's suggestion is ignored or 'rubbished'.

One of the great pleasures of teaching is that one does have an opportunity to impact in a positive way on pupils' lives and generally pupils do recognise that. Establishing an environment for both teachers and pupils where values are identified and shared, aims and objectives agreed and teaching methods approved encourages respect between staff and pupils. Some schools make this opening up of the needs of pupils, and the responses by teaching and other school staff, a formal process by a policy of hearing the 'Student Voice' through School Councils or Parliaments. Others informally ensure that all pupils, whatever their interests and talents, are recognised and built on through their project work in science or design & technology (see Chapter 6). STEM subjects draw on and are relevant to 'real life' and respect is not only important at the inter-personal level but also at the level of appreciating the contribution to STEM of other domains of knowledge. Respecting STEM subjects other than your own and appreciating their value and educational intentions is essential for STEM to flourish.

Conclusion

STEM is much more than one off projects, off-timetable activities to enliven the post-exam period or a thinly veiled excuse to entice young people into the manufacturing industries. Rather, the drawing together of the teachers of science, technology and engineering, and mathematics so that pupil work in one area can support and enhance their understanding in another is both efficient in classroom time and supports the way that we know young people learn. We have seen in Chapter 2 that whether teachers coordinate to support teaching across two areas, collaborate to work on a joint project, or integrate their work in a club or for a special project, staff need to look sideways at what colleagues are doing. If the STEM subject silos that have existed for so long in secondary schools are to be made much more 'porous' then a whole school approach is required that addresses the physical environment of the school, the learning needs of both girls and boys and the professional needs of their teachers. Sustaining the change is important.

> In times of change the learners will inherit the earth, while the knowers will find themselves beautifully equipped to deal with a world that no longer exists.
> (Eric Hoffer in Smith, 1996: 15)

Recommended reading

The physical environment

Andrew-Power, K., & Gormley, C. (2009) *Display for learning*. London: Bloomsbury.
Caviglioli, O. (2019a) *Dual coding with teachers*. Woodbridge, UK: John Catt Educational.

The pupils' learning environment

Dorantes-Gonzalez, D. J., & Balsa-Yepes, A. (2020) A neuroscience-based learning technique: Framework and application to STEM. *International Journal of Educational and Pedagogical Sciences*, 14(3), 197–200.
Weinstein, Y., Sumeracki, M., & Oliver Caviglioli, O. (2019) *Understanding how we learn: A visual guide*. Abingdon, UK: Routledge.

The teachers' professional environment

Duschl, R. A., & Bismack, A. S. (2016) *Reconceptualising STEM education*. New York: Routledge.
Jones, G. (2018) *Evidence-based school leadership and management: A practical guide*. London: Sage.

References

Adey, P. (2001) In need of second thoughts. *Times Educational Supplement*, 23 February 2001.

Barlex, D. (2000) Perspectives on departmental organisation and children's learning through the Nuffield Design and Technology Project. In J. Eggleston (ed.), *Teaching and learning design and technology*. London: Continuum.

Bereiter, C., & Scardamalia, M. (1996) Re-thinking learning. In D.R. Olson & N. Torrance (eds), *The handbook of education and human development: New models of learning, teaching and schooling.* New York: Blackwells.

Brighouse, T. (1991) *What makes a good school?* Stafford, UK: Network Educational Press.

Caviglioli, O. (2019b) *Dual coding with teachers*. Woodbridge, UK: John Catt Educational.

Desforges, C. (2001) Familiar challenges and new approaches: Necessary advances in theory and methods in research on teaching and learning. In *Nuttall/Carfax memorial Lecture, Cardiff 2000*. Southwell: British Educational Research Association.

Engeström, Y., Kaatrakoski, H., Kaiponen, P., Lahikainen, J., Laitinen, A., Myllys, H., Rantavuori, J., & Sinikara, K. (2012) Knotworking in academic libraries: Two case studies from the University of Helsinki. *Liber Quarterly*, 21(4), 387–405.

FutureLearn (2020) www.futurelearn.com/ (accessed June 5 2020).

Hargreaves, D. (2001) *Creative professionalism: The role of teachers in the knowledge society*. London: Demos.

Institute of Physics (2020) *Inclusive teaching*. https://education.gov.scot/improvement/Documents/UPDATEDtop-ten-tips-science.pdf (accessed June 5 2020).

Kaufer, D. (2011) *Neuroscience and how students learn*. https://gsi.berkeley.edu/gsi-guide-contents/learning-theory-research/neuroscience/#further (accessed June 5 2020).

Microsoft (2017) *Why don't European girls like science or technology?* https://news.microsoft.com/europe/features/dont-european-girls-like-science-technology/#sm.000093k65v2p9dm-fq8h1jr6pzzk8l (accessed June 5 2020).

OpenLearn (2020) www.open.edu/openlearn/ (accessed June 5 2020).

ORBIT (2020) www.educ.cam.ac.uk/research/projects/orbit/ (accessed June 5 2020).

Pamona College (2020) *Physics timelines*. https://research.pomona.edu/pt/ (accessed June 5 2020).

Quantum technologies (2020) *UK Quantum Technologies Programme*. www.stem.org.uk/quantum-technologies (accessed June 5 2020).

Royal Society (2011) *Neuroscience: Implications for education and lifelong learning*. RS Policy document 02/11, London: The Royal Society.

Seery, N. (2020) Pedagogy involving social and cognitive interaction between teachers and pupils. In J. Williams & D. Barlex (eds), *Pedagogy for technology education in secondary schools – Research informed perspectives for classroom teachers*. The Netherlands: Springer.

Smith, A. (1996) *Accelerated learning in the classroom*. Stafford UK: Network Educational Press.

Smith, A., & Call, N. (1999) *The ALPS approach*. Stafford UK: Network Educational Press.

Society for Canadian Women in Science and Technology(2020) *Women in STEM*. https://scwist.ca/resources/woman-in-stem-resources (accessed August 6 2020).

SSAT (2009) *Leading practice in STEM*. London: Specialist Schools and Academies Trust.

STEM Role Models https://medium.com/nevertheless-podcast (accessed June 5 2020).

Swailes, R. (2019) Choose it, use it, put it away. In O. Caviglioli (ed.), *Dual coding with teachers*. Woodbridge, UK: John Catt Educational.

Women in Mathematics (2020) *Famous women mathematicians*. www.amightygirl.com/blog?p=23086 (accessed June 5 2020).

11

Looking at STEM education in different countries

Introduction

In Chapter 1, we looked into the past to see how what we now call 'STEM education' has developed. In earlier chapters, we have set out the benefits to learners if their teachers 'look sideways' and take note of what is being taught in other aspects of STEM. In these final two chapters, we now look to the possible future of STEM education during the third decade of the twenty-first century. In this chapter we consider STEM across the world, with guest pieces written by experienced STEM educators describing the current position and the possible future situation of STEM education in the secondary schools in their countries.

STEM Across the World

In this section authors from Australia, Belgium, Brazil, China, Israel, Russia, Taiwan, and the USA write about STEM education in their particular countries. In each case, the piece is divided into three sections: the current situation, an exemplar and future developments. The authors, along with their countries are shown in Panel 11.1. Each piece has been extracted from a longer piece, all of which can be found at the website https://dandtfordandt.wordpress.com in the Papers section under STEM Papers.

STEM in Australia, David Ellis and John Williams

The current situation of STEM education in secondary schools

STEM projects have tended to be implemented more in lower secondary classes than in upper secondary because of subject demands. Lower secondary (Years 8–10) has more timetabling flexibility and a less rigid curriculum, whereas the upper secondary timetable is consumed with the preparation of students for their university entrance examinations at the end of Year 12.

PANEL 11.1 Authors from across the world who have written about STEM in their country

However, there are two factors that are leading to increased incorporation of STEM integrated activities in upper secondary classes. One is the increasing disillusionment with the university entrance examination system. While the examination scores remain a reasonable predictor of university success, it is increasingly recognised that they are not the best way to prepare for many of the professions, and so are now not the only mode of entry into university. Many university courses provide for entry by interview or portfolio pathways, which free up the Year 11–12 curriculum to incorporate, for example, integrated STEM activities that provide for discipline content learning and also have a portfolio outcome, which can be used as a component of university entrance.

The other factor that is leading to an increase of integrated STEM activities in the upper secondary years is the realisation that defined discipline content can be taught and learnt through carefully constructed integrated activity. It has been the attitude in the past, and remains in some contexts, that there is too much curriculum to cover in class, so no time can be 'given up' for STEM activities, resulting in STEM after school clubs or lunchtime projects. Teachers now realise that carefully co-constructed integrated activities can be quite an effective structure for the learning and application of disciplinary content. Consequently, some schools are enabling students to take fewer traditional Year 12 subjects, and rather enrol in an integrated STEM project-based subject, and not only achieve their university entrance goals, but better prepare themselves for further study in their chosen profession.

An exemplar of STEM education

Integrated STEM: A secondary school has decided to use a STEM project in Year 9 as the vehicle to achieve the school-wide Approaches to Learning (AtL) goals of:

- Self-Management: Organisational Skills – Plan short and long-term assignments; meets deadlines.
- Research: Information Literacy – Collect and analyse data to identify solutions and make informed decisions.
- Communication: Language Skills – Organise and depict information logically.

The design & technology department in the school leads the project, but works cooperatively with the science and mathematics departments. Each of these three departments agreed to contribute one class each week to the STEM project, which was then timetabled over three consecutive periods on one day each week for Year 9s. The departments agreed to this approach on the understanding that the STEM project would be co-designed in such a way as to enable the achievement of elements of each subject's Year 9 curriculum. So, the subjects are integrating in such a way that general goals of learning are being achieved, but also each discipline's curriculum content is being addressed. At the same time, students are working in an interdisciplinary context and learning how to work in teams.

The STEM project has a different focus each year. It is conducted in an open space, with access to specialist rooms. The equivalent of three classes (about 60 students) are timetabled at the same time, together with the three teachers (technology, mathematics and science) who act as resources for the groups of students working together.

Whole group presentations focus on the knowledge students need in order to develop solutions to problems, while at the same time addressing the content learning needs of each curriculum area. This is a very popular class in Year 9.

The future development of STEM education in secondary schools

The integrated version of STEM in Australian schools resides around the philosophical perspective of contextualised or situated learning (Putnam & Borko, 2000; Brown, Collins, & Duguid, 1989), and the benefits of integrating STEM education being considered 'best-practice' in the understanding of the relationships between disciplinary content and its relevance to the world around them (Burrows et al., 2017; Sanders, 2012). Schools that offer integrated or what has been labelled, in some cases, iSTEM (iSTEM, 2020), offer a 'stand-alone' elective course of study in a subject that integrates science, mathematics and technological concepts using pedagogical approaches such as project-based, problem-based, or inquiry-based learning (Kelley & Knowles, 2016). Space is made in the curriculum through the addition of iSTEM in a student's list of subject options. One popular example of an integrated approach are the 262 schools in NSW that are implementing the iSTEM course approved by the NSW Education Standards Authority in 2013. Developed in collaboration with local industries to address local needs, the curriculum 'presents maths and sciences to students in ways that challenge not only their understanding of these key subjects but also their ability to manage projects and work in teams' (Education Council, 2019, p. 17).

Differences in interpretation of what STEM education is, has also enabled a diversity of non-integrated approaches to STEM. Non-integrated STEM is still very much disciplinary based and may focus on one or a couple of disciplines (Blackley & Howell, 2015). According to Barlow and Ellis (2016), the genesis of STEM in Australia, emanating from the Office of the Chief Scientist, has presented a bias towards science to increase Australia's scientific and innovative competitiveness (Williams, 2011). Bias among the disciplines and non-integrated STEM has been facilitated by the availability of funding under a STEM label, yet is clearly oriented towards the specific funding goals. As an example, the 'Inspiring Australia – Science Engagement Programme' provides money for 'Students Science Engagement and International Competitions' (Australian Government, 2020), and has been utilised in schools to support science projects under the STEM banner.

STEM in Belgium, Didier Van de Velde

The current situation of STEM education in secondary schools

In the reformed Flemish secondary education (Flemish Government, 2018) we find the acronym STEM being applied to a group of tracks including pre-academic natural and 'industrial science' tracks as well as vocational education tracks. In middle school, besides compulsory math, science and technology, we find integrative STEM as an elective subject that allows young people to orientate themselves in a field of further secondary study. Besides this, the acronym STEM is more broadly used in government policies that want to encourage young people to consider a study track in higher secondary and post-secondary education that is in line with one or more components of STEM. These developments are sometimes combined with a pedagogical discourse that connects STEM with more active and integrative (and therefore more meaningful) learning. This approach aims to make the 'STEM subjects' more relevant and motivating (Flemish Government, 2015). We have seen an important number of secondary schools in recent years seize 'STEM initiatives' as a path to implement more integrative pedagogies.

In the Flemish secondary curriculum, the curriculum standards are grouped according to the European key competences (European Parliament, 2006). This enables the opportunity to better coordinate the learning progression in the various subjects: mathematics, natural sciences and technology. Besides this, a group of compulsory overarching curriculum standards have been developed that define 'STEM-practices' for all pupils in secondary education incorporating inquiry, design, problem solving and modelling. We find similar skills were set out in older curriculum documents, but these were often described from the position of either science or technology or mathematics, but not then from an overarching perspective. That now creates a new dynamic. These 'STEM-practices' are described as follows:

- applying a scientific method to develop reliable knowledge and to answer questions;
- analysing natural and technological systems using cross-cutting STEM-concepts;
- using instruments and tools with the necessary accuracy to observe, measure, experiment and investigate;
- working safe and sustainable with materials, chemicals, technological and biological systems;
- using measured values, quantities and units;
- investigating relationships between quantities in a quantitative way;
- developing scientific, technological, and mathematical models to visualise, investigate, solve and explain;
- designing a solution for a contemporary problem by using concepts and practices from STEM-disciplines;
- arguing choices from different perspectives in the design and use of technological systems and other STEM solutions;
- investigating interactions between STEM-disciplines and with society.

These 'STEM-practices' provide all subject teachers involved with a common set of goals with accompanying discourse that encourages coordination in learning STEM.

An exemplar of STEM education

Integrative pedagogies and STEM education

As technology education has embraced an integrative pedagogy focused around design, need-to-know resource tasks and case studies, we can observe that trend also in the natural sciences and mathematics, where more attention is paid to contexts, problem solving, modelling and systems thinking (Eurydice, 2011; Artique et al., 2012; European Commission, 2015). This reinforces arguments to support cooperation between subjects towards more meaningful STEM education.

Engineering design in the context of physics and math interactions: the prototype vehicle project

The 'prototype vehicle project' (STEM@school, 2018) offers an example of such a project designed to take about ten weeks to complete, with one hour/week of physics and three hours/week of engineering. The five hours/week of mathematics lessons were sometimes incorporated within both the physics lessons and engineering design on a 'just-in-time' basis when and where needed. Grade 9 students are tasked with designing and programming a prototype vehicle able to travel along a cascading array of green lights, the green wave, without having to stop at intersections. Completing the design task required learners to develop knowledge about kinematics, linear functions, and Arduino programming in the process of designing a mechanical prototype with drives and transmissions. To calibrate their system, students used a graphical representation of a function to map the controllers digital output value (representing the voltage on a controllable DC-motor) to the speed of the vehicle. And because the vehicle travels the green wave in a straight line within a run-on zone, both the steering and the non-linear acceleration phases need not be considered.

To address the sociocultural engineering parameters associated with such autonomous transportation systems, students were required to conduct a case study as part of the project. In doing so, they investigated the pros and cons of autonomous vehicles, starting first with an analysis of their own transport choices. By reflecting on a wide range of variables from traffic congestion to pollution, students generated transportation solutions designed to provide more sustainable forms of mobility. In this way the project aligned with technology education content and practices authentically situated within the broader sociocultural context of engineering design.

The pedagogical adjustments required to implement the STEM projects were imparting a new instructional paradigm on teachers where their concept of learning progression evolved from teaching math first, using that acquired knowledge in science, followed by application in technology, to a more integrative view where interdisciplinary interactions occur in a more natural way (Thibaut et al., 2018).

The future development of STEM education in secondary schools

The looser relationship between standards and subjects in the new Flemish curriculum makes it easier to combine subject matter in school practice. For instance, a gradually growing group of schools is working on a more integrated curriculum for natural sciences, geography (human and physical) and technology in the middle school. In Flanders, the legally validated curriculum is converted into a more manageable and more operational curriculum document for teachers, adding pedagogical recommendations. This allows school boards or networks of school boards (such as the network organisation of Catholic schools) to operationalise their 'freedom of education' that is enshrined in the constitution. In this way, the different roles of educational actors are defined: the 'what' is separated from the 'how'. In response to the new Flemish curriculum, curriculum guidelines for teachers have been developed supporting interdisciplinary interactions between the subjects natural sciences, geography and technology in lower secondary education.

Adoption of these more integrative pedagogical practices is not without its challenges. Recent literature review on the effects of subject integration confirms many critical points (Wilschut & Pijls, 2018). Most important is having a shared STEM mission among collaborating teachers and the school administration, providing an educationally supportive environment for all subject teachers involved (Van de Velde et al., 2016). The implementation process can be supported by a group of pioneers with teachers, coordinators and directors: a 'leading coalition' in the school. It is important that teachers and school leaders reflect on what they want to achieve and choose the appropriate resources. Teachers involved must have cooperation skills. Other critical elements include sufficient time for teachers to meet and co-plan the integration of inquiry and design-based learning approaches and a spirit of collaboration among all teachers involved. Teams having a strong collaborative spirit report better educational outcomes, which creates mutual dependencies among teachers and the potential to challenge established pedagogical beliefs among those involved. Also critical for successful STEM initiatives is motivating teachers to become involved and collaborate in integrative STEM education. As such, involvement of all school subjects becomes a relevant factor for building broad interdisciplinary school support.

Where cross-curricular cooperation in a complementary team runs smoothly, the experiences of teachers are often positive. They experience it as a meaningful way of professionalisation. According to educational research, if it is possible to develop such intensive partnerships, this dynamic also contributes to the innovative capacity of the school involved (Van der Bolt et al., 2006). Teachers involved in STEM projects report that the interdisciplinary approach, where inquiry is embedded within the design process, resulted in students more motivated to learn compared to traditional science instruction. Furthermore, participating math teachers, who at first were sceptical of the interdisciplinary approach and feared the potential loss of critical deductive thinking, were soon convinced of the motivational value of having students gather their own data for use in math courses. Specifically, math teachers realised that their students were better able to recognise the importance of different representations (tables, graphs, formulae) and the possibilities for modelling and predicting system behaviours (Van de Velde et al., 2016).

We can conclude that the international STEM dynamic has not missed its effects on Flemish secondary education. We can observe an effect on the proliferation of STEM-related study-tracks, a better aligned new curricula and growing pedagogical cooperation among STEM-teachers in schools. The new STEM curriculum will offer opportunities for more meaningful education and professional development of teachers in the future. More Flemish pupils have opted for general STEM education in secondary and higher education over the past decade, however STEM-vocational oriented secondary education is lagging behind.

STEM in Brazil, Vitor Mann

The current situation of STEM education in a secondary school

ORT School Brazil, a school where I had the privilege of study and where I have been work as a teacher since 2005, was created within the philosophy of 'learning by doing' ('hands on'), where the experience in the laboratories translates as a pedagogical reality. We are a school with more laboratories than classrooms, which characterises us as an environment of many discoveries, where our students develop a high degree of autonomy and creativity.

Today, in our junior high curriculum, we have three defined axes:

- General education (Portuguese, literature, writing, mathematics, history, geography, art, English and physical education);
- scientific and technological (physics, chemistry, biology, informatics and introduction to technology);
- Jewish education (Hebrew, Jewish culture and education);
- extracurricular activities (science club, robotics, programming, sports, theatre, dance, magic and board games).

The STEM philosophy has been part of our educational activity since 2010, when we started using this pedagogical proposal within our curriculum model, which led us to reconsider our didactic and evaluative methodologies. Coming from a conception of scientific and technological education with an experimental basis, a maker perspective, the adaptation to this philosophy occurred in a very natural and conscious way. The STEM philosophy helped us to be clear about our objectives and methodology, ratifying the three methodological approaches that guide our educational practice. In a very objective way, I can say that the application of this educational philosophy, as illustrated in Figure 11.1, has proved to be very efficient, combining three fronts of action:

- theoretical classes (information and contextualisation);
- practical classes (knowledge consolidation);
- author projects (construction of knowledge).

Theoretical classes are quite significant educational instruments and, today, due to their misunderstand application, they have gained a very negative status. Through

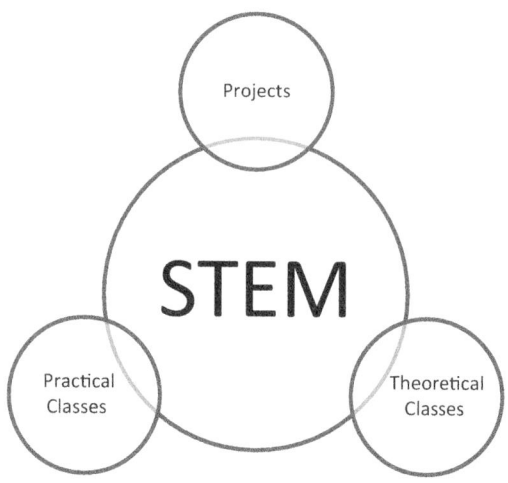

FIGURE 11.1 Three fronts of action in STEM

these moments of explanation, we have a unique opportunity to captivate and con-textualise the importance, be it academic or social, of the information presented. It is a critical moment, without which practical classes or projects cannot be developed.

Both in academic circles and in common sense, the theoretical classes have gained a pejorative meaning, justified by their passive and not creative approach. Therefore, these classes would not allow students an adequate environment for learning. Living in an Information Age, the expository and guiding classes would not appropriate in the daily lives of our students.

However, in view of our educational experience, we realise that the theoretical classes are moments of great pedagogical value. These classes are responsible for pre-senting the themes and concepts to be studied, allowing the teacher to contextualise future practical activities and projects. It is the perfect time for the teacher to 'catch' the students' attention and demonstrate his 'passion' for the contents, emphasising its importance for the young people's learning process.

When we gave up this first theoretical approach, we noticed that students were not motivated and even disinterested. A 'mechanic work' is produced in the devel-opment of the experiments and, many times, the projects do not reach the level of depth and relevance in which they were conceived. In this sense, we realise that when a first presentation of information, concepts and themes does not occur in an adequate and satisfactory way, the students' commitment to the activities is notori-ously impaired.

Theoretical classes do not allow students moments of autonomy and creativity, but they allow teachers to establish links between students and the content to be learned.

Practical classes, activities that follow theoretical classes, allow students to con-solidate knowledge, enabling them to experience controlled and oriented situations towards the construction of authorial knowledge. Faced with these activities, the teacher has a central role in the development of experiments, organisation of classes and presentation of problems. It is up to the teacher to build a controlled environ-ment, where after a safe and assertive way, all students will succeed in building their knowledge.

The practical classes, in laboratories or other creative spaces, allow students to have a concrete experience with certain level of autonomy. Despite all the previous work developed by the teachers, a significant flexibility in the learning processes is inher-ent to these activities. There are 'N' ways to conduct an experiment and interpret its results, which displaces the teacher from the comfortable position of owner of knowledge. It is a minimally controlled environment, considering that it is the role of teachers to ensure success in the learning process, but it is clear that practical activities enable students to follow their own paths for an effective transformation of informa-tion into knowledge.

In this context, the development of reports, as well as the discussion of results, is essential tools for a satisfactory appropriation of knowledge. The practical classes demand well-developed scripts (previously defined by the teachers), offering stu-dents a healthy environment for experiencing concrete situations and build their own knowledge. It stands out as an environment of little creativity, but that allows a real experience of autonomy and independence.

Finally, the projects are activities that demand high creative capacity from the stu-dent, as well as an effective social change in the role of the teacher. In this context, the teacher has the role of tutor, losing his or her hegemonic position of authority. An environment of cooperation is created where, effectively, the knowledge to be learned results from the interest and individual ability of the students. It is an environment of very little control, but where very complex knowledge and relationships can be built in a very natural way.

Understood as an educational tool (learning methodology), we can affirm that the project approach is the most complete pedagogical experience. It offers students a high level of autonomy and creativity, requiring a high degree of commitment and dedication. There is no possibility to produce a project without motivated and involved students to develop it. There is no place for imposition here, only cooperation and guidance

An exemplar of STEM education

Using a concrete example, the theme 'plastics and recycling' is pertinent in all disciplines, each developing skills and competences within its curricular scope. Theoretical classes use resources such as articles and films to mobilise students, seeking to commit them to the topic in question. Associated with these activities, practical classes constitute a series of activities and experiments that aim to enable students to develop basic knowledge inherent to the theme. A practical activity related to the density of plastic materials, when concomitant with a theoretical context, allows students to build complex relationships (concepts and skills), which will be required in the academic future of these young people.

In contrast, the projects and their holistic approaches enable students to build the skills and competencies necessary for the formation of a citizen. It is knowledge that is not restricted to the academic environment, but that needs a solid and efficient academic formation to be achieved. In the case of plastics, a possible project would be the 'production of biodegradable plastics from natural sources', a job without curricular or disciplinary limitations, but which requires good training to be conducted in the correct way. A student cannot develop a bioplastics project if he does not understand and knows certain concepts such as: what is a plastic, what are its impacts on the environment, what is degradation and biodegradation, what are natural resources and development sustainable?

For the construction of projects, it is necessary to consolidate a complete and complex theoretical/curricular base. Therefore, for students to have 'freedom' it is necessary that they have a good orientation and contextualisation. A certain degree of freedom is not reached without good academic formation.

The future development of STEM education in secondary schools

The projects approached considered and ratify the complexity and holistic vision inherent of knowledge, enabling students to have a more natural and sensitive learning process. In this perspective, there are no labels, limitations, definitions or categorisation, it is resulting in the consolidation of a more realistic and comprehensive view of the knowledge built. The exercise of autonomy and creativity are inherent in this methodology, leaving the teacher with a very significant role of guide and evaluator.

Contradictorily, here we find that a project-based approach, while very creative and interesting, does not allow students to consolidate basic knowledge. The high degree of independence and autonomy, as well as the limited control of teachers, do not guarantee students basic competences that will be demanded of them in future evaluations and selection systems. We can affirm that the development of projects without a theoretical curriculum base previously established and idealised, results in the formation of notoriously unprepared students.

Therefore, understanding this perspective and seeking to consolidate the formation of complex and complete subjects, going forward our curricular format comprises:

Curricular courses (mandatory)

- 40% theoretical classes;
- 40% practical classes;
- 20% projects.

Extracurricular courses (optional)

- 100% projects.

STEM in China, Yang Chunling and Ke Shan

The current situation of STEM education in secondary schools

The introduction of STEM education could be seen in the field of science and technology education from 2001 in China. Since 2012, research on STEM has begun to flourish, reaching a climax in 2016 (China STEM Education White Paper, 2017). Since 2016, many regional governments have released policy papers emphasising the overall planning and systematic implementation of STEM education in their regional schools. In 2017, Chinese National Institute of Education Sciences promulgated the '2017 China STEM Education White Paper' emphasising that STEM education is in line with China's economic development and talents cultivation. In February 2017, the new standard for the primary science curriculum was launched and explicitly stated that STEM content should be integrated into the primary school curriculum. In September 2017, the Ministry of Education issued 'Outline of Curriculum Guidance for Comprehensive Practical Activities in Primary and Secondary Schools', which stipulates that integrated practical curriculum must be integrated into the school curriculum. Also, in 2017, many universities including Peking University and Tsinghua University opened Maker and STEM education admissions (Ren Jing, 2018).

School education in China has been dominated by subject teaching. Mathematics, physics, chemistry and biology have been important subjects in Chinese secondary schools and have received widespread attention from schools and students. Relatively weak areas in Chinese schools are practical activities in technology and science, and the lack of the cultivation of students' practical skills and innovative skills. In recent years, the Ministry of Education has implemented a variety of educational reform strategies to improve quality education. Take promoting STEM education for example, the government hope to develop student's integrated knowledge of the science subjects and skills to solve real-life problems through the integrated STEM approach.

An exemplar of STEM education – 'Eco-Intelligent Scientific Research Station' project in Beijing Zhongguancun High School

Project background

Beijing Zhongguancun High School is a public school catering for students aged 12 to 17. This STEM project is based on the academic situation of junior students and their field investigations at a few science and technology investigation stations around Beijing. The project integrates knowledge in real-life situations, which engages the students to design solutions applying relevant knowledge of various disciplines. This integrated approach breaks through disciplinary boundaries and improves students' abilities to solve practical problems (Su & ZhengFu, 2019).

The implementation of the project

Based on the project-based learning process, the project course emphasises that the students are driven by the real problems they find to design an eco-intelligent scientific research station that meet the ecological characteristics of a specific area, and the students complete the whole process from design to model making.

The project lasts 15 weeks and is designed and implemented according to the teaching process of determining the theme, identifying problems, designing plans, collecting materials, designing and reporting project works. The more detailed project process and project objectives are as follows:

Stage one: Each student group identify the problems need to be solved through collecting and reading related literature, studying subject knowledge, sharing ideas with other groups, and analysing problems with scientific methods.

Stage two: Each group design a plan for building an eco-intelligent scientific research station. In this design process, the students learn and master basic scientific design ideas and techniques. They use knowledge of physics, geography, mathematics, information technology and other subject knowledge to solve real-world problems. Group work develops their teamwork spirit and their communication, negotiation and expression skills.

Step three: Each group make a model for their designed research station and the making process involves using 3D printing, laser cutting, turning control, woodworking and metalworking. This process cultivates the students to think like a real engineer and improves their abilities to transform their design ideas into practical models. Table 11.1 shows the complete implementation process of the project:

Project evaluation

In this STEM project, the scheme of each group is different. In addition, the course involves multi-disciplinary integration, and the approaches for completing the project are diverse. Therefore, the school uses a variety of course evaluation methods include the following categories:

Formative evaluation. The formative evaluation runs through every step of the project implementation. The evaluation focuses on the students' learning process in the design and production of the research station, not the result. Evaluation methods include student self-evaluation, peer evaluation and teacher comments.

TABLE 11.1 Project implementation

Week	Content	Project Procedure
1	Identify research themes and problems	• Conduct topic discussions based on the questions raised by the students within each group. Before the discussion, the students have visited different scientific research stations located in different places such as wetlands, deserts, grasslands, seashores, forests, etc. They raise their questions based on their real experience. • Select the area of the research station to be established. Design a model of a more complete intelligent research station based on the current status of the station in the selected region.
2	Develop project plan	• Determine the background, purpose, and significance of the project. • Project content o The characteristics of houses in different regions o The stability factors of houses o The adaptability of houses to the ecological environment (such as how to adapt to alternative environments) o Intelligent facilities of the house o The aesthetic factors of the house • Formulate the stages and weekly plan for the project. • Identify the conditions required for the project. • Division of labour among team members. • Clarify the necessary group work system.
3	Modify the project plan	Adjust the project plan based on group self-evaluation and the evaluation from other groups and the teacher comments. Post the final plan on the display board, and each group can share their ideas and the progress of their work.
4	Collect related literature	Master the method of literature collection. Combining the literature with their field trips to the research stations, the students investigate the status and problems of the selected stations.
5	Design the house – part I	Based on the information collected and group discussions, each group draw the sketch of their design of the house of the research station considering the function of the house for scientific research purpose and the aesthetics of the building. Then each group present their sketch and share ideas.
6	Design the house – part II	Modify the design of the house combining the stability factors of the house in different environments. Then each group present their sketch and share ideas.
7	Make a model of the house based on the design completed in weeks 5 and 6	Make a model of the house to validate the design combining aesthetics and robustness and make improvements to the design accordingly.
8	Design the house – part III	Modify the design of the house adding the factors of energy, ecological adaptation and environment protection. Then each group present their sketch and share ideas.
9	Design the house – part IV	Modify the design of the house to make it an intelligent house. Based on the academic situation of different students, the intelligent design requirements are divided into three levels as follows: • All students learn basic programming in class, which enables them to design a scientific research station with basic intelligence • Students with technical expertise can perform high-level design • Students who are particularly interested in this project can participate in additional courses after class to learn about functional design and improve the intelligence level of their research stations

(Continued)

TABLE 1.1 (Continued)

Week	Content	Project Procedure
10	Finalise the design	Based on the design and modification of the previous weeks, modify the design and determine the final design based on the discussions of the group members and the suggestions of other groups and the teacher.
11	Complete the engineering drawing of the research station	Draw the engineering drawing of the station using related computer software. For the specific use of the software, the students can ask the teacher for help or search the internet for instructions.
12/13	Make the model	Choose the production method according to the selected materials. It can be completely handmade, or it can use 3D printing, laser cutting, woodworking and other technologies. It is recommended that students who have learned how to build models be the team leader. If students encounter problems, they can ask other teachers for advice.
14	Write project report	The report can include an explanation of how to make a good 'Eco-Intelligent Scientific Research Station' model and should also include a reflection on the project process, as well as empirical research materials such as design drawings and working photos of student groups.
15	Submit project work	Submit group work and share it in class and with other classes of the year group.

Demonstrative evaluation. Through group display and personal presentation, evaluate the students' participation in the whole project.

Overall, the 'Eco-Intelligent Scientific Research Station' STEM project is a comprehensive STEM course that integrates interdisciplinary knowledge and develops students' multiple skills. Such a STEM project basically reflects the understanding and requirements of Chinese secondary schools for STEM education as current stage.

The future development of STEM education in secondary schools

'China STEM Education 2029 Innovation Action Plan' (see STEM Education Research Centre, 2018, 2019) (The Plan hereafter) was launched in May of 2018 opening up a new stage in the systematic development of STEM education in China. The Plan aims to train a group of future-oriented innovative talents for China, improve students' scientific inquiry ability, innovative awareness and the ability to solve complex problems. As Wang Su, Director of STEM Education Research Centre of National Institute of Education Sciences, pointed out, to achieve the goal of building China into an innovative country, China needs to train and reserve talents in advance. It is hoped that through the efforts of the next decade, there will emerge more innovative talents with international competitiveness in China.

The Plan explains the connotation of STEM education and how to promote STEM education regarding the challenges for China's introduction of STEM.

In the Chinese context, the implications of STEM education are explained from five dimensions:

1   STEM education should be included in the national innovative talent training strategy.

2   STEM education is a national lifelong learning activity.

3   STEM education is a coherent curriculum group across disciplines and segments.

4   STEM education is a carrier for cultivating the comprehensive quality of all students.

5   STEM education is an educational innovation practice involving the whole society.

At the launching ceremony of The Plan, Wang Su explained that The Plan emphasises the universality of participating institutions, calling for more social forces to collaborate in STEM education innovation; advocates that STEM education can benefit all students, especially students of special groups; hopes to cultivate the ability of innovative thinking and scientific inquiry, and change the evaluation method and innovation training mode by focusing on the measurement of the learning process.

The following seven aspects are proposed to promote the development of STEM education in China:

- Promote the top-level design of STEM education at the national level to facilitate the full implementation of STEM.
- Improve the STEM education curriculum and teaching system, and promote the effective connection of STEM content in all grades.
- Promote the professional development of STEM teachers.
- Establish the corresponding standards and evaluation system for STEM education.
- Promote the formation of a STEM education environment that the whole society values, and build an integrated STEM innovation ecosystem.
- Promote the unified thinking and understanding of the whole society, that is, to establish the strategic position of innovative human resources, mainly scientific and technological talents and innovative entrepreneurs, in the future development of the country.
- Summarise the experiences of effective STEM implementation cases and promote the successful model.

The first phase of The Plan focuses on STEM innovative talent strategy research, current situation research, and international comparative research. Meanwhile, a benchmark programme will be launched to lead schools to conduct research on STEM courses, teaching methods and evaluation to explore effective STEM education practice models and through a group of STEM pilot schools to cultivate STEM seed schools and STEM seed teachers.

The regional governments have been driving The Plan since it was launched, and more than 70 schools from different cities have become STEM pilot schools, more than 200 schools have been chosen as seed schools and over 70 teachers have been selected as seed teachers to share their experiences in implementing STEM. According to the research conducted by Fang HaoYing et al. (2019), 35 pilot schools being surveyed provide extensive STEM courses to the students.

STEM in Israel, Osnat Dagan

The current situation of STEM education in secondary schools

Junior high schools

Since 1995 there has been an integrated syllabus for science and technology in junior high school that reflects the STS (Science and Technology in Society) approach. Books and other learning materials for this integrative subject were developed with teacher guides and many hours of in-service teacher training materials were developed and delivered in order to implement the program's rationale and aims. This syllabus was updated several times over the years according to the Ministry of Education emphases and changes. The last update occurred in 2019 when meaningful learning aspects and twenty-first-century skills were adapted to this syllabus.

In 2011, the MoE launched a special program for excellence: 'Scientific and Technological Reserves' for grades 7–12. This six-year program aims to raise the number of students excelling in STEM subjects in high school. In order to fulfil this aim, excellent students in junior high school study additional hours in math, physics and computer sciences. In high school, they receive extra lessons as needed. Although the aim of this program is excellence in STEM, only three of the STEM subjects are covered (math, physics and computer sciences) and the integrative aspects of STEM are missing.

A new program aiming to motivate students to learn STEM subjects and enhance studying twenty-first-century skills is in development. A multi-sector committee agreed to prepare a program for long-term sustainable change in STEM excellence and decided to focus on junior high school while strengthening the infrastructure of knowledge, skills, capabilities and motivation. The main concepts of this program are to implement the knowledge base and skills of each discipline: sciences (chemistry, biology and physics), technology and engineering and the integration between them (Sheatufim, 2019).

Makers' spaces were established in various local authorities to enhance the innovation and entrepreneurship in some primary schools, but mostly for junior high school pupils.

Senior high schools

There are 32 tracks in high school technology education. These are divided into three categories: (a) science-based technology tracks, (b) technology tracks and (c) vocational tracks. It is only in the science-based technology tracks that some of the science, technology and engineering subjects are integrated (Dugger, 2010 – category b), for example, biotechnology, electronics, mechatronics, environmental sciences, scientific engineering tracks. In fact, the scientific engineering track is the only one that truly integrates STEM (Dugger, 2010 – category c).

The students study in technology tracks for three years. In their first year, they have to study either physics or technology sciences. During the second and the third years, they have to choose topics from their chosen track and develop projects in which they design and make solutions to everyday problems and present them to the external examiners. These projects implicitly encompass STEM ideas and concepts.

The scientific engineering track

The scientific engineering track is the only one that integrates all STEM subjects explicitly. This unique track actually existed prior to the STEM era, allowing students to study by way of analogies. Analogies that mean different things in different disciplines are chosen and all its meanings across those disciplines are studied. Examples include pace, objects and fields, resonance, waves, etc. These analogies are taken from sciences such as physics, biology and chemistry as well as from engineering as electronic systems, technology, mathematics, algorithmics, bio-medical systems, robotics and more. Analogy as an integrating concept between STEM disciplines is one approach to integrative STEM teaching-learning. This track requires a high level of abstraction ability and is intended for excellent students.

TO"V (Technician and Matriculation)

Another program is TO"V (Technician and Matriculation). This track is for students from grades 9 to 12 and provides them the opportunity to qualify as technicians at the same time they complete the requirements for their matriculation certificate. At the end of this program, students can complete an additional two years and graduate as Practical Engineers with the option of continuing studies towards an academic degree. They study each discipline in isolation. They study everything required for matriculation as well as for the Technician's diploma and receive additional hours for mathematics, science, English and language (Hebrew or Arabic as first language). The goal of this program is to register each year at least 2,500 students who will successfully complete the entire program and receive both their Technician's diploma and their matriculation certificate, completing everything by the end of 12th grade (Kearney, 2016).

iSTEAM

ORT Israel (an educational network that manages science and technology-oriented high schools in Israel) has been developing and implementing a new Project Based Learning (PBL) curriculum pilot based on innovation, science, technology, engineering, art and mathematics (i-STEAM). This i-STEAM-PBL curriculum will empower young students with essential knowledge, skills and values relevant to the twenty-first century. It aims to bridge the gap between the knowledge acquired in school and real-world knowledge (Choresh, 2016). This project began as a pilot project in six schools in 2014 and ORT Israel hopes that its success will lead to its adoption across the Israeli education system. Nobel Laureate, Professor Dan Shechtman, a member of the i-STEAM program steering committee, has succinctly expressed the program's aims as follows: 'Hi-tech, creativity and entrepreneurship go hand-in-hand. The idea is to teach every child in Israel entrepreneurship, just like you teach mathematics, physics, chemistry and English' (Shechtman, 2018).

I-STEAM is an integrative program with the following characteristics: (a) an interdisciplinary approach and theme; (b) PBL with an ICT-rich pedagogy; (c) innovative and inventive thinking methods; (d) inspiration, entrepreneurship and career development relating to the real world and high-tech industries and (e) examining moral dilemmas of science and technology, based on culture, heritage and values.

The iSTEAM curriculum emphasises the encouragement of emotional involvement and motivation among students. In addition to the skills of project management and construction of knowledge through independent learning, students experience active collaborative work. They structure ongoing exploration in which they deal with current challenges in science, engineering, technology, arts, and combinations thereof. The students' work plan consists of ten stages: exposure, initial thinking, definition of a problem/need, presenting ideas and receiving feedback, studying, gathering information, interim evaluation, writing a Wiki entry, preparing the final product, presenting the project (Choresh, 2016). At the beginning of the pilot project, massive teacher training was delivered, focusing on two main elements: (a) iSTEAM integrative learning, the power of PBL and ICT skills; (b) educational management aspects such as how to coordinate an integrative subject.

An exemplar of an integrative STEM education M.Ed. degree at Beit Berl College

The M.Ed. degree in integrative STEM education was developed at Beit Berl College in collaboration with four faculties from various STEM disciplines (computer sciences, environmental sciences, physics and technology) (Dagan et al., 2019).

The main aims are to expand and enrich the teachers' understanding of the different STEM-based fields, introduce them to new integrative fields found in industry and academia, and provide them with the necessary foundations to implement integrative STEM education using cutting-edge teaching and learning techniques.

Rationale: Understanding the uniqueness of each STEM subjects and their common attributes enables a deeper understanding and better application of the relevant knowledge and transfer principles and methods from one area of knowledge to another. The program emphasises the development of skills that enable learning, research and application of problem-solving methods in work teams comprising students from diverse academic backgrounds, where presentation and feedback also take place. Problem-solving processes are central to the program and will be expressed in all courses. The program also raises social and ethical aspects inherent in these disciplines.

The organising principles are: integrativity and relevance, implementation of project-based learning (PBL), evaluation of students, partnerships and integration of women.

The M.Ed. program lasts two years and has both a thesis and a non-thesis track (38 and 42 academic credit hours, respectively). The program consists of compulsory courses (22 credits), two seminars (8 credits), elective courses (4 credits), and a final project (8 credits) to be implemented in educational settings. The program includes interdisciplinary courses such as 'Biomimicry' and 'Biosphere Research', and education and pedagogy courses, such as 'Education for Values in Science and Technology'. The two intended research seminars are: 'Reforms and Changes in Scientific' and 'Technological Education and their Implications for Teaching and Learning', based on theoretical research methodologies, and 'Action Research in Relation to STEM Project Implementation', which involves methods of action and self-research on a case study conducted in the program.

Three core courses will be conducted in the PBL method and enhance teamwork: (1) 'Investigating Authentic Projects from the Academic World and Industry', which adopts a reverse engineering approach to analyse areas of knowledge and methodologies used in the design and development of the project; (2) 'Developing an

Integrative STEM Project', which aims to solve a real problem and develop a product; (3) 'Developing an Integrative STEM Education Project' to be implemented in an educational framework, and research this implementation. In these core courses, students will work in teams from various disciplines. Through the second core course, 'Developing an Integrative STEM Project', students will experience both the design process and the integration of disciplines. This experience will assist them in the third course, to develop integrative STEM curriculum materials and implement them in secondary schools.

The focus of evaluation will be on team-learning processes, with an emphasis on the process, including in relation to the expression of 'soft skills', and not only on the product. Formative, summative, qualitative and quantitative assessment will all be used.

This unique and challenging program will prepare and train teachers to teach STEM while working in integrative teams with their colleagues at school to assist their learners to work in the same way in their STEM projects. This program is now in the approval process and will be launched in the near future.

STEM in Russia, Sergey Gorinskiy

The current situation of STEM education in secondary schools

At the turn of the twentieth and twenty-first centuries, the 'Technological Literacy' concept aiming to provide school students with ability to use, manage, assess and understand technology became the core of changes in technological education. This model was developed as a logical response to the transformation the industrial society into an information society during the period referred by many economists as the fifth *Kondratieff cycle*. The explosive growth of IT companies and the IT market demanded, first of all, the corresponding growth of qualified *users* of new technologies.

Nowadays, in line with changes in technology, the economy, the labour market and international educational trends, Russian schools gradually introduce other approaches aiming to change pupils' attitudes towards technology from *users* to *creators*. To reach this aim it is important to teach and study technology and engineering in close connection with science and mathematics.

However, for most of the schools, STEM remains 'terra incognita', science and mathematics subjects are studied as theoretical disciplines that are not connected enough with engineering, technology and real life in general. Today, the implementation of STEM education approaches in Russian schools faces the following problems:

- In the last decade, teaching high school students is largely aimed at obtaining good results in the Unified State Exams, which are the obligatory tests taken by all Russian school leavers. The results of these exams are acknowledged by universities. Currently, there is a situation where both high school students and teachers are motivated to successfully pass tests in individual subjects, and not to carry out interdisciplinary projects, since they are not taken into account when entering universities. Moreover, the results of the Unified State Exams are an important criterion for determining the quality of teachers' work.

- At the national level, there is no long-term program for the development of STEM education in schools. Such programs are being developed and tested just in some schools, for example in schools-participants of the international educational network World ORT, which is one of the leaders in the field of STEM education.

■   Unreadiness of most of the teachers to go beyond the scope of their subject, lack of qualifications and motivation to develop and implement interdisciplinary lessons and projects.

Given the above problems, in the near future the development of STEM education in Russian secondary schools can be considered both within the framework of formal school education and non-formal education and training. These two systems are not isolated and can complement each other.

An exemplar of STEM education in secondary schools

The Moscow school # 1540 (Moscow ORT Technology School) – one of the leading educational institutions in Russia in the field of development of STEM education – demonstrates a model with external integrators. The school participates in the educational project 'Engineering Class at a Moscow School', supported by educational authorities, leading Moscow universities and hi-tech companies.

Traditionally, ORT schools pay special attention to the study of engineering and new technologies. In the Moscow ORT Technology School students study robotics, electronics, 3D modelling, programming, web design, IoT (Internet of Things), and experimental physics within the frameworks of the compulsory program, additional lessons and extracurricular activities (including Engineering Immersion).

The school principal, Dr Tatyana Khotyleva, mentioned the project-based approach as the key element of STEM methodology. Children begin to participate in project activities in preschool, at the age of four to five years. Starting from the age of six years, at the stage of preparation for school, participation in project activities becomes mandatory for all children. This activity is organised in the Lego technology lessons. Work is carried out only in small groups; each group selects, develops and presents its own projects.

At various stages of schooling, all children participate in three types of project activities: socially significant projects, creative projects and STEM projects. When they reach high school, students create serious technological projects, including changing the internal environment of the school.

For example:

■   Thanks to the 'Smart Light' project, a biodynamic lighting system was introduced in the school.

■   The purpose of project 'Biotech – Green House' is creation in the school buildings vertical landscaping zones, control of humidity, temperature and chemical composition of soil using microcontrollers and mobile applications.

■   Reorganisation projects for school premises and spaces are being developed by students. For example, a project 'Maker-place' in a high school building aims to create an engineering zone designed to implement all stages of school engineering projects: from finding ideas for solving problems, design, and manufacturing.

The future development of STEM education in secondary schools

To consider the future of STEM education in Russia, we have to take into account the traditions of *polytechnical school education* during the Soviet period and *technology education* in Russian secondary schools since 1992–1993. The idea of Soviet *polytechnical education* was not to give children just the sum of different technical skills but ensuring for the secondary school's leavers the ability to freely choose a profession. In 1952, *labour training* lessons were introduced in Soviet schools. Workshops, training and experimental labs were created at schools. In high school, classes are held in engineering, electrical engineering and agriculture. In 1993 the *subject area technology* was introduced instead of subject *labour training* into the Basic Curriculum of secondary schools of the Russian Federation. Technology became the main practice-oriented subject area in the school curriculum, which can be described with the words design and manufacture. It must be emphasised that the experience of teaching design & technology in England and Wales had a great influence on the formation of technological education in Russian secondary schools. Pilot projects in two regions of Russia (Nizhnii Novgorod and Novgorod the Great) were implemented with the support of the British Council. With the direct participation of British experts, the first technology textbooks for Russian secondary schools were prepared and published.

Thus, subjects related to the modern concept of STEM have always been an important part of Russian school education; however, the goals and approaches to teaching STEM subjects periodically undergo significant changes, responding to the challenges of the economy and changing needs of the labour market.

STEM in Taiwan, Kuen-Yi Lin and Yu-Jen Sie

The current situation of STEM education in secondary schools

In recent years, Taiwan has been committed to the national 12-year basic education reforms in the curriculum outline and syllabus for different subjects announced in 2014. This was followed by successive announcements of the specific syllabuses for science, technology and mathematics disciplines in 2018. The plan was to implement the new syllabuses in phases beginning from 2019 (Ministry of Education, 2014), year by year. However, policies related to STEM education were not clearly defined in the published curriculum outline and syllabuses for the various disciplines. There was also no systematic planning and implementation of STEM education in the core literacy and learning priorities of the relevant syllabuses. Taking the syllabus of the science and technology discipline announced in September 2018 as an example, (See Figure 11.2) the syllabus contents contain the related connotations of integrating STEM or STEAM (incorporating arts). However, its planning philosophy, core literacy, and learning focus were mainly guided by the science and technology discipline, rather than the STEM disciplines. As such, the outline of the national 12-year basic education curriculum planned by the Ministry of Education (MoE) did not contain any systematic or complete plan for STEM education.

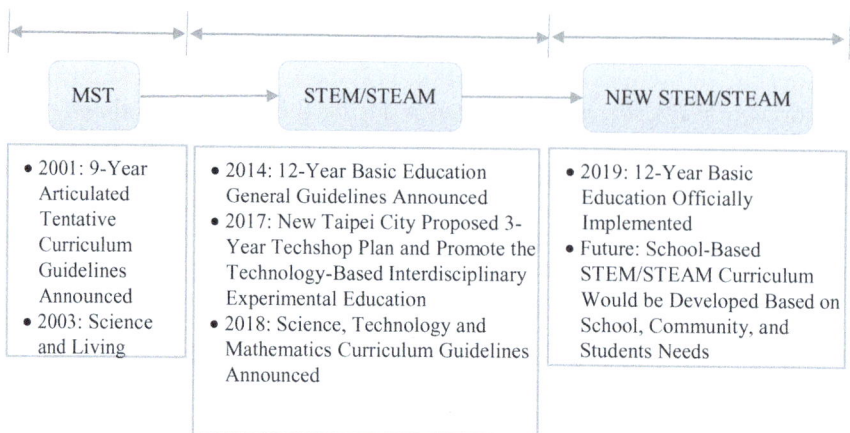

FIGURE 11.2 Development of STEM education policies promoted in Taiwan
Source: Lin et al. (2018)

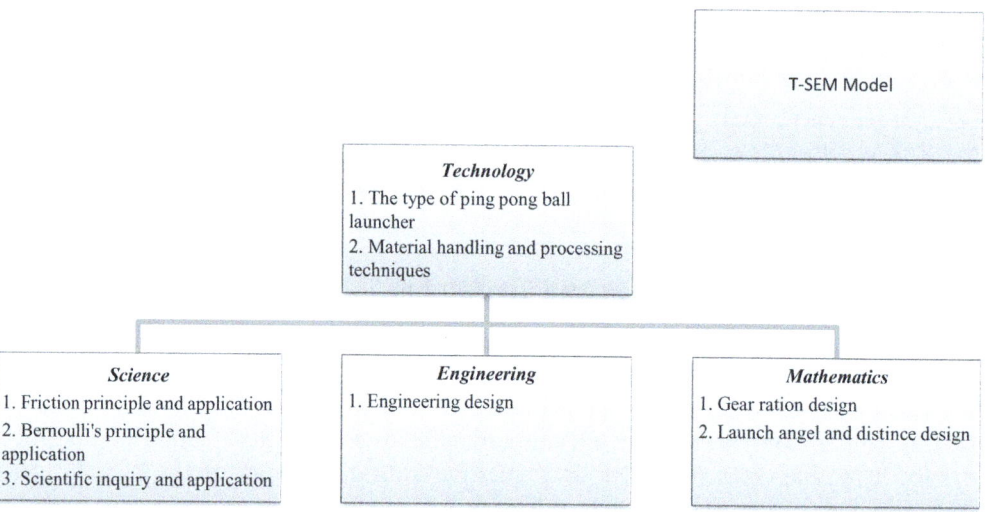

FIGURE 11.3 The related STEM knowledge and competency of the ping pong ball launcher

An exemplar of STEM education in secondary schools

In Taiwan's national 12–year basic education curriculum outline for the science and technology discipline, STEM education is mainly designed for the high school stage. A STEM practical activity for students in the first grade of high school known as 'Designing a ping pong ball launcher' is used here to illustrate the implementation of integrated cross–disciplinary STEM teaching. Here, the T-SEM model (Figure 11.3) will be used as the basis to illustrate the way the activity makes associative integration with science, engineering, and mathematics.

Context design

Ping pong is a kind of sport played in pairs for all ages. People who love playing ping pong may not devote themselves to the game since they could not find a partner. In this regard, many different types of electric ping pong ball launchers are sold in the market but these are costly and unaffordable for many people. To help ping pong enthusiasts practise the sport alone, please design an electric ping pong ball launcher. This launcher must be able to serve straight and spin balls to meet the players' needs.

The related STEM knowledge and competency

The design and production of the ping pong ball launcher involve knowledge of science, technology, engineering, and mathematics, as shown in Figure 11.3. This will enable students to effectively understand the role of STEM disciplines in the thematic activity of creating a ping pong ball launcher, which will then guide them to integrate and apply the relevant knowledge to complete the task. The introduction of STEM knowledge in the curriculum can be supplemented with examples of its daily life applications. This will help them to understand and apply conceptual knowledge (i.e. extract their prior experiences in daily life). When teachers teach the scientific principle of friction in daily life applications, take the ping pong ball launcher as an example; students can understand the relationship between the launcher and friction.

The future development of STEM education in secondary schools

Williams, Jones and Buntting (2015) believe that clear goals and objectives should be addressed when implementing STEM education, which may shed light on the current status of promoting STEM education in Taiwan. Since the goals and systematic policies of STEM education have not been clarified and formulated, there might be more difficulties during the actual implementation of STEM education in the future. Therefore, educational authorities should carefully examine authentic problems revealed in various educational stages in Taiwan. For example, most Taiwanese students pay much attention to knowledge acquisition and have great performances in many international tests, such as the Programme for International Student Assessment (PISA) and Trends in International Mathematics and Science Study (TIMSS). However, their performance on learning attitudes and interests is limited. Based on the results above, identifying alternative ways of enhancing students' learning and career interests should be viewed as an important goal of promoting STEM education in Taiwan.

STEM in the USA, John G. Wells

The current situation of STEM education in secondary schools

Initiatives launched throughout the Obama administration made STEM education a national priority, and culminated with support in the final budget providing an investment of $3.0 billion dedicated to STEM education programming across more than a dozen Federal agencies (Handelsman & Smith, 2016). Following in these footsteps, under the current administration in 2018 the National Science and Technology Council (NSTC) released a five-year Federal strategic plan intended to ensure that all Americans would have equal access to high quality STEM education (NSTC, 2018). The report calls for an integrative approach to teaching STEM concepts beginning 'at an early age – elementary and secondary school – because they are the essential prerequisites to career technical training, to advance college-level and graduate study, and to increasing one's technical skills in the workplace'. To further ensure achievement of this strategic plan the fiscal budget submitted by the White House in 2019 (OSTP, 2019) proposed allocating $200 million every year on STEM education, with an additional $20 million in grant funding to advance career and technical education programs. As with the previous administration, this strategic plan is continuing to emphasise building a stronger transdisciplinary foundation for STEM literacy, increasing diversity and equity within STEM education, and developing new education models intended to prepare students for the twenty-first century STEM workforce.

A decade of making STEM education a US national priority has resulted in substantive changes to schooling in America. Top among those are new national education standards embracing technological/engineering (T/E) design as a central instructional strategy, innovative K-12 schooling models that incorporate instructional spaces specifically tailored to support design-based learning experiences, and new designerly pedagogies for teaching STEM content and practices in more integrative, authentic and meaningful ways. The new or newly revised education standards prepared by the K-12 education associations for science (NSTA), technology and engineering (ITEEA), and mathematics (NCTM) each now emphasise their unique claim to STEM education, though consistent among all is the use of T/E design-based learning for teaching their respective disciplinary content and practices.

The success of STEM education reform in the US has fostered a fundamental shift in pedagogical practice calling for new instructional strategies that intentionally target acquired disciplinary knowledge in response to the cognitive demands inherent to design based learning. A pedagogical shift such as this imposes on teachers the need to reflect on their current teaching practices, analysing what they actually do when they teach and why, in order to employ new integrative STEM education strategies. Perceived this way, the national focus on STEM education sustained in the US over the past two decades is not so much reflective of educational reform writ large, but more an example of *reformed education* (Wells, 2019a) at the school and classroom levels. Such *reformed education* has produced a plethora of preK-16 programs uniquely effective in teaching and learning developed for their specific student populations, and which are demonstrated best using good classroom examples of STEM education.

An exemplar of STEM education

Recent calls at the Federal level for advanced, systemic approaches to STEM education in the US are resulting in novel educational models that can seamlessly connect student learning between middle school, high school, and post-secondary university programs. A large rural public education system in southwest Virginia is currently implementing just such an approach to STEM education through a university and secondary school partnership.

The STEMbot project utilises a programmable precision biotechnical robotics system as a platform around which to develop open-ended design-based instructional modules and other technological systems such as 3D printers to create a powerful, flexible learning platform. Specifically, STEMbot immerses middle school, high school, undergraduate and graduate students in designing, implementing and evaluating authentic design-based biotechnical applications. Design-based learning modules challenge students to program the sophisticated system to autonomously monitor and control growth conditions, measure the impacts of variables on living organisms, and design new data collection and analysis tools as a means of developing student knowledge of STEM content and practices. Students must solve open-ended design problems such as programming the STEMbot to calculate plant growth by capturing digital images and counting pixels, or 3D-printing new robotics tools that determine pH or light intensity. This multi-generational approach to STEM education promotes learning across all grade bans where students acquire knowledge from each other while honing their creative and critical thinking skills through T/E DBL experiences. For example, as students complete design problems in these learning modules and master the competencies (Standards of Learning – SOL) of the school system's three-year course sequence, they demonstrate an understanding of scientific and engineering practices – asking questions and defining problems, exploring dependent and independent variables, and offering simple solutions to design problems (Life Science SOL LS.1). Through their hands-on experiences with the precision biotechnical robotics system, they deepen their understanding of the carbon, water and nitrogen cycles (LS. 4) and chemical energy processes like photosynthesis (LS. 5). Students also design projects that combine hardware and software components to collect data (Computer Science SOL 6.5) and break down problems into parts to facilitate design (CS 7.4). They construct programs to accomplish a task as a means of scientific exploration (CS 8.1) and systematically test and refine programs (CS 8.3). Students also apply real-world mathematics skills using 3D Cartesian coordinates (Math SOL 6.8) to program STEMbot and represent data from practical situations in a variety of ways (Math 6–8 SOL: Probability and Statistics strand). Finally, students will report on their collaborative, small-group learning activities (English SOL 8.1) through interactive presentations (SOL 8.3).

The STEMbot project is an example of a multi-generational educational approach designed to teach the critical-thinking skills necessary for the twenty-first century workforce. University and public-school system partnerships such as the STEMbot will attract and motivate a wider range of students into the STEM fields. More importantly, T/E design based educational approaches like STEMbot create the type of powerful cross curricular opportunities between technology and engineering education and core academic subject areas that lead students toward high-demand career pathways.

The future development of STEM education in secondary schools

Improving student understanding by connecting content and practices through curricula that integrates science, mathematics and technology (SMT) has been a national priority in the US for well over half a century. This is evidenced through early 1980s publications such as *A Nation At Risk*, *Science for All Americans*, and *Benchmarks for*

Science Literacy, which at the time gave rise to a focus on the integration of SMT as an instructional priority (Wells, 2019a). The unmistakable intent reflected in these documents was to envision the teaching of SMT as an integrative endeavour, as clearly articulated by presenting the concept of science as 'the union of science, mathematics, and technology that forms the scientific endeavor' (AAAS, 1989: 25) and 'the ideas and practice of science, mathematics, and technology are so closely intertwined that we do not see how education in any one of them can be undertaken well in isolation from the others' (AAAS, 1993: 321–322). These documents refuted the traditional silo approach to teaching SMT content and practices, and contend that integration is best achieved by following a learning approach where design is a central instructional strategy, such as that being practised in technology education (NRC, 1996: 135–138).

STEM education in the US is still an acronym often misinterpreted and/or misunderstood. To ensure integration remains the educational priority, practitioners must recognise STEM for what it truly is – it is simply an acronym for science, technology, engineering and mathematics. Specifically, what is important for practitioners to recognise is that STEM is *not* a separate discipline or a meta-discipline. Nor is it a field of study, a curriculum, or a single subject taught in schools. STEM is an acronym originally intended to convey the concept of transdisciplinary integration. It is a concept of integration intent on moving education beyond the traditional siloed, mono-disciplinary approach, to one that embraces experiential learning where students seamlessly integrate disciplines within authentic and relevant technological and engineering design-based learning scenarios (Wells, 2019b). The results of a landscape analysis of STEM education published by Stimmer and Froschl (2019) provided insights from key stakeholders in STEM education and related fields regarding trends and future directions of education both formal and informal. Among those insights, two points are of particular importance to those responsible for classroom implementation of STEM education. The first is that implementing STEM education at any level is not easy. Aside from the educator needing more content and pedagogical knowledge than was provided in their preparation program, there is a lack of sufficient funding, limited quality professional development, and schools structured to support siloed teaching of disciplines. Any one of these obstacles would deter even the most determined educator from attempting to implement STEM education in their classrooms. The second is that new pedagogical approaches are necessary for teaching STEM education, particularly for T/E design-based learning. Unless confronted by cognitive dissonance, teachers will continue to teach the way they were prepared. Maintaining the fidelity of science, technology, engineering, and mathematics pedagogies when implementing STEM education requires a thorough understanding of those pedagogical practices.

Such understanding requires quality professional development that provides opportunities to experience new pedagogies, time enough to practise those pedagogies, and follow-on support during classroom implementation.

Acknowledgements

John Wells wishes to acknowledge and thank Cheryl Morgan, Amanda Feldes, and Mark Hainsworth for providing descriptions of how implementation of Integrative STEM Education occurs in their educational settings.

Recommended reading

Freeman, B., Marginson, S., & Tytler, R. (2019) An international view of STEM education. In A. Sahin & J. Mohr-Schroeder (eds), *STEM education 2. 0: Myths and truths – what has K-12 STEM education research taught us?* Leiden, The Netherlands: Koninklijke Brill NV.

Nuangchalerm, P., Prachagool, V., El Islami, R. A. Z., & Abdurrahman, A. (2010) Contribution of integrated learning through STEM education in ASEAN countries. *Journal of Progressive Education*, 10(1), 11–21.

Murphy, S., MacDonald, A., Danaia, L., & Wang, C. (2019) An analysis of Australian STEM education strategies. *Policy Futures in Education*, 17(2), 122–139.

Suter, L. E., & Camilli, G. (2019) International student achievement comparisons and US STEM workforce development. *Journal of Science Education and Technology*, 28, 52–61.

References

American Association for the Advancement of Science (AAAS) (1989) *Science for all Americans.* Washington, DC: AAAS.

American Association for the Advancement of Science (AAAS) (1993) *Benchmarks for science literacy: Project 2061.* Washington, DC: AAAS.

Artique, M., Dillon, J., Harlen, W., & Léna, P. (2012) *Learning through inquiry. Resources for implementing inquiry in science and in mathematics at school.* The Fibonacci Project. Internet: www.fondation-lamap.org/sites/default/files/upload/media/minisites/action_internationale/inquiry_in_science_education.pdf (accessed June 6 2020).

Australian Government. (2020, 14 February) *Grants to sponsor students to take part in STEM events.* www.business.gov.au/Grants-and-Programs/Sponsorship-Grants-for-Student-Science-Engagement-and-International-Competitions#inspiring-australia (accessed June 6 2020).

Barlow, J., & Ellis, D. (2016) Are the T and the E dimensions being recognised in the Australian STEM education discourse? In H. Middleton (ed.), *Creating contexts for learning in technology education* (pp. 8–14). Adelaide: The Design and Technology Association of Australia.

Blackley, S., & Howell, J. (2015) A STEM narrative: 15 years in the making. *Australian Journal of Teacher Education*, 40(7), 102–112. doi:10.14221/ajte.2015v40n7.8

Brown, J. A., Collins, A., & Duguid, P. (1989) Situated cognition and the culture of learning. *Educational Researcher*, 18(1), 32–42.

Burrows, A. C., Garofalo, J., Barbato, S., Christensen, R., Grant, M., Kinshuk, S. . . . Tyler-Wood, T. (2017). Editorial: Integrated STEM and current directions in the STEM community. *Contemporary Issues in Technology and Teacher Education*, 17(4), 478–482.

Choresh, C. (2016) The ISTEAM program – case study: 'Steaming' forward to a multidisciplinary approach, innovation, entrepreneurship, and a start-up culture. *MIT LINC 2016 Conference*, Cambridge, MA. pp. 6–70. https://linc2016.mit.edu/files/2016/08/LINC-Proceedings-2016-1.pdf (accessed June 6 2020).

Dagan, O., Ragonis, N., Goldman, D., & Wagner, T. (2019) Integrative STEM education – A new M. Ed. program: Development, objectives, and challenges. *PATT*, 37, 125.

Dugger, E. W. (2010) *Evolution of STEM in the United States.* Paper presented at the 6th Biennial International Conference on Technology Education Research in Australia. http://citeseerx.ist.psu.edu/viewdoc/download?doi=10.1.1.476.5804&rep=rep1&type=pdf (accessed June 6 2020).

Education, Audiovisual and Culture Executive Agency (EURYDICE) (2011). *Science education in Europe: National policies, practices and research.* http://eacea.ec.europa.eu/education/eurydice/documents/thematic_reports/133EN.pdf (accessed June 6 2020).

Education Council. (2019) *STEM school education interventions: Synthesis report*. www.education-council.edu.au/site/DefaultSite/filesystem/documents/Reports%20and%20publications/Publications/STEM%20Education%20Initiatives%20Synthesis%20Report.pdf (accessed June 6 2020).

European Commission Directorate-General for Research and Innovation. (2015) *Science education for responsible citizenship. Report to the European Commission of the expert group on science education*. https://ec.europa.eu/research/swafs/pdf/pub_science_education/KI-NA-26-893-EN-N.pdf (accessed June 6 2020).

European Parliament, Council of the European Union (2006) *Recommendation on key competences for lifelong learning*. https://eur-lex.europa.eu/legal-content/EN/TXT/?uri=celex%3A32006H0962 (accessed June 6 2020).

Fang, H. Y., Laing, H., Xie, Y. S., Zhou, S., & Zhou, X. L. (2019) Case study of the first pilot schools based on the 'China STEM Education 2029 Innovation Action Plan'. *China Academic Journal Electronic Publishing House*. www.cnki.net (accessed June 6 2020).

Flemish Government (2015) *STEM-kader voor het Vlaamse Onderwijs*. Brussel: Departement Onderwijs en Vorming (publication in Dutch). www.vlaanderen.be/publicaties/stem-kader-voor-het-vlaamse-onderwijs (accessed June 6 2020).

Flemish Government (2018) Decreet tot wijziging van de Codex Secundair Onderwijs van 17 december 2010, wat betreft de modernisering van de structuur en de organisatie van het secundair onderwijs. Stuk 1469 Nr. 3 van 28 maart (publication in Dutch). www.vlaamsparlement.be/parlementaire-documenten/parlementaire-initiatieven/1245244 (accessed June 6 2020).

Handelsman, J., & Smith, M. (2016) *STEM for all*. White House Office of Science and Technology. https://obamawhitehouse.archives.gov/blog/2016/02/11/stem-all (accessed June 6 2020).

iSTEAM, ORT Israel (2020) *Integrated STEM*. https://en.ort.org.il/category/isteam/ (accessed June 6 2020).

Kearney, C. (2016) Efforts to increase students' interest in pursuing mathematics, science and technology studies and careers. *National Measures Taken by 30 Countries – 2015 Report*. Brussels: European Schoolnet.

Kelley, T. R., & Knowles, J. G. (2016) A conceptual framework for integrated STEM education. *International Journal of STEM Education*, 3, 1–11. doi:10.1186/s40594-016-0046-z

Lin, K. Y., Hsiao, H. S., Chang, Y. S., Chien, Y. H., & Wu, Y. T. (2018) The effectiveness of using 3D printing technology in STEM project-based learning activities. *Eurasia Journal of Mathematics, Science & Technology Education*, 14(2), em1633. https://doi.org/10.29333/ejmste/97189.

Ministry of Education. (2014) *General outline of twelve-year basic education curricula*. www.naer.edu.tw/files/15-1000-7944,c639-1.php?Lang=zh-tw (accessed June 6 2020).

National Institute of Education Sciences 中国教育科学研究院 (2017) *China STEM education white paper*. Beijing: NIES

National Research Council (1996) *National science education standards*. Washington, DC: National Academy Press.

National Science and Technology Council (NSTC) (2018) *Charting a course for success: America's strategy for STEM education*. www.whitehouse.gov/wp-content/uploads/2018/12/STEM-Education-Strategic-Plan-2018.pdf (accessed June 6 2020).

Office of Science and Technology Policy (OSTP) (2019) *Progress report on the federal implementation of the STEM education strategic plan*. www.whitehouse.gov/wp-content/uploads/2019/10/Progress-Report-on-the-Federal-Implementation-of-the-STEM-Education-Strategic-Plan.pdf (accessed June 6 2020).

Putnam, R., & Borko, H. (2000) What do new views of knowledge and thinking have to say about research on teacher learning? *Educational Researcher*, 29(1), 4–15.

Ren Jing. (2018) *Learning to do research: A case study on student's experience of STEM learning at middle schools in Shanghai*. http://cdmd.cnki.com.cn/Article/CDMD-10269-1018820941.htm (accessed June 6 2020).

Sanders, M. E. (2012) Integrative STEM education as 'Best Practice'. In H. Middleton (ed.), *Explorations of best practice in technology, design, & engineering education.* 2 (pp. 103–117). Brisbane: Griffith Institute for Educational Research.

Sheatufim. (2019) *A recommendation for systematic method for excellence in STEM education in middle schools.* Position paper. Israel (in Hebrew).

Shechtman, D. (2018). ETF – European Training Foundation. www.etf.europa.eu/en/news-and-events/news/etf-awarded-winning-israeli-curriculum-transfers-philippines (accessed June 6 2020).

STEM Education Research Centre, National Institute of Education Sciences 中国教育科学研究院STEM教育研究中心 (2018) *China STEM Education 2029 Innovation Action Plan.* Beijing: NIES.

STEM Education Research Centre, National Institute of Education Sciences 中国教育科学研究院STEM教育研究中心 (2019). *China STEM Education Survey Report.* Beijing: NIES.

STEM@school (2018). *Learning modules for 9th grade: Autonomous driving car.* www.stematschool.be/en/our-learning-modules (accessed June 6 2020).

Stimmer, M., & Froschl, M. (2019). *Voices from the field: A snapshot of STEM education today.* New York, NY: FHI 360.

Su, W., & ZhengFu, L. (2019) *STEM education.* Beijing: Educational Science Publishing House.

Thibaut, L., Ceuppens, S., De Loof, H., De Meester, J., Goovaerts, L., Struyf, A., Boeve-De Pauw, J., Dehaene, W., Deprez, J., De Cock, M., Hellinckx, L., Knipprath, H., Langie, G., Struyven, K., Van de Velde, D., Van Petegem, P., & Depaepe, F. (2018) Integrated STEM education: Conceptualizing an instructional approach for secondary education. *European Journal of STEM Education,* 3(1), 02. https://doi.org/10.20897/ejsteme/85525

Van de Velde, D., Van Boven, H., Dehaene, W., Knipprath, H. & De Cock, M. (2016). *Pré-university STEM education: How do teachers perceive the implementation ?* Publication in Dutch: Doorstroomgericht STEM-onderwijs: hoe beleven en percipiëren leraren in de tweede graad van het secundair onderwijs de opstart? Brussels: Impuls – Acco. www.acco.be/nl-be/items/TAB000086/Impuls-Jaargang-2016-2017---enkel-abonnement (accessed June 6 2020).

Van der Bolt, L., Studulski, F., Van der Vegt, A., & Bontje, D. (2006) *De betrokkenheid van de leraar bij onderwijsinnovaties, een verkenning op basis van literatuur.* Research review commissioned by the Ministry of Education, Culture and Science (NL). www.rijksoverheid.nl/documenten/rapporten/2006/07/04/de-betrokkenheid-van-de-leraar-bij-onderwijsinnovaties (accessed June 6 2020).

Wells, J. (2019a) STEM education: The potential of technology education. In M. Daugherty & V. Carter (eds), *The most influential papers presented at the Mississippi Valley Technology Teacher Education Conference.* Council on Technology and Engineering Teacher Education, 62nd Yearbook (pp. 195–229), Muncie, IN: Ball State University. http://ctete.org/wp-content/uploads/2019/05/2019-CTETE-Yearbook-Mississippi-Valley-Conference-21st-Century.pdf (accessed June 6 2020).

Wells, J. (2019b) Technology education and designerly ways of knowing: The pedagogical goal of design is understanding. *Proceedings of the 2019 International Symposium on Technology Education* (pp.15–34). College of Education, Capital Normal University, Beijing, China.

Williams, P. J. (2011). STEM wducation: Proceed with caution. *Design and Technology Education: An International Journal,* 16(1). https://ojs.lboro.ac.uk/DATE/article/view/1590

Williams, P. J., Jones, A., & Buntting, C. (2015). *The future of technology education.* Singapore: Springer.

Wilschut, A. & Pijls, M. (2018). *Effecten van vakkenintegratie: een literatuurstudie.* Kenniscentrum Onderwijs en Opvoeding Hogeschool van Amsterdam. https://pure.hva.nl/ws/files/5112546/Effecten_van_vakkenintegratie.pdf

Future visions for the STEM curriculum

Introduction

This chapter is in three parts and builds on the future ideas of STEM education in different countries we saw in Chapter 11. The first part, 'Big issues and STEM education', describes the major issues facing the world with which STEM education needs to engage. The second part, 'STEM and disruptive technologies – an opportunity to future gaze', discusses the consequences of new and emerging technologies and how these might be addressed in STEM education in schools. The third and final part, 'Your vision', considers four scenarios and the future visions of STEM within each scenario.

Big issues and STEM education

Barlex and Steeg (2017) have argued that one of the Big Ideas that underpin design & technology is 'critique' and that engaging learners with critique may be achieved through the lenses of justice and stewardship. In this section, we argue that the big issues facing the world can also be viewed through these lenses, and that learning in the STEM subjects can be used to engage learners with this critique. We will consider each in turn.

Justice

There are many kinds of justice, but for our purpose it is the idea of social justice that is important. This concerns the relationship between the individual and society as measured by the distribution of wealth, opportunities for personal development and social privileges. In a just world, all people should be able to live in freedom from hunger and fear and have shelter from harm. They should have opportunities to pursue happiness and make the best of their lives. Currently, there are many situations in the world where people are hungry, afraid and lack shelter from harm. The plight of migrants fleeing from war zones and finding themselves living in makeshift camps in squalid conditions is an obvious example. The lives of subsistence farmers are also a cause for concern: a family or community growing just enough food for them to be able to eat with little if any surplus. Any disruption of this endeavour quickly leads to hunger and starvation. In 2015, about 2 billion people (slightly more than 25 per cent of the world's population) in 500 million households living in rural areas of developing nations survived as subsistence farmers, working less than 2 hectares (5 acres) of land (Rapsomanikis, 2015). This problem is also likely to become worse. The Food and Agriculture Organisation of the United Nations (2009) warned that the world population will have reached over 9 billion by 2050, from its current population of some 7.4 billion. This will place a significant burden on food production. For example, the report warns that it is estimated that by 2050 developing countries' net imports of cereals will more than double from 135 million metric tonnes in 2008/2009 to 300 million in 2050.

There are many subjects in the curriculum in which teachers can raise awareness of situations in which there has been or is a lack of social justice (history, geography, English literature, religious studies for example), but teachers of the STEM subjects can do more than just raise awareness. They can ask their students to consider how the knowledge, understanding and skills learned through such subjects can be deployed to alleviate such situations.

In the case of subsistence farming…

The science teacher might teach about the needs of plants as follows:

- fertile soil in which to grow, for some soils fertilisers might be needed;
- appropriate weather conditions to supply sunlight and water at temperatures that do not harm the plants. In adverse conditions additional water, protection from sunlight and cold might be required;
- protective measures against pests and disease that affect yields;
- drainage to prevent the soil becoming waterlogged and preventing growth;
- appropriate planting to maximise yields and enable harvesting.

This can easily be related to the nature of the soil, availability of water, the climate and weather in places where subsistence farming is the norm and the vulnerability of folk dependent on this is established. Adding to this that the impact of climate change is likely to make conditions in these places even worse, then the case for taking action to achieve social justice is very strong.

The mathematics teacher might teach about the scale of the problem. Just how big is 2 billion? That's two with nine noughts after it, 2,000,000,000. What does this mean? The populations of the ten largest cities in the UK and the US at the moment are shown in Table 12.1

The total population of these cities is 40,386,441. Let us call this 40 million. Simple arithmetic shows that the number of people having to exist on subsistence farming at the moment is five hundred (500) times bigger. The numbers are staggering and the extent of human misery when such farming fails can scarcely be imagined. As with science teaching, mathematics teaching makes the case for taking action to achieve social justice very strong.

And it is, of course, the taking of action, or intervening, that is the role of technology and engineering. A word of caution is necessary here. Asking young people

TABLE 12.1 Population data

City in UK	Population*	City in US	Population**
London	8,907,918	New York	8,398,748
Birmingham	1,155,717	Los Angeles	3,990,456
Glasgow	612,040	Chicago	2,705,994
Liverpool	579,256	Houston	2,325,502
Bristol	571,922	Phoenix	1,660,272
Manchester	554,400	Philadelphia	1,584,138
Sheffield	544,400	San Antonio	1,532,233
Leeds	503,388	San Diego	1,425,976
Edinburgh	488,050	Dallas	1,345,047
Leicester	470,865	San Jose	1,030,119
Total	14,387,956	Total	25,998,485

* Source The Geographist, 2020
** Source Wikipedia, 2020

to devise better ways for subsistence farmers to farm is almost certainly inadvisable. Although struggling in their endeavour, subsistence farmers bring generations of knowledge, understanding and skill to bear on the problem and it would be arrogant for young people in more favourable circumstances to tell them how to do this demanding task. Mishak Gumbo (2020) has written very convincingly about the science and technology that is embedded within the practices of indigenous peoples, particularly with regard to farming and cooking.

The case of disaster relief through designing emergency kits that might be dropped by parachute to help those in distress is less contentious with the possibility of the container providing elements that might be used to construct temporary shelters as well as including food, water, water sterilisation tablets, bedding, clothing and simple communication devices. An additional problem to be solved here is the provision of instructions in pictorial form that do not require the use of words.

Rosa Lyster, writing in the *London Review of Books* (2020) paints a bleak picture of water supply (see Panel 12.1).

A consideration of the infrastructure that might conserve, purify and distribute water would provide an interesting STEM topic in which science, mathematics and design & technology teachers would each have a significant part to play. The writing of Claes Classander and Jonas Halstrom on teaching technological systems (2020) will help teachers provide ways of thinking about both the problems and possible solutions.

Stewardship

In developing an appreciation of stewardship, it is important to move beyond the standard evaluation of designed outcomes, which usually limits itself to answering the question, 'Did it do what it was supposed to?' A designed outcome might well do

In five years' time, two thirds of the world's population is going to be living in a state of 'water stress', according to the UN. Either we won't have enough or it will be dirty or we won't be able to access it without difficulty. Thirty-three cities are currently suffering 'extremely high' water stress, according to the World Resources Institute, which is another way of saying that they are using most of the water they have. This will only get worse as the effects of climate change intensify. Rising temperatures will encourage the flourishing of bacteria and other pathogens. Rising sea levels will salinate freshwater sources, rendering them unsusable. More drought means more hunger, but it also means more violence, according to the growing body of research that indicates an 'overt' correlation between acute water stress and violent conflict (recent studies have also pointed to the strong connection between resource depletion and violence against women). More flooding means more damage to already compromised sanitation infrastructure, as well as contamination of the remaining supply. In ten years' time, India will have half the water it needs, as will Zimbabwe, although in its case ten years is an optimistic timeframe, given the unwavering severity of the drought there. Forty per cent of Beijing's water supply is currently too polluted to use, and Mexico is draining its aquifers 50 per cent faster than they can be replenished.

PANEL 12.1 The state of water supply across the world

what it was supposed to do, but this interrogation must be extended to include the following questions:

- Is what it is supposed to do worth doing?
- To what extent does it contribute to a future worth wanting?
- What might be the unintended consequences of wide scale use?
- To what extent will these consequences compromise the wellbeing of Planet Earth and the creatures that live there now and in the future?

It has long been realised that we cannot continue to consume the Earth's resources at the current rate (Leonard, 2010). Engaging in stewardship will require us to teach young people about different economic models, the deleterious impact of the current linear economy and the need to move towards a circular economy. Underpinning this teaching will be life cycle analysis, which in the case of a linear economic model is termed 'cradle to grave' analysis, but in a circular economy model is termed 'cradle to cradle' analysis. The circular economy model bases itself on the way nature operates in that all material flows are cyclical and the waste from one life form becomes the feedstock for another in nature. Nothing is wasted, with the overall driver for all the many millions of cycles being energy from the sun. A much-simplified version of this is shown in Figure 12.1.

The Ellen MacArthur Foundation (2020) has dedicated itself to making the case for circular economies to replace the current linear economy and persuading businesses

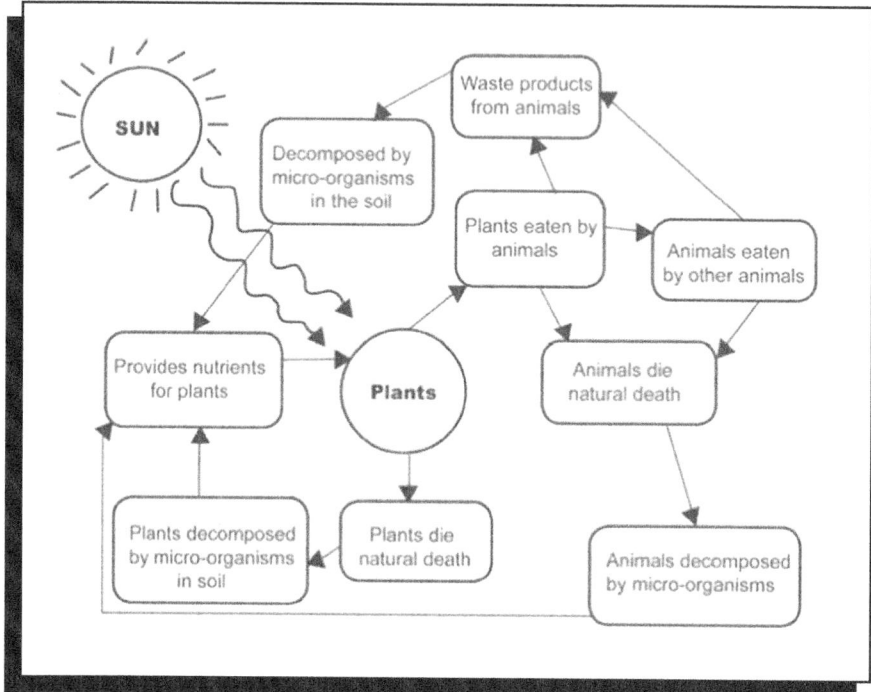

FIGURE 12.1 A simplified version of nature's circular economy

to develop and adopt them. An infographic developed by the Ellen MacArthur Foundation is shown in Figure 12.2. There is no doubt that moving from a linear to a circular economy is a huge challenge, and one important part of meeting this challenge must be to educate young people about the circular economy so that they become informed citizens who are advocates for meeting this challenge. In addition to teaching *about* the circular economy, teachers can engage young people in re-designing products to be suitable for a circular economy so that once their useful life is over they may be returned to the manufacturer for disassembly and the materials and components re-circulated.

An intriguing aspect of a circular economy is that the aspect of personal ownership is challenged in that products that were bought and then discarded in a linear economy will instead be leased from manufacturers who have the responsibility to re-circulate the materials and components.

As circular economies are adopted, they will play a large part in meeting the challenge of climate change, but the situation caused by the emission of greenhouse gases is too urgent to wait for this to happen. Hence teaching young people about the impact of climate change and how this is likely to affect the planet is also an important part of STEM education dedicated to stewardship. Teaching how new and emerging technologies may be used to replace those technologies that are responsible for greenhouse gas emissions will be important and, as with the circular economy, this is important in giving future citizens the ability to be advocates for their support as in many cases (e.g. wind generated power) (Merrick, 2018), government investment as well as private investment will be necessary. A particularly interesting example of young people advocacy is taking place is in the US where a group of young

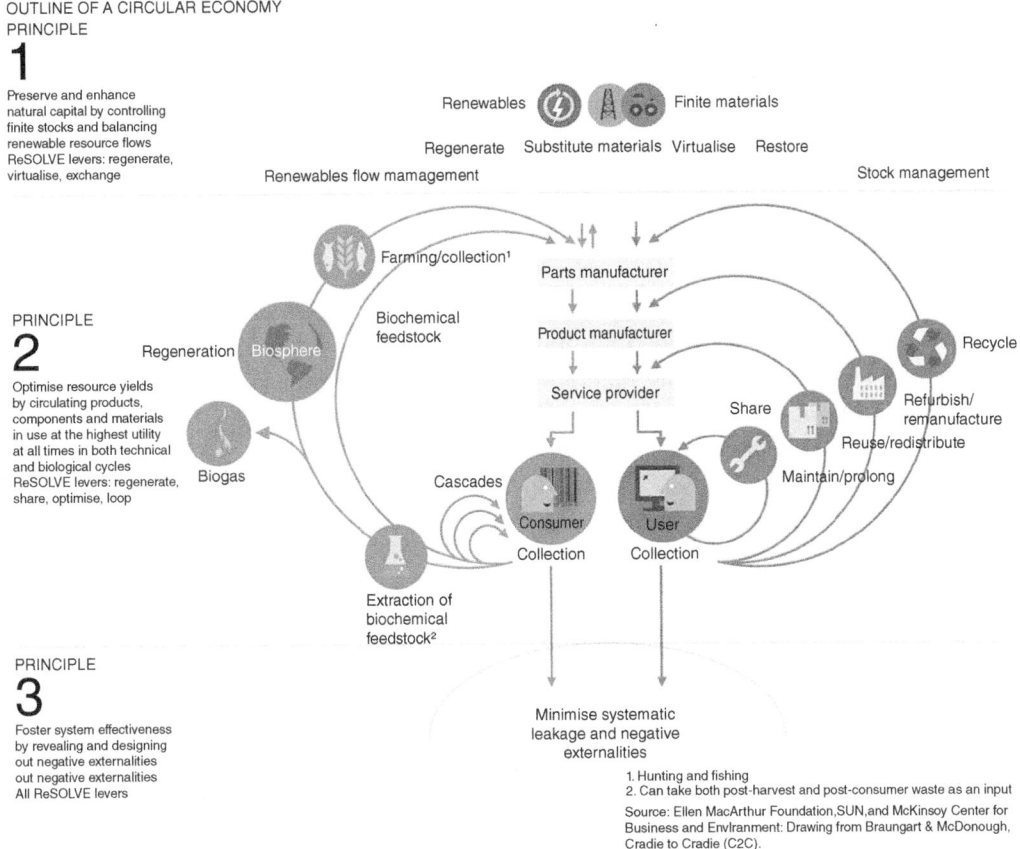

OUTLINE OF A CIRCULAR ECONOMY

PRINCIPLE 1

Preserve and enhance natural capital by controlling finite stocks and balancing renewable resource flows ReSOLVE levers: regenerate, virtualise, exchange

PRINCIPLE 2

Optimise resource yields by circulating products, components and materials in use at the highest utility at all times in both technical and biological cycles ReSOLVE levers: regenerate, share, optimise, loop

PRINCIPLE 3

Foster system effectiveness by revealing and designing out negative externalities out negative externalities All ReSOLVE levers

Renewables   Finite materials

Regenerate   Substitute materials   Virtualise   Restore

Renewables flow mamagement   Stock management

Farming/collection[1]

Parts manufacturer

Biochemical feedstock

Product manufacturer

Regeneration   Biosphere   Recycle

Service provider

Refurbish/remanufacture

Share

Biogas   Cascades

Reuse/redistribute

Maintain/prolong

Consumer   User

Collection   Collection

Extraction of biochemical feedstock[2]

Minimise systematic leakage and negative externalities

1. Hunting and fishing
2. Can take both post-harvest and post-consumer waste as an input

Source: Ellen MacArthur Foundation,SUN,and McKinsoy Center for Business and Envlranment: Drawing from Braungart & McDonough, Cradie to Cradie (C2C).

FIGURE 12.2 A circular economy as envisaged by the Ellen MacArthur Foundation
Source: © Ellen MacArthur Foundation

people is taking the president of the United States to court in the action Juliana v United States (2020). Their complaint asserts that, through the government's affirmative actions that cause climate change, it has violated the youngest generation's constitutional rights to life, liberty and property, as well as failed to protect essential public trust resources.

In addition, lifestyle choices will play a part in the fight against global warming and discussion of the move towards reducing significantly if not eliminating the eating of meat and the consumption of dairy foods is something that STEM teachers must address in their teaching. Many teachers would agree that the compelling case being made by Greta Thunberg (2019), Extinction Rebellion (Harding, 2020), Greenpeace (2020) and significant commentators such as Sir David Attenborough (Shukman, 2020) for governments and individuals to act *now* is a call to stewardship and one that STEM educators are morally bound to endorse.

Several of the pieces written for Chapter 11 indicate that secondary schools are beginning to take seriously the role of STEM education in confronting the Big Issues facing the world. The importance of 'looking sideways' and collaborating with colleagues across the STEM subjects will become even more significant as understanding and resolving these issues requires knowledge, understanding and skill from all the contributory subjects.

STEM and disruptive technologies: An opportunity to future gaze

Barlex, Givens and Steeg (2020) have identified and discussed nine disruptive technologies that they consider suitable for inclusion in the secondary school curriculum and are highly relevant to STEM education. These are listed with brief descriptions in Table 12.2. Their potential for disruption is justified by the extent to which they meet the criteria for being disruptive as identified by the McKinsey Global Institute (2013). These are:

- Disrupt the status quo – they will overturn existing hierarchies and may (or may not …) lead to different and more democratic hierarchies.
- Alter the way people live and work – they may increase or decrease employment opportunities, change the knowledge and skill sets required for employment, impact on education and alter relationships.
- Rearrange 'value pools' – they influence existing and new commercial activity in ways which redistribute financial gain, generally towards those who are deploying these technologies.
- Lead to entirely new products and services – they will provide types of products and services that have not previously existed.

Enabling young people to appreciate the current and potential role of such technologies in our society is part of developing their technological perspective, which provides insight into 'how technology works'. This informs a constructively critical view of technology and helps avoid alienation from our technologically based society. It should enable consideration of how technology might be used to provide products and systems that help create the sort of society in which young people wish to live. This is an aspect of STEM education that transcends programs justified merely on vocational grounds. The vocational justifications inevitably involve relatively small numbers of young people and do not address an entire cohort. Given the likely impact of such technologies on our way of living it is important that as many young people as possible are engaged; hence as we saw in Chapter 1 and many earlier chapters, we argue that this aspect of STEM should be seen as part of general education for all. Given the nature of the particular technologies, it is likely that some will be more suitably taught in particular STEM subjects than others because of teacher specialisms.

A possible distribution building on likely STEM teacher expertise is shown in Table 12.3. Note that there is the potential for overlap, hence some duplication which might well encourage 'looking sideways'.

One might imagine that teachers specialising in engineering might be able to teach about those technologies taught by design & technology and computer science teachers. In terms of what might be taught, it is important that the idea of *disruption* is clearly conveyed. Barlex, Givens and Steeg (2020) have identified three categories of disruption, incidental, intentional and cultural as follows:

Incidental disruption may be seen as the result of a new and emerging technology that was developed, as are most technologies, with the intention of solving a particular problem and/or providing financial gain for those who invest in the technology. It was not conceived or implemented with the express intention of causing disruption; but disruption, as defined by the McKinsey Global Institute, happened,

TABLE 12.2 Disruptive technologies to teach in the secondary school

The technology	The description
Additive manufacture (AM)	AM involves fabricating physical objects in successive thin horizontal layers, according to digital models derived from CAD designs, 3D scans or video games. Such printing can take place at different scales from nano structures to complete buildings and may involve a wide range of materials: human tissue, electronics, and food as well as traditional industrial product materials.
Artificial intelligence (AI)	AI can be categorised at three different levels. First is 'narrow' AI that specialises in one area (e.g. the AI that plays games such as chess or go better than humans). Some AI are used in collaboration with humans, in the judicial system, for example, The second and third levels are concerned with more general ability. 'General' AI can perform as well as a human across the board (i.e. it is AI that can perform any intellectual task that a human can). Such AI is yet to be developed. Third is 'super intelligent' AI (i.e. an AI that performs better than human brains in practically every field).
Augmented reality (AR)	Augmented reality (AR) is a live, direct or indirect view of a physical real-world environment whose elements are augmented (or supplemented) by computer-generated sensory input such as sound, video, graphics or GPS data as discussed in Chapter 9.
Big data	Big data is data that exceeds the processing capacity of conventional database systems. The data is too big, moves too fast, or doesn't fit the strictures of standard database architectures. It is collected by large corporations and governments (and, increasingly, open data from 'citizen' scientists) and using big data analytics it can give insights into the behaviour of potential consumers and citizens.
Programmable matter	Imagine a product made up of fine-grained computing elements (in much the way that you are made up of cells). The way these elements are programmed, including their response to physical stimuli, can affect the physical properties of the bulk object, such as shape, texture, colour, conductivity, transparency and so on. This is programmable matter. Currently, the smallest programmable elements are ~10 mm-sized, but there are active research projects aimed at driving this size down.
Internet of Things (IoT)	The Internet of Things (IoT) is the networking of physical objects (i.e. things that have embedded electronics, software and sensors), which are connected to one another over the internet and can exchange data. This allows extensive communication between the physical and digital worlds, enables remote control of devices across the internet and produces vast amounts of big data. The successive roll outs from 3G, 4G and now 5G each offering increased download speed and reduced latency increases the significance of the IoT.
Neuro-technology	Neuro-technology is concerned with technologies that inform about and influence the behaviour of the brain and various aspects of consciousness. Current neurotechnologies include various means to image brain activity, stimulation of the brain by magnetism and electricity, measuring the electrical and magnetic brainwave activity, implant technology to monitor or regulate brain activity, pharmaceuticals to support neurotypical brain function, and stem cell therapy to repair damaged brain tissue. Recently, measurements of brain activity have been used to control real world artefacts.
Robotics	A robot may be defined as 'a machine that carries out a physical task autonomously using a combination of embedded software and data provided by sensors'. This definition embraces relatively simple robots such as the Roomba vacuum cleaner to extremely complex robots such as the Google self-driving car.
Synthetic biology	Synthetic biology is the process of designing and creating artificial genes and implanting them in cells. In some cases, all existing genes have been removed; in others, the new genetic sequences are introduced into the DNA of existing cells. It is far more than simply borrowing existing genes from nature. Synthetic biology is the process by which completely new life forms (i.e. life forms that have never previously existed) are created, as we saw in Chapter 8.

TABLE 12.3 Distribution of disruptive technologies in STEM subjects

The technology	Taught in design & technology	Taught in science	Taught in computer science
Additive manufacture (AM)	✓		
Artificial intelligence (AI)			✓
Augmented reality (AR)	✓		
Big data			✓
Programmable matter	✓	✓	
Internet of Things (IoT)	✓		✓
Neuro-technology	✓	✓	
Robotics	✓		
Synthetic biology		✓	

nonetheless. The development of the original car mass production system by Henry Ford in 1913 falls into this category.

There are some technologists who develop products with the **deliberate intention of disruption**. This is the case for Ken Gabriel who managed the development of an automated cancer therapy treatment from a starting point that used 17 different machines, took up to 22 days to develop the therapy from the blood, and cost up to $450,000 per treatment. He set his engineers the target of producing an automated system within a single piece of equipment within a ×10 framework; that is, it was to cost ten times less and work ten times faster. He described this work as 'intentionally disruptive' (Gabriel, 2019).

Cultural disruption is perhaps the most thought provoking. The philosopher Christopher J Preston has written at length in his book *The Synthetic Age* (Preston, 2018) about the way our development and deployment of technologies in recent years is fundamentally changing our relationship with Planet Earth, with nature and with what it might mean to be human. He identifies the following:

- the production of nanomaterials, the like of which cannot be produced in nature;
- the use of AI to solve immensely complex problems beyond the scope of ordinary humans; and
- the use of synthetic biology and neuro-technology to augment humans to the point where we become a new species, no longer homo sapiens. Yuval Harari (2014) coined the name 'Homo Deus', to describe humans with almost god-like powers to capture this change.

On a perhaps more mundane level, but no less significant, is the ubiquity of mobile phones, which through access to a 'library in our pockets', gives us vast amounts of information and can be seen as changing the way we *think*. Preston believes that this possible trajectory presents humankind with an enormous problem. He suggests that we need to consider carefully whether we should choose to take this path. In particular, it needs to be noted that the way that new and emerging technologies play out has to be put into the context of the impact we are having on all other life forms on the Earth and on the behaviour of the planet itself. The recent reports from the Intergovernmental Panel on Climate Change (IPCC, 2018) and the UK Committee on Climate Change (UKCCC, 2019) plus the warnings given by the eminent broadcaster Sir David Attenborough (Shukman, 2020) should leave us in no doubt that the disruption caused by our deployment of technology goes far beyond the commercial disruption first envisaged by Clayton Christensen (2012) and its reconfiguration by McKinsey.

While this chapter focuses deliberately on social and economic disruptions as embodied in the McKinsey criteria, we should recognise that some new or emerging technologies, singly or in combination, may impact on the biosphere on a global scale in ways that may result in profound social and economic disruption.

Preston (2018: 173) makes an eloquent and compelling plea for the involvement of *all* citizens in deciding what technologies to develop and how they should be deployed. He writes:

> Making big choices is always hard. Making irrevocable choices for the whole planet is unprecedented. But at this point, we have changed too much to stand back and do nothing. We need to look at as many of the various options as we can, talk about them, argue about them, investigate and research them as thoroughly as possible. Conducting this discussion thoughtfully, fairly and inclusively is perhaps the worthiest, and certainly the most important political task of our time. It is also one that we can no longer shirk.

In addition to learning about the nature of disruption, it is important that young people are given the opportunity to explore what such disruption might involve. Two ways of doing this have been developed by Barlex (1995, 2017) and although they were developed in the context of teaching design & technology, they are sufficiently general that they can be applied to disruptive technologies being taught by other STEM specialisms. The first is a winners and losers analysis. This involves using the target chart shown in Figure 12.3. The disruptive technology being considered

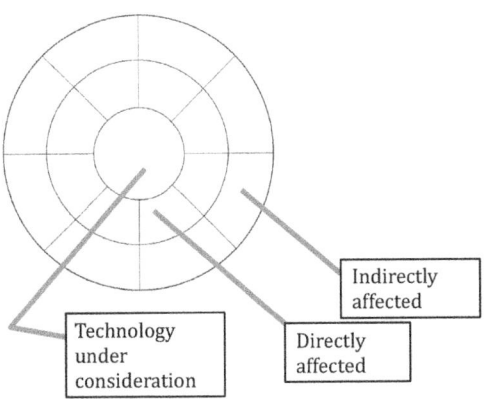

FIGURE 12.3 Winners and losers target chart

is placed in the centre of the chart. Consideration of the immediate impacts of this technology gives rise to list of groups and individuals who will be affected by these impacts. Those who are directly affected are written in the inner circle, those indirectly affected are written in the outer circle. Then these entries are classified as either winners (those who benefit from the technology) or losers (those for whom the technology causes some form of detriment). Given one colour to the winner segments and a different colour to the loser segments gives an immediate visual picture of the winner loser balance, although care must be taken in the interpretation of this as it does not, of itself, give any indication of the size of the groups involved or the severity or otherwise of any impact. If time is short and you as the teacher decides that you need to 'cut to the chase', you can always produce a filled-in version of the chart and use that to stimulate discussion about the disruption that might be caused by the technology under consideration. An effective way to stimulate discussion around a completed chart is to assign some learners the role of 'winners' and other learners the role of 'losers'. The task then facing the learners is for the winners to justify why it is permissible for them to win at the losers' expense, and the losers to argue for some form of recompense from the winners. In these discussions, in whichever of the individual STEM subjects they take place, there will be ample opportunity for the *learners* to look sideways and bring knowledge and understanding from across their different subjects to bear on the content of their discussions.

The second is scenario building and exploration. Ideally, one would want learners to build scenarios for themselves but they will not find this easy and any techniques will require specific teaching in terms of the technique itself and the understanding of specific concepts on which using the technique relies. A general approach often used to build scenarios is to identify two sets of so called 'critical or significant uncertainties' and to use these as axes to create four quadrants such that there is a particular scenario located in each quadrant (see Figure 12.4.). Each of these can be fleshed out into a human story, which can be explored from various critique perspectives.

Learners might use scenario building to explore, for example, the way relationships between humans and robots might play out by means of some scenario developments. The initial task is to identify the critical uncertainties for the axes needed

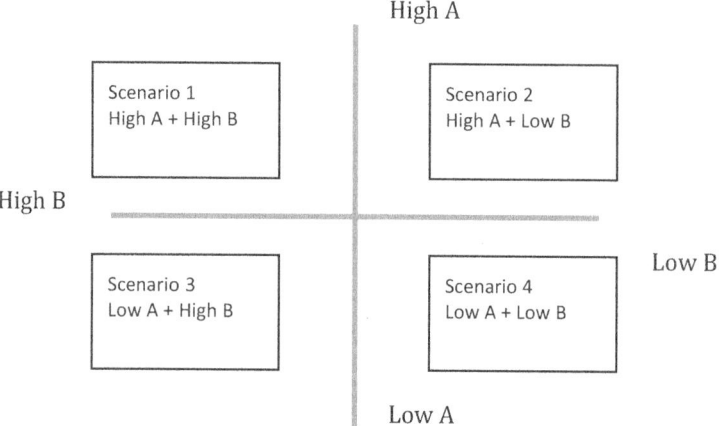

FIGURE 12.4 A general representation of the critical uncertainties A and B to create four scenarios

to create the four scenarios. Whether humans are prepared to accept robots into their life and work (as co-workers, substitute workers or helpers) seems a crucial factor. In this example, some people may feel uneasy about a robot presence at work or home especially if they see the robot as somehow 'messing with nature' in that technology has created a sentient being that is not a human but a machine. Others will welcome a robot presence on the grounds that robots tackle tasks they do not wish to do and that they can provide companionship. Hence the Y-axis in the scenario development concerns 'acceptance'. Related to acceptance is the possibility that as robots become more sophisticated, to the point where they become first person conscious and moral agents in their own right, they might be granted rights or not. Such rights would curtail the way in which humans could treat robots preventing them, for example, being seen and treated as disposable once they were deemed no longer fit for purpose. Hence the granting of rights to robots forms the X-axis. The resulting four quadrants are shown in Figure 12.5. Each of the quadrants provides the basis for learners to write a brief descriptive piece in which they, their friends or members of their family are the main protagonists in working out their relationship with robots according to the constraints of the particular quadrant. This requires imagination and empathy, and it would probably be worth talking with teachers in the English department about how best to support such writing and also to ensure that you have appropriate expectations and are not fobbed of with writing well beneath the standard that your learners are able to achieve. It might also pay to let the learners move away from the standard essay format and add illustrations or even move into 'graphic novel' format. Considering each scenario from a 'winners and losers' perspective might be worthwhile. However, the more complex the demands of the writing the longer it will take to achieve, so it will be important to match expectations to both time available and curriculum significance. In the end, it is important that learners ask of each scenario whether it represents, what Shanon

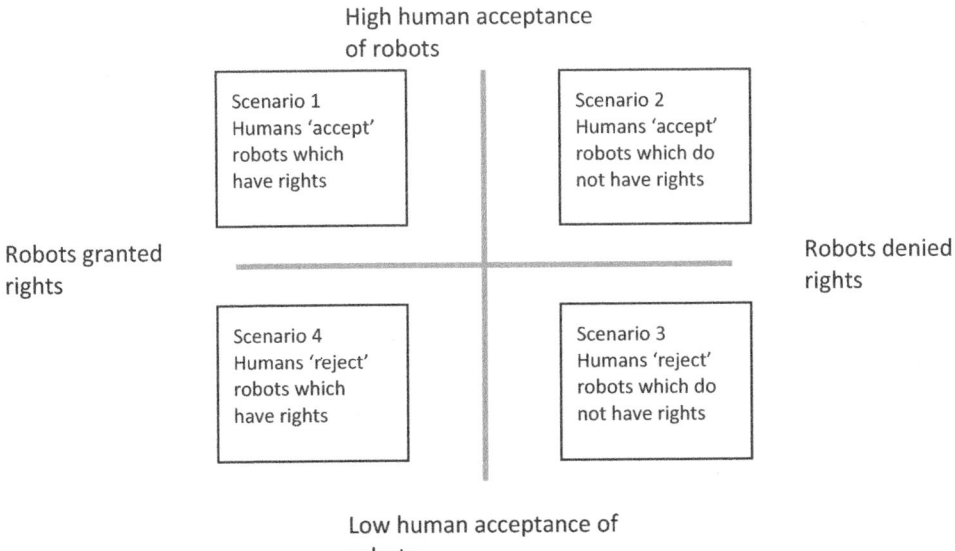

FIGURE 12.5 Four scenarios from axes of uncertainty concerning human acceptance of robots and the granting of rights to robots

Vallor so aptly refers to, as a future worth wanting. For an intriguing exploration of the possible human robot relationship *Machines Like Me* (McEwan, 2019) provides considerable food for thought.

Your vision

In building scenarios for possible STEM futures, we will identify two sets of so-called 'critical or significant uncertainties' and use these as axes to create four quadrants such that there is a particular scenario located in each quadrant. The two sets of critical uncertainties that we will use are as follows:

■   The extent to which STEM subjects operate in isolation as opposed to the extent to which they operate collaboratively; these are labelled S.T.E.M. (taught in isolation) and STEM (taught collaboratively) at the extremes of the vertical axis.

■   The extent to which the subjects are seen as vocational education or general education; these are labelled Vocational and General at the extremes of the horizontal axis (see Figure 12.6).

Note that the scenarios operating in each quadrant are not mutually exclusive, and it would be possible for two or more of these scenarios to operate simultaneously in a single secondary school. It is also important to realise that these scenarios do not necessarily represent what will happen; they are a product of the chosen uncertainties and choosing different uncertainties would result in different scenarios but they do give us the opportunity to explore possible futures from various perspectives and consider what the consequences of such futures for STEM education.

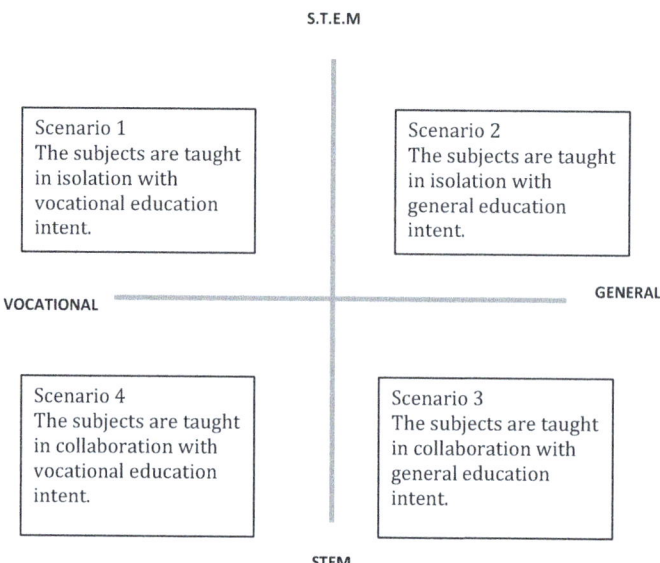

FIGURE 12.6   Four scenarios from axes of uncertainty concerning isolation/collaboration and vocational/general education

Scenario 1

This scenario manifests itself through teaching the subjects more or less in 'silos' particularly valuing those subjects that are gatekeeper subjects providing qualifications that enable progression into further study. While this reinforces the status of science and mathematics, it does little for design & technology or engineering as these do not function as gatekeeper subjects.

Scenario 2

This scenario appears on the surface to be very similar to Scenario 1 but it is likely that the make-up of the classes taking design & technology and engineering options will be different. This is because at 'option time', when subjects are chosen by learners aged 14+ for further study, design & technology and engineering will have been promoted on general education grounds for all learners as opposed to appealing just to those the learners who wish to pursue a particular trade or technical career pathway. Hence it is likely that classes will include learners who have shown aptitude in academic subjects as well those who have been successful in 'practical' subjects. As in Scenario 1, the subjects will be taught in isolation.

Scenario 3

As in scenario 2 the make-up of the classes post 14+ is likely to include learners who have shown aptitude in academic subjects as well those who have been successful in 'practical' subjects. However, in this scenario definite attempts will have been made by the teachers to 'look sideways' and utilise the learning in some STEM subjects to support the learning in other STEM subjects. In some situations, there will be timetable suspensions, so-called drop down days, or even weeks, in which learners collaborate in tackling an open task, sometimes of their own choosing, and in so doing use their learning from across the STEM subjects. In such cases, the learners will probably need to be supported by a team of teachers from different STEM subjects.

Scenario 4

In this scenario teachers will 'look sideways' and utilise the learning in some STEM subjects to support the learning in other STEM subjects. Given the vocational intent, they are likely to involve their learners in multi-disciplinary events that promote STEM careers such as TeenTech, which was mentioned in Chapter 8.

Matthew James of Lewis Girls School in Wales was asked to develop a curriculum that would try to combine science, technology, engineering and maths into one project/topic-based learning activity, which the school would put onto the timetables of Year 7 and 8 (ages 12 and 13) pupils called STEM. This would be in addition to usual separate lessons in science, design & technology, textiles, food technology and ICT. He decided to try and do two separate projects, one in Year 7 and one in Year 8, each running throughout the year on one lesson a fortnight. The teaching focus for Year 7 was 'structures' and pupils were engaged with designing and making frame structures (to act as a support for a wind turbine) and shell structures (to be part of monocoque chassis framed toy racing car). Each was tested against a range of performance criteria.

The teaching focus for Year 8 was 'energy' and involved pupils designing and making a range of alternative energy power sources and using them to light the LEDs in a model of their school. The pupils were able to test the efficiency of their power sources under various conditions. In both projects there was ample opportunity for pupils to be taught science, design & technology and mathematics, and to use what they learned in their designing and making. Matthew's STEM projects would seem to fit well within Scenario 3.

Paul Gittins is an interior designer who works with Space Zero (2020), specialists in designing environments for learning, and he has been thinking about the way the built environment can support learning in the STEM subjects. Interestingly, his ideas support learning within individual subjects as well as enabling significant collaboration and multidisciplinary work. He calls the whole assembly a 'STEM hub'. As his preliminary concept sketch shows (see Figure 12.7), he envisages three separate teaching areas for science, mathematics and technology surrounding a central open area that houses a stage for presentations, four display areas and two pods dedicated to design activity. Between the mathematics and science areas is a bank of raised seating to give clear viewing of the stage area. Between the science and technology areas are a staff collaboration area and two preparation rooms. Such a design has considerable implications for those who have to teach there. It would, of course, be possible to retreat into the separate subject teaching areas and ignore the opportunities afforded by the central open area, but by creating a central inspirational hub with separate

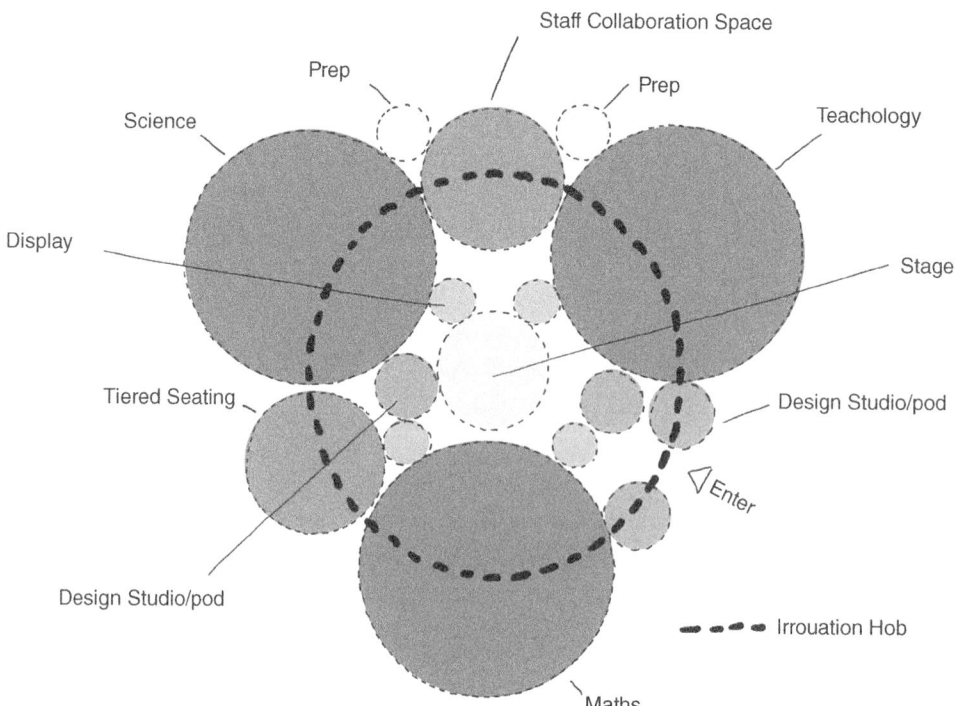

FIGURE 12.7 Paul Gittin's concept sketch for a STEM hub

spaces for each of the subjects, specialisms become both connected but can remain independent and dedicated to their individual subject. This would promote the possibility of combined teaching and then the hub can be used to launch projects to all subjects and careful planning of adjacencies will allow separate subject exploration in their specialist teaching environment. Paul envisages that at the start of such combined project-based teaching pupils would use the central area in much the same way as we use an Apple store – browsing the displays to access relevant information. He thinks that this space would bring to view and mind the industries and products of the world making STEM more conscious and present in the school not just a pedagogy and spatial organisation principle. So, in Paul's vision of a STEM learning environment it would be possible to operate in any of the four STEM scenarios outlined above.

In Chapter 1, Mike Watts painted a picture of the qualities required from a 'STEMer' – a young person well versed in the STEM subjects and able to apply their learning in the world outside school. However, he has lamented the inadequacy of the curriculum of many schools to develop such individuals as we spend 85 per cent of our lives tackling the tasks life throws our way as opposed to the 15 per cent of our lives that we spend in formal education. Tackling such real life tasks in school requires learning 'on the hoof' and has been called complex learning. Mike explains:

> An important characteristic of complex learning (Wintrup, Wakefield, & Davis, 2015) is that learners deal with a range of (often complicated) materials incorporating a considerable number of interacting elements. The learning tasks involved are deemed to be authentic because they provide the driving force for learning – through which learners integrate knowledge, skills, and attitudes to solve real-life problems. Not only must they interact with science, technology, engineering and mathematics on a daily basis but modern life is predicated on engagement with each element of STEM. Very few have the inclination to return to 'the simple life' (whatever that is) – and even then it is intrinsically the domain of homo fabricus, the tool-maker, who is alive and well even in simplicity. It behoves us then, to use the 15% to set the foundations for life in 85%.

So, we should ask, 'Which of the scenarios developed above provides the foundation for life in the 85%?' The main contender would seem to be Scenario 3.

In conclusion

The scenarios teachers find themselves in will, to some extent, be determined by the teachers' personal visions of STEM education. If they regard subject integrity and identity in very high esteem and see collaboration with other subjects as a threat to this, then they will find themselves in Scenarios 1 and 2. On the other hand, if they do not see collaboration as threatening integrity and identity, then they will be prepared to 'look sideways' and find themselves in Scenarios 3 and 4. And any personal vision will to some extent have to be modified in the light of the contexts in which teachers find themselves. In a school in which there is established and successful practice of individual subjects working in isolation with vocational intent, it is difficult to adopt a different position, and arguing for change may put a teacher at odds with colleagues. But, even in these situations, it is possible to 'look sideways' in your own practice and show the benefits this brings to the young people you teach.

Whatever your vision, we find ourselves returning to the ever-important idea of conversation – with colleagues in your own discipline, colleagues from other disciplines, senior leaders in your school, students and their families, the wider school community and within and across the various professional bodies that represent and engage with STEM education. These conversations alone will be insufficient to implement your vision but without them we believe that however attractive and worthwhile that vision, it will not become a reality.

Looking sideways at what your colleagues are doing and talking to them about what *you* are doing is vital. Such conversations are the starting point for change.

Recommended reading

The following deal with some of the issues raised in this chapter:

Berners-Lee, M. (2019) *There is no Planet B*. Cambridge: Cambridge University Press.

Bhagwat, S. (2019) *Eating for the environment*. www.open.edu/openlearn/nature-environment/eating-the-environment/content-section-0?active-tab=content-tab (accessed September 30 2020).

Harari, Y. (2014b) *Homo Deus a brief history of tomorrow*. London: Vintage.

McEwan, I. (2019b) *Machines like me*. London: Vintage.

Nourbakhsh, I. (2013) *Robot futures*. Cambridge, MA: MIT Press.

O'Neil, C. (2016) *Weapons of math destruction*. Great Britain: Allen Lane.

Preston, Christopher J. (2018b) *The synthetic age*. Cambridge, MA: The MIT Press.

Thunberg, G. (2019) *No one is too small to make a difference*. London: Penguin.

Williams, P. John & Stables, K. (2017) *Critique in design and technology education*. Singapore: Springer.

References

Barlex, D. (1995) *Nuffield design and technology student's book*. Harlow: Longman.

Barlex, D. (2017) Disruptive technologies. In P. John Williams and Kay Stables (eds), *Critique in design and technology education*. Singapore: Springer.

Barlex, D., Givens, P., & Steeg, T. (2020) Teaching about disruption: A key feature of new and emerging technologies. In A. Hardy (ed.), *Learning to teach design & technology in the secondary school*, Edition 4. Oxon: Routledge.

Christensen, C. M. (2012) Disruptive innovation. In M. Soegaard & Dam F. Rikke (eds), *Encyclopedia of human-computer interaction*. Aarhus, Denmark: The Interaction-Design. org Foundation. www.interaction-design.org/encyclopedia/disruptive_innovation.html (accessed April 29 2020).

Classander, C., & Jonas Halstrom, J. (2020) Making the invisible visible: Pedagogies related to teaching and learning about technological systems. In P. John Williams & David Barlex (eds), *Pedagogy for technology education in secondary schools*. Singapore: Springer.

Ellen MacArthur Foundation (2020) *A circular economy*. www.ellenmacarthurfoundation.org (accessed April 29 2020).

Gabriel, K. (2019) Interview with Jim Al-Khalili on *The Life Scientific*, 12 March 2019. www.bbc.co.uk/programmes/m00035tc (accessed April 29 2020).

Greenpeace (2020) *Earth stewardship.* www.greenpeace.org.uk/about-greenpeace/ (accessed April 29 2020).

Gumbo, M. (2020) Teaching technology in 'poorly resourced' contexts. In P. John Williams & David Barlex (eds), *Pedagogy for technology education in secondary schools.* Singapore: Springer.

Harari, Y. (2014) *Homo Deus a brief history of tomorrow.* London: Vintage.

Harding, J. (2020) The arrestables. *London Review of Books,* 42(8). www.lrb.co.uk/the-paper/v42/n08/jeremy-harding/the-arrestables

Intergovernmental Panel on Climate Change (IPCC) (2018) *Global warming to 1.5°C.* www.ipcc.ch/sr15/ (accessed April 29 2020).

Juliana v United States (2020) *Securing the legal right to a safe climate.* www.ourchildrenstrust.org/?utm_term=0_4094e87487-8a53d2a846-116192553 (accessed April 29 2020).

Leonard, A. (2010) *The story of stuff.* London: Constable & Robinson Ltd.

Lyster, R. (2020) Where water used to be. *London Review of Books,* 42(7) www.lrb.co.uk/the-paper/v42/n07/rosa-lyster/diary (accessed April 29 2020).

McEwan, I. (2019) *Machines like me.* London: Vintage.

McKinsey Global Institute (2013) *Disruptive technologies: Advances that will transform life, business, and the global economy.* www.mckinsey.com/mgi (accessed 30 September 2020).

Merrick, R. (2018) *Wind and solar power investment crashed after government cut funding, show new figures.* www.independent.co.uk/news/uk/politics/wind-power-solar-investment-drop-uk-government-funding-environment-figures-budget-a8162261.html (accessed April 29 2020).

Preston, Christopher J. (2018) *The synthetic age.* Cambridge, MA: The MIT Press.

Rapsomanikis, G. (2015) *The economic lives of smallholder farmers: An analysis based on household data from nine countries.* Rome: Food and Agriculture Organisation of the United Nations. www.fao.org/3/a-i5251e.pdf (accessed February 26 2019).

Shukman, D. (2020) *Sir David Attenborough warns of climate 'crisis moment'.* www.bbc.co.uk/news/science-environment-51123638 (accessed April 29 2020).

Space Zero (2020) *Rethinking learning environments.* https://spacezero.co.uk/about-us/ (accessed April 29 2020).

The Food and Agriculture Organisation of the United Nations (2009) *How to feed the world in 2050.* Rome, Italy: FAO. www.fao.org/fileadmin/templates/wsfs/docs/expert_paper/How_to_Feed_the_World_in_2050.pdf (accessed April 29 2020).

The Geographist (2020) *1000 largest cities and towns in the UK by population.* www.thegeographist.com/uk-cities-population-1000/ (accessed April 29 2020).

Thunberg, G. (2019) *No one is too small to make a difference.* London: Penguin.

UK Committee on Climate Change (UKCCC) (2019) *Net zero: The UK's contribution to stopping global warming.* Report to Government. www.theccc.org.uk/publication/net-zero-the-uks-contribution-to-stopping-global-warming/

Wikipedia (2020) *List of United States cities by population.* https://en.wikipedia.org/wiki/List_of_United_States_cities_by_population (accessed April 29 2020).

Wintrup, J., Wakefield, K., & Davis, H. (2015) *Engaged learning in MOOCs: A study using the UK engagement survey.* York: Higher Education Authority.

Index

Page numbers in *italics* refer to content in *figures;* page numbers in **bold** refer to content in **tables.**